Structures of Complexity

Rupert Riedl

Structures of Complexity

A Morphology of Recognition and Explanation

Springer

club of vienna

WIEN KULTUR

Rupert Riedl (Deceased)
Konrad Lorenz Institute for
Evolution and Cognition Research
Altenberg, Austria

Translated by Michael Stachowitsch, University of Vienna, Austria

This translation was funded by the *Vienna Municipal Department 7* and the *Club of Vienna*

ISBN 978-3-030-13063-3 ISBN 978-3-030-13064-0 (eBook)
https://doi.org/10.1007/978-3-030-13064-0

This Springer imprint is published by the registered company Springer Nature Switzerland AG.
The registered company address is: Gewerbestrasse 11, 6330 Cham, Switzerland

Foreword 1

Today, humankind faces challenges that go far beyond the uncertainties and quirks of nature, to which we have had ample time to adapt. The problems we now face arise from the interplay between two sources of complexity, that of the natural system and that of the technical world we ourselves have created. "We are intervening in a world that we have not yet understood" was one of Rupert Riedl's most common pronouncements. Importantly, the cognitive pathway runs in the counter-direction to the developmental pathway. Riedl's insights and methodological approaches push the traditional boundaries of science and are widely applicable. This is because the journey along the two pathways in pursuit of underlying causes means tackling both the "hierarchy of disciplines" and the evolutionary hierarchies. No solution can be reached without illuminating the causes behind the developmental pathway.

Responsible, scientifically founded advice to decision-makers requires going beyond the easily recognizable symptoms to understanding the causes behind the problems we face. Exponential growth as a fundamental principle has accelerated every aspect of human endeavor, and science itself has helped fuel the process. This growth has diverted funds and reduced the time available for dealing with underlying processes. We spend too much time plausibly describing visible phenomena and deriving short-sighted measures. This is a dead end. We need to query the causes behind the individual disciplines and promote interdisciplinarity to better shape our world. Technology has enabled us to transcend our innate sensory (spatial and temporal) perception, has abetted ever-narrower specialization, and has enticed us to intervene at all levels. Nonetheless, short-circuiting the "polluter pays" principle, i.e., eliminating the feedback of an action on the perpetrator, as new technologies or the media tend to do, scuttles vital learning processes. A case in point is the obligatory interaction between material and formal causes of adjoining tiers in the hierarchy of complexity, or the purposes and drives that traverse all tiers. In the technical disciplines, the two internal (formal and material) causes apparently suffice for most researchers. The purposes and drives, in contrast, are left to other disciplines despite their effects and feedbacks on every human intervention. This shifts an ethics of responsibility into one dominated by opinion—a recipe for responsibility-free conduct. Today's dominant corporate logic demonstrates that the structure of thought has been fully translated into real structures of action, circumventing the barriers imposed by democratic institutions.

Evolution has equipped us astoundingly well to survive in the complex world in which we arose, making us the dominant higher-developed species on the planet. New methodologies and technologies have boosted our capacity to realize ideas that far exceed direct, intuitive perception. The unforeseen repercussions have hit us unprepared. We have replaced sensible observation of changes in the real world with the virtual reality modeled in subdisciplines. Neglecting "external effects," a routine approach in economics, is foreign to evolution and leads to "collateral damage" in the technical applications by some sciences. These threaten the existence of human civilization. Fragments of Rupert Riedl's thoughts and insights have increasingly cropped up in publications, conferences and symposia dealing with the development of new technologies and with our efforts to understand them. The present book compiles this vision in condensed form. As in his other highly recommended works, this volume is excellently illustrated: as a biologist, Rupert understood that comprehending complex structures is best mediated by relying on our visual capabilities.

This translation was prompted by internationally active colleagues from a range of disciplines who have recognized that the fundamental perspectives and insights outlined in this book are widely unknown in today's English-speaking community, with its quantitative bent. International collaborations in various sectors of technology and economics have also increasingly underlined the need for an English translation of *Structures of Complexity*.

Thanks to funding from the Vienna Municipal Department 7, the Club of Vienna was able to support the translation into English. Special thanks go to the translator, Dr. Michael Stachowitsch, a student of Rupert Riedl's, and to his daughter, Dr. Barbara Schweder, who provided editorial input. We are also grateful to Rupert's family, who consented to and supported publication, and to Springer for taking on this project.

January 2019

Hermann Knoflacher
President Club of Vienna, Gugging
Vienna, Austria

Foreword 2

When I first met Rupert Riedl 33 years ago through my wife, he was already a renowned scientist. Back then, he gave us a copy of his book *The Strategy of Genesis* with the dedication "fondly remembering our computer discussions." As a software engineer and freshly minted computer artist, I was convinced that evolution and number structures were deeply intertwined. I was truly impressed that a zoologist knew so much about highly abstract phenomena of patterns.

Fifteen years later, Rupert, who in the meantime had become a friend, gave me a copy of his latest book *Strukturen der Komplexität* (*Structures of Complexity*). It changed my life. Up until then, I, like many others, was certain that the future of science lay in ever-further specialization. Rupert, however, felt that morphology had unjustifiably slipped from the focus of scientific endeavor. The developments of the past 10 years have brilliantly proven him correct. Even the insight that entropy and the monetary system are tightly interrelated is already addressed in that volume.

It saddens my heart that Rupert Riedl's important contribution to the history of information processing and to the development of computer art—along with modern concepts such as digitalization, the Internet of Things, or crypto-economy—is so poorly recognized. One potential explanation is that his most important contribution has only now been translated into English.

I am very proud to have been able to contribute to making the translation and new edition of such essential reading for the twenty-first century a reality.

February 2019

Peter Kotauczek
Burg Hartenstein
Weinzierl am Walde, Austria

Preface

Some people are naturally attracted to complex phenomena. They simply creep up on us in our daily lives or we actively seek them out. That must be what happened to me. This was no doubt promoted by my "right brain hemisphere preference," the pleasure afforded by a synoptic worldview, and the artistic world of my sculptor father—aided and abetted by the tenets of morphology and by my teachers *Ludwig von Bertalanffy* and *Konrad Lorenz*.

Where is my experience anchored? I started out with the systematics and microscopic anatomy of marine invertebrates and then published the *Fauna und Flora der Adria* (*Fauna and Flora of the Adriatic Sea*), later expanded to encompass the Mediterranean, followed by *Biologie der Meereshöhlen* (*Biology of Marine Caves*) and, finally, a book examining the Mediterranean as an ecosystem (*Gärten des Poseidon*, *Gardens of Poseidon*). Each endeavor sought to interlink thousands of species. The complex interrelationships I recognized in those efforts led to my books *A Systems Theory of Evolution* (*Systemtheorie der Evolution*) and *The Strategy of Genesis* (*Strategie der Genesis*). I soon recognized that the thought processes behind grasping complexity—involving differentiated and recursive causality—were poorly understood. The impression was that we were simply projecting our thought patterns onto natural patterns of order. This prompted me to develop a "naturalized theory" of cognitive processes in a series of further volumes, whose contents contribute to the discussions in this book.

What are the new aspects? I admittedly rely here on some of the above experience and have updated selected illustrations with proven didactic merit. Basically, however, I present all the new knowledge that has enriched science as a whole, thanks to the integration of anatomy, systematics, evolutionary theory, and epistemology.

This approach builds on juxtaposing the terms rational and ratiomorphic, cognition and explanation, and systems of thought and conceptual structures, as well as on distinguishing between structural and class hierarchies. Doing justice to the structures of complexity benefits from perceiving these phenomena in the form of twin pyramids comprising standard building blocks and individual components. If we wish to adapt to complex systems, we must recognize the cognitive dualisms behind our understanding of causes and effects, and we need to perceive and differentiate the suggestions triggered by observation and explanation. This is because the explanatory pathways run counter to the cognitive pathways, with both recapitulating the developmental pathway of complexity in this world.

This book's perspective is rooted in biology and is complementary to the treatments proffered by physicists and mathematicians. For one, I come from biology. Equally important, modern biologists are instilled with the complexity of their discipline (2 million species and 500,000 system categories multiplied by dozens of specific characters)—many millions of individual facts making up a single, enormous, interrelated constellation. This complexity generates a greater body of experience than other scientific disciplines. Moreover, we biologists have accessed the cognitive processes defining our species in the framework of "evolutionary epistemology"—and have learned to relate these to external reality.

My aspiration is to be able to apply this body of gained experience to any complex system. I have time and time again had the opportunity to investigate this issue under a variety of perspectives. This is honed with an appreciation for the difficulties in conveying this message based on the reception given to my books and on classroom feedback. Our makeup, our faculties, are unprepared for unraveling complex matters, and this inability is reflected up into the structure of our universities. Importantly, we ourselves are complex: we live based entirely on complexity and our survival depends on it. This warrants an attempt at providing an overview and summary.

Altenberg, Austria
March 2000

Rupert Riedl

Contents

1 The Issues Tackled Here: An Introduction 1
1.1 The Topic at Hand 1
1.1.1 Research into Complexity Today 1
1.1.2 Complexity: Its Characteristics and Its Meaning 3
1.1.3 Why, of All Things, Structures? 7
1.2 On Methods 8
1.2.1 Morphology, Systems Theory and Gestalt 8
1.2.2 Structuralism and Functionalism 9
1.2.3 On cognition, Explanation and EE 10
1.2.4 Biology as the Conceptual Framework 11

2 The World and Cognition as a Problem 15
2.1 What Appears Reasonable to Us 15
2.1.1 What Arose with Consciousness? 16
2.1.2 The Conceivable Validations of Perception and Knowledge Gain 17
2.1.3 Forms of Perception Versus Communication 18
2.2 How Knowledge Is Gained 26
2.2.1 The Levels of Cognition 26
2.2.2 What All This Can Tell Us About the World 31
2.2.3 The Purpose Served by Such Knowledge 32
2.3 The Nature of Our Knowledge 33
2.3.1 Construction and Reality 33
2.3.2 Emergence, Notions and Language 36
2.3.3 Perceiving (Cognition) and Explaining 39

3 The Systems of Cognition 51
3.1 Conditions of Perception 51
3.1.1 Perception Means Problem Solving 52
3.1.2 Fundamentals of Association and Conditioning 52
3.1.3 The Transition to Cognitive Processes 53

3.2 Processing Consecutive Coincidences 54
3.2.1 The Composition of the Algorithm 54
3.2.2 Wherein the Seeds of Success Lie 55
3.2.3 Wherein the Deficiencies Lie 55
3.2.4 How to Overcome the Deficiencies 56
3.3 Processing Simultaneous Coincidences 57
3.3.1 The Composition of the Algorithm 57
3.3.2 The Reasons for Success 64
3.3.3 The Deficiencies of the Program 65
3.3.4 The Path to Overcoming These Deficiencies 71
3.4 On Structures and Classes 73
3.4.1 The Evolution of Memory 74
3.4.2 Fields of Similarity 75
3.4.3 On Structural and Class Hierarchies 78

4 Structuring the Perceived 91
4.1 A Theory of the World 92
4.1.1 The Hierarchic Structure of Things 92
4.1.2 On Transformation and Emergence 96
4.1.3 The Broadest Parameters 100
4.2 The Order of Things 106
4.2.1 The Process of Reciprocal Enlightenment 106
4.2.2 The Three Fundamental Types of Complex Similarity 117
4.2.3 The Four Fundamental Forms of Complex Order 120
4.3 The Principles of Morphology 126
4.3.1 The Theorem of Homology 128
4.3.2 Type and Bodyplan 144
4.3.3 A Theory of the Phene and Character 152
4.4 The Principles of Systematics 165
4.4.1 The Weighting Problem 166
4.4.2 Optimizing the Class Concepts 172
4.4.3 The Nature of the Natural System 177

5 The Systems of Explanation and Understanding 181
5.1 The Conditions and Our Faculties 182
5.1.1 The Preconditions 182
5.1.2 The Hypotheses of Causes and Purposes 183
5.1.3 Common Sense and Intuition 186
5.1.4 The Psychology of Explaining and Understanding 188
5.2 Changes in Cultural History 189
5.2.1 The Beginnings in Our Culture 191
5.2.2 Antiquity and the Middle Ages 192
5.2.3 The Modern Era 194
5.2.4 The Concepts of Understanding Today 196

5.3 The Conditions of Explaining 198
5.3.1 On Causal Explanations............................. 199
5.3.2 The Double Pyramid of Explanation 209
5.3.3 The Three Pathways: Assuming, Explaining and Emergence.................................... 212
5.4 The Forms of Understanding............................ 213
5.4.1 On Teleological Explanation 214
5.4.2 Understanding Actions............................. 215
5.4.3 Understanding in the Humanities.................... 217

6 The Structure of the Explained and Understood................. 219
6.1 The Path to a Dynamic Explanation of the World 219
6.1.1 Origin of the Inorganic............................. 220
6.1.2 The Origin of Organisms 222
6.1.3 Paradigms of the Origin of Reason 223
6.2 The Order of Causes 225
6.2.1 The Reduction of the Concepts of Causality 226
6.2.2 The Causes and Growth of Historicity................ 232
6.2.3 Four Interactions, Four Forms of Causality............ 238
6.3 The Principles of Explanation 240
6.3.1 Explanatory Models in the Inorganic Realm 240
6.3.2 Evolutionary Theories in the Organic Realm........... 247
6.3.3 The Explanatory Models in the Organic Realm 265
6.4 The Principles of Understanding 286
6.4.1 The Explanatory Models of Human Behavior........... 287
6.4.2 Explanation of Artefacts Exhibiting Genealogies 290
6.4.3 The Explanation of Institutions in Civilization 299

7 Overview and Outlook 305
7.1 On the Unity of the World and of Cognition 306
7.1.1 On Our Faculties, Language and Culture 306
7.1.2 The Systems of Cognition and the Structures of the World 307
7.2 On Naive and on Nasty Deceptions........................ 308
7.2.1 Can Explanation Replace Cognition? 309
7.2.2 On the Origin, Nature and Controllability of Losses...... 310

Bibliography 313

Index... 323

Chapter 1
The Issues Tackled Here: An Introduction

Our worldview has been negligently compartmentalized and simplified. At the same time, we have allowed the world around us to become so complex that we are increasingly less able to comprehend it (Riedl and Delpos 1996a). This is abetted and formalized by the largely analytical approaches of the neatly subdivided disciplines in the natural sciences. It is also promoted by human society, which rewards those who most skillfully intervene in nature. The result is, sad to say, that we tend to confuse graspable circumstances with the real world itself.

The definitional nature of our logic and languages may well have set the stage for this. The rationalistic bent of modern culture has further channeled our thought processes into simplifications, adding insult to injury.

1.1 The Topic at Hand

The above situation requires delving into the issue of complexity and focusing on holistic perspectives, on interdisciplinarity and on synoptic approaches. Although these concepts have become modern catchwords, much still remains to be done to remove the many hurdles facing this new movement.

This calls for (Sect. 1.1.1) reviewing the research landscape, (Sect. 1.1.2) discerning the features of complexity, and finally (Sect. 1.1.3) outlining the importance of focusing on structure.

1.1.1 Research into Complexity Today

Complexity was a staple for classical biologists, in contrast to the 'exact sciences'. In physics, the traditional approach was to circumvent the complexity of this world and focus on the remnants that proved to be mathematically representable. The

R. Riedl, *Structures of Complexity*,
https://doi.org/10.1007/978-3-030-13064-0_1

successes of this reductive method are legendary and led to the assumption of 'immutable laws', to the expectation that every natural phenomenon could be reduced to such laws. This stylized the methodology applied by physics into a fundamental paradigm for all natural sciences, reducing biologists to narrators of ephemeral stories. It also further marginalized the humanities.

The first step is to compare the position of (a) biology with (b) the shift in focus of inorganic scientists to complexity, (c) their current situation and (d) the strategy underlying my approach.

(a) *Biologists* have increasingly endeavored to fill the divisive trenches. One motivation was that biology itself faced the prospect of being cleaved into a causalistic physiology and a 'hermeneutic' morphology. This was accompanied by the recognition that this artificial methodological divide was the very site at which inorganic scientists and scholars in the humanities could initiate a dialogue (Riedl 1985).

(b) Importantly, however, the *inorganic sciences* are also undergoing change. This is prompted by their successes in tackling biological questions and by the attendant necessity of addressing complexity. The exact time point of the shift is difficult to determine—perhaps as far back as Schrödinger's 'What is Life' (1957).

The result, however, is indisputable. Irreversibility—the historicity of the complex world—was discovered, and the inorganic realm was no exception. We recognized the phase transitions in every form of evolution and the limits of predictability. We acknowledged 'inner' conditionalities, ordering parameters and the buildup of order by exporting entropy. The resulting insight is that complex systems cannot be entirely reduced to their constituent components. All these concepts converge in the paradigm of biology and are treated here.

The available literature is voluminous. New disciplines have arisen—non-equilibrium thermodynamics, synergetics, chaos research and fractals to name a few—each yielding its own series of monographs. This heroic movement, which transcends classical physics, has made headway both on the scientific and popular level. Ebeling and Feistel (1994), Gell-Mann (1994), Lewin (1993), Nicolis and Prigogine (1987) or Ebeling et al. (1998) are examples. For a key symposium, see Schweitzer (1997).

(c) Nonetheless, the overall *situation* is perceived as being unsatisfactory. The study of complexity is said to be 'in a crisis' (Horgan 1995). Herbert Simon felt that the potential of mathematical solutions might be overestimated. Insufficient consideration is being given to emergence phenomena in the phase transitions. Will we fail to overcome reductionism after all, expecting to ultimately be able to reconstruct complex entities based on their components? Many of these concepts crop up in the following chapters.

Of course, every attempt to overcome classical physics is again anchored in physics itself. Which is all very legitimate. Unsurprisingly, the exceptional insights afforded by modern molecular research have spawned the rationale that we must continue to work from that perspective. In my opinion, surmounting

this paradigm calls for an epistemological shift and for refocusing on those paradigms that have already proven themselves in biology.
The necessity for pursuing the epistemological approach created a movement whose intellectual history is outlined by Mainzer (1994). That effort still predates a 'naturalized epistemology', foremost 'evolutionary epistemology' as advocated by Lorenz, Mohr, Oeser, Riedl, Vollmer and Wuketits. The latter approach tackles the origins of human reason and examines the inherent difficulties that history has imposed on arriving at an understanding of complexity. This is the point of departure for this book.

(d) With regard to *the paradigms*, the paradigm of physics is not the cornerstone for my approach. While the insights provided by physics cannot be contradicted, they do require a new superstructure. My arguments are founded in the paradigm of biology. Why? Because, as I will outline in detail, biology is the hub that can re-link the inorganic realm with the humanities and social sciences. The answer lies in reconciling cause-and-effect physiology with hermeneutically operating morphology.

Importantly, any shift in perspective must build on a new set of tools. A new method and terminology along with a new form of portrayal are required to do justice to the topic. As the term 'synoptic' implies, observation and perception are central. The task involves attaining a 'combinational overview', something we are quite well equipped to do (Riedl 1987a). This calls for 'pictorial' abstraction because such depictions—both of natural forms and cognitive forms—provide immediate and convincing insights. Synthesizing the overarching principles then establishes the gateway to the structures of complexity.
The task is twofold: it involves investigating the structures of extra-subjective reality and, equally importantly, the structures of our thought patterns and their history. Our forms of thought (at least those vital for survival) were shaped by correctly processing the structures in our surroundings. Where the two match, they often serve us better than we think. Where they prove to be overtaxed—something we like to ignore—they become potentially treacherous stumbling blocks. This describes the issues tackled here.

1.1.2 Complexity: Its Characteristics and Its Meaning

We refer to structural and functional interrelationships as being complex and group them according the gradual permutations of certain features, be they natural objects, artefacts, notional forms or thought processes. They can be complicated, but complication is not their defining feature.

This calls for (a) formulating a definition of complexity and then examining (b) its manifestations, (c) its significance and (d) how to approach it technically.

(a) Sharply honing a definition of complexity would be misleading. This is because complexity is a wide-ranging and manifold condition in our world, and also consistently polymorphic. Arriving at an adequate definition would require incorporating a lengthy series of features.

The better approach is therefore to (a1) distill its general features, which also encompass other states, and to (a2) define the series of specific features characterizing all gradients.

- (a1) *Complexity contains forms of order.* This means complexity is variously removed from physical equilibrium, from thermodynamic chaos (the equal distribution of matter and its temperatures). Creating and maintaining order must comply with the conditions governing open systems: they are traversed by matter and energy and achieve their ordered structure by exporting entropy. They can also be referred to as dissipative (German: *zerstreuende*) systems because, at the very least, they release heat. Order in complex systems can broadly be understood as law multiplied times application (LxA; details in Riedl 1975). This is equally valid for a local cultural ordinance (complex law times rare application) as it is for the universal law of gravity (simple law times countless applications).

  At the same time, every hot object emits heat. Although every piping system releases frictional heat, it cannot be said to be complex even though material and energy flow through it. Crystallization also releases heat and leads to a high degree of order without fully satisfying the conditions of complexity.
- (a2) Complexity contains gradients of features. Gradients because the features themselves can be expressed in highly differentiated form. Experts refer to system properties or, more broadly, to the product of self-organization processes (versus minimal 'outside organization'). Three groups are differentiated here (i–iii):

  (i) *Historicity* is a main feature. This refers to historical uniqueness and encompasses three sub-characters: *irreversibility*, *phase transitions* and *emergences*. The first means that the developmental processes can neither be traced back along the same path nor be repeated. Rather, they have gone through phase transitions, each of which in itself is unique and typically neither accessible nor repeatable. In fact, these transitions lead to the emergence of new features that are not contained even in traces in the old system, making them unpredictable in practice.

  Celestial bodies are the largest and most long-lived objects with historicity, followed by the oceans and continents, the kingdoms of organisms, biocoenoses, languages and cultural artefacts—all fundamental physical processes. Lasers can serve as an example. A rubidium crystal, excited by an energy source, will emit a ray of light. The direction that light takes, however, is unpredictable thanks to the 'parliament of molecules' (Haken 1978); this is an event marked by shortest historicity.

  Of course, quite complex systems with very short historicity also exist. Examples include snowflakes, a rapidly emptied compost heap, or a piece of prematurely scrapped cultural legislation. These stand in contrast to complicated states such as metal shavings, a rubbish dump or a jabbering crowd, none of which are complex.

(ii) *Hierarchic organization* is a second feature and is related to *polymorphy*.

We find the most complex hierarchies, with up to 18 tiers (Riedl 1975), in higher organisms. Step by step—from atoms and molecules up to organs and individuals—each new tier exhibits phase transitions and manifests new emergent qualities, accompanied by yet another descriptive terminology. Cultures, languages and artefacts follow suit. The simplest systems, in turn, are textbook examples from physics. An example is Bénard cells (compare Nicolis and Prigogine 1987): heating a thin water layer from below and cooling it from above creates convection cells in which heated molecules rise up centrally and sink down along the outer margins.

One feature deserves mention here: redundancy. Although it does not determine hierarchy, it does influence the patterns of order in a hierarchy. Redundancy refers to the repetition of nearly identical structural elements. Examples include the hydrogen molecules of a sun, the molecules making up genetic material, the brain cells of an organism, the leaves of a tree, the spruce trees in a forest, the ties of a railway track, or the number of books in a particular edition.

Polymorphy alone, however, by no means determines complexity—despite being very high in a rubbish heap or landfill. The situation differs in hierarchies. Even the simplest hierarchy of structural elements or functions leads us to expect complex conditions. This points to the third group of features.

(iii) *System conditions in the narrower sense*. Complex systems are always embedded in a broader setting. This affects the modalities of that system's origin, maintenance and change, but it does not do so alone. The internal workings of a system gain autonomy and follow novel trends, constraints and degrees of freedom. And they are considerably more stable than changing environmental conditions because their fate can be intimately tied to such a system.

Gradients in expression can be expected here as well. The top tier again encompasses higher organisms, which pass along certain features over billions of years, followed by languages, all of which contain and separate nouns and verbs. Even architectural styles and false theories can persist for centuries, independent of the milieu, based on autonomous internal principles. Additional catchwords pertaining to complexity in this connection include feedback, multilateral and recursive causality, internal conditionality and stability principles.

This, however, already goes beyond the general definition of complexity and touches upon the level of research, of discussion and controversies.

(b) *Complexity is omnipresent*. In principle, a chemical bond or even a heavy atom already meet the definition of a polymorphic structure and of a functional interrelationship. Complexity is missing only when the component elements have not yet been linked or when we ourselves have disassembled or disrupted the link. In such cases, the historicity is blurred.

The definition excludes any accumulation of unrelated items such as those found in fresh garbage piles. The same holds true when polymorphism is lacking, such as in a pile of sieved sand. A sandy beach, however, does represent a complex system. Pure redundancy, such as that characterizing a dot matrix, should also be excluded. This is also valid for an absolute lack of redundancy, such as found in the garbage heap, or in an ahistorical randomness such as shattered glass. The amount of such disjunctivity and chaos that civilization produces is astounding.

(c) *The relevance of complexity* for living organisms is a reversal of the above-mentioned constellation. We live exclusively on complexity. In one sense that sounds trivial because we ourselves are complex systems. A single quantum of light might be sufficient to activate a melanocyte in our skin or to trigger dramatic responses after hitting our retina. In short, any environmental factor that stills our hunger, satisfies our urge for physical activity, elicits affection or gratifies our sense of esthetics is complex.

Our whole existence is anchored in complexity. To turn a phrase coined by Schrödinger: we feed on complexity, we live on its degradation.

At the same time, we are destined to create complexity. Complexity rules supreme in all things of value that cultures produce, whether in agriculture or animal husbandry, artefacts, organizations or ideas. All this forms a precondition for our survival, compensating for the degradation of order that nourishes us.

(d) *Our technical treatment of complexity*. The issue of complexity calls for special methodologies, terminologies and forms of depiction. The upcoming chapters are devoted to developing such terms and depictions. The first step (preceded by a preamble) is to broadly outline the methods.

This approach requires a few words about (d1) the synoptic approach and (d2) the stance behind it.

- (d1) *The ambition of the synoptic approach* stands in opposition to the traditional scientific method, in particular to that of the natural sciences. Subjectively, this approach tends to be relegated to the arts or, alternatively, the arts are touted as being the 'spice of the sciences'. It pays to be objective.

  The sciences are commonly interpreted as being analytically oriented, but their terminologies already prove to be as synthetic as their systems and intellectual frameworks. Categorizing the sciences as being analytical is mostly based on their particularistic product. These a purely manners of speech. The fact remains that the sciences tend to subdivide our world into parts rather than pulling it together. This is patently evident in the successive dissection into disciplines and sub-disciplines, each with its own, unconnected scientific language and sub-language. This all runs counter to the vital necessity of comprehending the world and human thought as an entity, as a whole.

  The synoptic approach, in turn, appears to be largely synthetic (and speculative to boot). What, one might ask, can be synthesized if not analytically gained building blocks? Again, it's the product that creates that impression.

The product itself is largely synthetic. The allegation of speculation also proves to be a mere preconception. This is because the cognitive process is characterized by a natural cycle of alternating synthesis and analysis, referred to in this book as inductive and deductive processes. And induction necessarily contains speculative elements: it is a necessary heuristic principle in all sciences.

Of course, a synoptic approach is always riskier for researchers than pure analysis. It is more risky intellectually, it is disparaged as a minority phenomenon, and it receives little funding because its products promise less profit or political influence. Nonetheless, foregoing this approach poses high risks for society's understanding and treatment of the world.

- (d2) We are actually ideally equipped for *synoptic tasks*, specifically for gestalt perception. The process is automatic and features an innate sensibility. It guarantees our survival by summarizing and sorting complex, highly polymorphic shapes and forms. Classical examples include the recognition of faces, species or styles. Such perceptions unerringly steer us through our complex world.

  Of course, this talent appears to be unevenly distributed. The same holds true for the often contrasted talent for mathematics and logic. The explanation remains unclear. Many researchers assume a brain hemisphere preference: an individual can rely either more on his or her left, analytic-deductive dominated brain half or on the right, synthetic-inductive half (see Chap. 4, Sect. 4.3). Another explanation is our subordination to modern educational approaches, which apparently favor left brain hemisphere problem-solving (compare Chap. 7, Sect. 7.2). In principle, synoptic tasks involve being motivated to synthesize and to draw comparisons (and to trust that this is useful).

1.1.3 Why, of All Things, Structures?

There are apparently no functions without structures, at least in the macro-realm. Only in quantum physics do particles and their functions (the energy state of a wave) in some sense begin to coalesce. And the rule equally states: no structures without functions, or at least no structures without any effect whatsoever. If that is the case, then why start here with structures?

A second rule of thumb is that we discover phenomena through structures and explain them through functions. Although this needs to be challenged, the rule does contain a kernel of truth. Gestalt perception is required to experience structures, but not to experience function. We automatically perceive shapes and form, whereas explaining the accompanying functions requires a rational framework.

This insight serves to structure the main parts of this book: Chaps. 3 and 4 juxtapose our innate method of cognition with the constructional method of explanation presented in Chaps. 5 and 6.

This structure also mirrors another circumstance. It turns out that the explanation for a phenomenon can change without the matter itself changing. Vice versa, however, a new insight into a phenomenon immediately prompts a new explanation.

When dealing with complexity, the cognitive process is the prerequisite and more reliable element.

1.2 On Methods

A few additional terms and concepts help anchor the methodologies applied in this book. They all have accepted definitions and stem from biology, or from psychology and sensory physiology as biologically oriented disciplines. And they all exhibit a holistic character along with the cognitive ambition to integrate the processes of perception and of thought. The authors of selected 'key works' are briefly introduced in order to characterize the disciplines; later chapters provide more detailed discussion.

The task here is—building on terms rooted in biology (Sect. 1.2.1)—to present (Sect.1.2.2) those dealing with structure and function. Then, in Sect. 1.2.3 a discussion of cognition and explanation is used to (4) develop the framework for the overall argumentation.

1.2.1 Morphology, Systems Theory and Gestalt

These schools of thought—as differently as they may have unfolded—all arise from the aura of biology.

(a) Morphology is a term stemming from biology or, more precisely, from 'comparative anatomy'. It can be traced back to the physician Karl Friedrich Burdach, was developed by Goethe, further advanced by Oken and Owen, and came to dominate classical biology as a whole. This is reflected in a remarkably rich literature, albeit one whose theoretical underpinnings were largely formulated outside the English-speaking world.

Morphology serves as the first methodological approach. Firstly because no other discipline can boast more experience in cognition and greater achievement in comparative tasks. Two million species multiplied times an average of at least ten unique anatomical features have yielded over 20 million terms (names)—five times the vocabulary of all the major languages combined. Secondly, morphology—as the name implies—is the science of form (gestalt) or, more precisely, the interpretation of form. It is therefore 'epistemological' and helps resolve the interplay in analytical-synthetic processes. It provides the framework for all practical endeavor in comparative anatomy and phylogenetics and is the cornerstone of every comparison involving complex systems.

Structuralism, which is closely related to morphology, is treated further below.

(b) *Systems theory* has also developed from biology. It can be traced back to my teachers von Bertalanffy and Paul Weiss in Vienna and deals with the causal relationships in complex systems, in particular their interdependencies. In contrast to morphology, which has only minimally influenced the cultural sciences, systems theory has permeated almost every science. This includes the study of cognitive processes.

(c) The term '*gestalt*' comprises more than today's colloquial 'form', 'structure' or 'pattern'. It comes from the German word '*gestellt*' and describes the act of forming. Accordingly, it incorporates the viewer him- or herself, i.e. the person transforming perceptions into gestalt. This recursive concept has been absorbed virtually unaltered by other languages. It also harbors a theoretical component that spread from Austria and southern Germany, initially through Ehrenfels, Koffka and Wertheimer, in the form of *gestalt theory*. This became an important concept in the early twentieth century. The theory goes a step beyond the field of psychology but remains rooted in the phenomenon of perception.

In the upcoming topics, these concepts form a troika for delving into the synoptic tasks of cognition and the structuring of theories.

1.2.2 Structuralism and Functionalism

The methods of morphology, because they are supported by gestalt perception, were soon no longer scrutinized: morphology was once again practiced intuitionally and taken to be equivalent to comparative anatomy. Its theory remained rooted mostly in the German-speaking world, but its subject matter blossomed and became indispensable. While this was insufficient to trigger a true renaissance, it did usher in a thematic substitute.

The task here is to juxtapose (a) structuralism with (b) its counterpart (functionalism) and to recognize the (c) relationship between the two.

(a) *Structuralism* originated from the French linguistic tradition and was expressed by Lévi-Strauss (1968) before being picked up by the developmental psychologist Piaget (1973) and ultimately reaching contemporary, English-speaking authors.

Structuralism presages a relationship with morphology. It cites Geoffrey Saint Hillaire and the English authors Owen, Gregory Batson, D'Arcy Thompson and Waddington, refers once again to holism, transformation, self-regulation, organization and order, and espouses two important views. First (i) it demonstrates that, beyond the functional explanations provided by Neodarwinism, additional 'inner principles' must be at work that help understand the product of evolution. Second (ii) it highlights that—beyond the diachronic, explanatory approaches to the problem—synchronous, 'descriptive' approaches must be postulated in order to better comprehend the phenomenon of evolution.

(i) The former view involves terms from holism and systems theory, superimposed on Neodarwinism.
(ii) The latter view approximates the distinction between cognitive and explanatory approaches, two concepts that form the main body of this book and that are juxtaposed in Chaps. 3 and 4 versus Chaps. 5 and 6. The proponents are mainly authors from the 1980s such as Ho (1984), Hughes and Lambert (1984), Rieppel (1990) and Webster and Goodwin (1984). The methodological difference between the two approaches has never been clarified. One might think that manners of speech, 'ways of seeing', are involved. 'Rational morphology' may come to mind, although 'rational' is a misnomer because the word also means 'reasonable', 'practical', or even 'purposeful'. Defining the methods turns out to be the key issue, and this book is devoted to that task.

(b) The term *functionalism* encompasses the fundamental paradigm behind mainstream natural sciences per se, namely ongoing reductionistic causalism. In the case of evolutionary theory, this spawns the expectation that random mutations and environmental selection alone provide a satisfactory explanation.
(c) We clearly need to consider *the interplay* between the two interpretational directions. The term 'functional structuralism' has even been introduced to reflect this. Functions are naturally attributed to structures. Equally, in the macroscopic realm—whether it involves physics or cultural products—functions are never perceived without the attendant structural elements. Only when descending into the realm of microphysics do the borders between functions and structures (waves and particles) become more fluid. I caution against mixing the two perspectives: even their methods differ fundamentally. This is a core issue of this book.

1.2.3 On cognition, Explanation and EE

EE—evolutionary epistemology—underpins the theoretical framework espoused here. The theory of evolution takes on a core role because EE itself can be understood as a satellite theory of the evolutionary perspective.

Accordingly, (a) EE can help to differentiate the processes of (b) cognition and (c) explanation.

(a) Evolutionary epistemology studies the hereditary basis of the human psyche, of our social and—more interesting in the present context—our cognitive performance or faculties. It holds that this performance is the product of our adaptation to extra-subjective reality. It also incorporates the history of human organization. This theory was anticipated by Ernst Haeckel, brought to paper by Konrad Lorenz and then further developed in the 1970s by Lorenz, Campbell, Vollmer and myself. Chaps. 2 and 3 are devoted to this approach.

Evolutionary epistemology is also closely related to another biological discipline, namely comparative ethology, which itself again presupposes the theory of evolution. EE has also become integrated into numerous other disciplines in the natural sciences and humanities (Riedl and Delpos 1996b), ranging from mathematical theory and physics to law and political science. It has also granted me numerous insights into the methods of science (Callebaut 1993) and, importantly, the recognition that the methods of cognition and explanation differ considerably from one another.

(b) *The process of cognition* is 'ratiomorphic', i.e. it resembles reason but is clearly not rational in the strict sense. Rather, it operates on a largely preconscious level and is directed at recognizing rule-based or 'lawful' simultaneity. This has been only poorly studied, and its outputs are therefore experienced as being gained intuitionally (see Chaps. 3 and 4). Clarifying this conundrum is doubly useful. First, it clears out the misconception that the method is unscientific (based on a perceived ignorance about its structure) despite its recognized, fundamental role and high reliability. Second and most important, this method is eminently suited for dealing with complex phenomena.

(c) Although *the explanatory process* also has a ratiomorphic basis, it ultimately operates consciously. It is directed at detecting and unravelling lawful successions of events: it is considered to be well studied, its outputs are experienced as rational constructions, it is downright paradigmatically held to be scientific, and it alone is considered to be acceptable in the framework of the so-called exact natural sciences. Chaps. 5 and 6 compare and relativize the explanatory and the cognitive process, especially because the former depends on the latter and is itself less suitable for deciphering complexity.

1.2.4 Biology as the Conceptual Framework

In retrospect, biology has clearly delivered most of the tools required for tackling complex systems. This reflects biology's unparalleled experience in dealing with complexity, ultimately under three conditions. These conditions need to be set in relation to their consequences.

Biology experienced (a) a schism of methods early on. It also strove (b) to elucidate the cognitive processes and (c) to rebut the allegation of 'biologism'. This effort exposed (d) the gradients behind the methodological schism, yet without (e) intermixing the methods themselves.

(a) The *methods of biology* lie at the crossroads between those of the inorganic and cultural sciences. Biology's physiological disciplines, down to molecular biology and biophysics, operate in a causalistic, explanatory manner. Their underlying (yet unattainable) ambition is to trace even the most complex phenomena back to the laws governing matter. In contrast, anatomists and systematists operate morphologically and comparatively. Their 'hermeneutic' approach follows—as this book will demonstrate—a recursive method of 'reciprocal enlightenment or illumination', an approach that characterizes the humanities as well.

At this point, biology starts to unravel, revealing thinly veiled misunderstandings. This is the impetus to untangle the misleading schism along with its methodological implications. This second ambition is to dismantle the schism between the exact natural sciences and the humanities. Taking the task of deciphering complexity seriously means not shying away from the ultimate litmus test—addressing the complexity of the sciences themselves.

(b) *Cognitive patterns and natural patterns* must, asserts EE, be complexly interrelated. This is ultimately an epistemological issue involving the nature of knowledge.

Clearly, many of us have become diligent citizens of our planet and equally capable researchers without having delved into questions of cognition. Nonetheless, we can all benefit from recognizing the process behind grasping a situation and gaining knowledge. The heart of the matter is the degree to which the structure of our thought processes mirrors that of extra-subjective reality.

(c) The term *biologism* refers to a very specific type of allegation. It is directed against a worldview holding that mental and social phenomena are attributable entirely to biology. This critique is both imprecise and unfair. Of course, all laws governing 'deeper' layers permeate the successively 'higher' ones. This makes them necessary but by no means sufficient to comprehend and explain the higher tiers.

The laws of physics and chemistry indisputably operate at the organismic level. They prove to be vital for all life processes. At the same time, their action alone does not define life. Perception, activity and needs are new, superimposed qualities. Every level must be viewed on its own. In many cases, no traces of the newly emerged qualities in a particular tier can be detected in the constituents of the preceding tier. Logic, religion and literature, for example, have no roots in the animal kingdom. Nevertheless, the laws of biology are essential for the existence of humaneness, for social and cultural traits. They are necessary—yet at the same time insufficient—to perceive and explain humanity.

Putting the above to pen may seem almost trivial, but the central role played by biological methods and biological insights make this perspective very helpful indeed. Overall, it is fair to say that biology has attained a new status.

(d) *Three gradients* differentiate the two methods along the entire spectrum of complexity in the sciences.

This context is evident in (d1) the conventional arrangement of the sciences, in (d2) the degrees of complexity in the inorganic realm, and in (d3) the methodological overlap.

- (d1) The *conventional arrangement* of the sciences represents a gradient. For every science, that gradient extends from a typical or core manifestation to some irrelevant or inapplicable endpoint. This is equally valid for the causalistic method of inorganic chemists and physiologists as it is for the hermeneutic, comparative approach of morphologists and humanities scholars (Fig. 1.1).

Here, 'conventional' disciplines are understood as being those traditional university subjects that adhere to the particular horizons of complexity defining their objects. They develop theories about a specific cross-section of the world. The concept of 'longitudinal theories' will be introduced later to refer to efforts that, like the theory of evolution, chaos theory or systems theory, seek to unite all the levels under a particular point of view. Some of these will prove useful here, yet belong to another type.

The modus operandi of physics is causalistic. All phenomena are attributed to the four physical interactions (weak and strong nuclear forces, electromagnetism, gravity). Such energy transformations indisputably underlie forms of human endeavor such as the arts. At the same time, explaining a school of art, such as that of Raphael, based on nuclear forces and gravity would miss the point. Rather, the analysis would start by comparing artistic creativity and invoking gestalt perception. In contrast, the analysis of elementary particles in physics would benefit little from introducing gestalt perception—even if shapes and forms are visible in the bubble chamber.

A 'physics of culture' or an 'atomism of cultures' would do little justice to physics as a discipline. A 'culture of the inorganic realm' or a 'comparative culture of atoms' would be equally senseless (Fig. 1.1).

When viewed from such inapplicable endpoints, this seems so self-evident as to render the entire matrix trivial. Nonetheless, the gradients of complexity of the

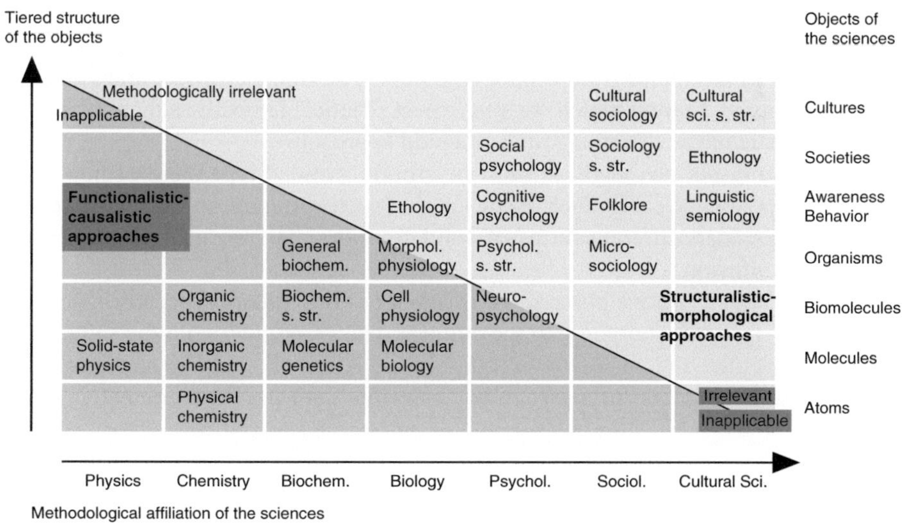

Fig. 1.1 Arrangement of conventional sciences based on their level of complexity and the degree to which the causalistic versus morphological method is applied. The irrelevant or methodologically inapplicable ends of the matrix are indicated. Morphological treatments of the objects of physics are irrelevant (methodologically inapplicable). In causalistic treatments of cultural objects, complexity is irrelevant (s. str, sensu stricto)

objects treated by these disciplines (Fig. 1.1) reveal a transitional field for the causalistic and the morphological approaches. Biology takes a central position in this field, where the methods either conflict with or augment one another.

- (d2) The *degrees of complexity in the inorganic realm* are also instructive. Starting with structural chemistry and proceeding to mineralogy, geology, geomorphology and physical geography, the structurally perceptible features gradually increase in size until, ultimately, they dominate entirely. In that same sequence, the causalistic perspective wanes and the structural perspective waxes.

  In mineralogy, insights may be prompted by gestalt perception, but many of the structures can still be causally attributed to the shapes of the component molecules and ultimately to the laws governing chemical bonds. In physical geography, however, at the other end of the series, gestalt perception dominates. Clearly, the form and position of today's continents can be causalistically attributed to the distribution of masses in the Earth's crust, to fluid and tension forces. Nonetheless, factoring all this in would yield less new insight than a good map of Africa.
- (d3) This leads to the issue of *methodological overlap*, to the possibility of examining the same object both causalistically and structurally. This will be treated more in-depth later and is an essential perspective for a deeper understanding.

  It suffices here to state that this possibility arises when the prerequisites for both gestalt perception and for causality are fulfilled. On the one hand, an object must be sufficiently differentiated to warrant a comparison within a 'field of similar forms'. This degree of complexity is already evident at the level of mineralogy. On the other hand, making useful statements requires that the sector we seek to explain causally not be too far removed from elementary conditions. In principle this can extend up to the complexity level of cultural sciences, as demonstrated by examples from economic theory. The limits of a sector are defined by the bounds beyond which we can expect practicable results either only from a causalistic or only from a gestalt-oriented approach.

(e) *Intermixing the two methods* must be strictly avoided regardless of how well they supplement each other. Both involve such different approaches and are subject to such different forms of validation that any amalgamation can only cause confusion.

From the cognitive perspective this is a curious situation because complex things—from the inorganic to the cultural realm—comply with the same laws that governed the make-up of our brains. After all, natural scientists and humanities scholars still prove to be 'crossable', at least experimentally.

Chapter 2
The World and Cognition as a Problem

From the biologist's perspective, the phenomenon of perception as well as the problem of cognition need to be examined in a manner that philosophers will find unusual. This is because, for biologists, cognition is already relevant in animals, whereas philosophers set their sights on humans, on the semantics and syntax of our culture. Cognitively, this is accompanied by a subliminal pursuit of truth. It turns out, however, that consciousness followed by deliberate, critical reflection were prerequisites for making cognition and perception into phenomena that unmask the deficits behind knowledge and truth.

This calls for distinguishing (Sect. 2.1) what appears reasonable to us, (Sect. 2.2) how knowledge is gained and (Sect. 2.3) what kind of knowledge we in fact possess.

2.1 What Appears Reasonable to Us

Raising the question why human reason harbors such unreasonable streaks reveals that two different types of reason are involved. The first simply refers to our clear-mindedness, setting us apart from 'dumb animals'. Ever since Kant, philosophy also understands this level as being characterized by the development of concepts and the intellectual capacity to recognize relationships and draw conclusions. Irrationality, in contrast, encompasses the unreasonable behavior that impedes success and reduces the quality of life—key concerns in this book.

Making the world and cognition into a problem is quintessentially human. Fertility figurines fashioned tens of thousands of years ago and funeral rites that date back 40,000 and 60,000 years show that this characteristic, metaphysical problem must have originated very early on. Even back then, it was coupled with the existential question of where we come from and where we are going.

R. Riedl, *Structures of Complexity*,
https://doi.org/10.1007/978-3-030-13064-0_2

For some, this issue may have lost urgency because it doesn't seem to reach much beyond the perceptible natural world we encounter on a daily basis. Nonetheless, open questions do remain. And pondering questions that go beyond personal knowledge is clearly a mark of every thinking person. If our cosmos arose from a Big Bang, which I posit as still being the most acceptable theory, where then did all that energy come from (especially without the pre-existence of space and time)? Such questions can be raised without any expectation of a definitive answer.

This means that metaphysics cannot be sidestepped. Pursuing 'speculative metaphysics' promises little gain because it strives to derive the world from the highest of supposed principles. 'Inductive metaphysics' (Hartmann 1964), however, provides a way forward because it reveals the preconditions that we accept when we—in every branch of research—extend our query from the known to the unknown. Metaphysics therefore accompanies us, consciously or not, in our inevitable ambition to ask 'off-limit' questions (for example what caused the Big Bang?). Nonetheless, we need to accept that it is not the most dependable of guides.

The problems surrounding how we understand the world are of a more fundamental nature. They begin with the contradictions between what our hereditary 'cognitive apparatus' suggests and what conscious reflection concludes. Accordingly, the problem hovers somewhere between experience and reason, empiricism and rationality. This calls for taking a position.

The first step is to deal (Sect. 2.1.1) with consciousness, (Sect. 2.1.2) with the validation of knowledge gain and (Sect. 2.1.3) with the difference between forms of perception and language. This provides the basis for differentiating our questions and solutions.

2.1.1 What Arose with Consciousness?

How did early humans handle the many puzzles that surrounded them in everyday life? Who could be behind all the hardships and indignities our ancestors faced and with what entities did one have to arrange oneself or beseech? Wasn't there some sort of intent—much like their own intentions and the intentions of those around us—which bore responsibility for all these adversities?

This gave rise, either by revelation or rumination, to the gods, which stood apart from the recognizable natural objects. These were initially conceived as being demons, then endowed with all the good and especially all the bad characters of humans, finally transformed into loving fathers, whereupon humans discovered their own god-like nature. The development of this worldview, however, was marked early on by critical voices, for example by those of pre-Socratic thinkers at the very roots of our culture (compare Capelle 1968).

A relevant sentence by Anaximander (611-545 BC) has been preserved. "I write what I believe to be the truth, because the lore passed on from the Greeks appears to me to be too much and too absurd." And there we have it: the problem of truth has been raised. It came to take on many iterations and has accompanied us ever since.

This issue, with its many inconsistencies and contradictions, can also be formulated as a dilemma (see Sect. 2.1.3).

(In the philosophy of the modern age, Kierkegaard elaborated this into a problem of human existence, Nitzsche and Dilthey into a 'life philosophy', Sartre into a type of nihilism, Heidegger and Jaspers into an 'existential philosophy'. Its influence on literature was considerable, on the sciences minimal. This need not be pursued further here.)

For our topic, the above development raises the question of how we actually perceive things and gain knowledge. In the language of philosophy: "How do the defining characters of the object transpose themselves onto the subject?"

2.1.2 The Conceivable Validations of Perception and Knowledge Gain

In humans, a relationship of some sort must exist between extra-subjective reality and our senses and thought processes. Interestingly, cultural history reveals only few attempts to develop such a validation into a well-reasoned system.

As an introduction, I juxtapose the (a) transcendent, (b) the transcendental and (c) the evolutionary methods.

(a) The oldest attempt was developed by Plato (427-347 BS) in his '*theory of ideas*". In brief, he assumes that, beyond the physical world, there are principles in which both the objects constituting extra-subjective reality and the 'soul' of the subject 'participate' in. These ideas behind all things are mirrored in our concepts. This is referred to as a 'transcendent' principle, i.e. one standing 'above and beyond' the physical world. This has survived until today in the traditions of idealistic philosophy and Christianity.

    Aristotle (384-322 BC) presented a more worldly interpretation. He assumed that the 'particles' that make up our senses are similar to those of the outside world, enabling a match. Whether this represents a complete theory of cognition is open to debate. Nonetheless, this assumption principally underlies the work of all natural scientists.
(b) Kant (1724–1804) developed a theory based on the possibilities of gaining experience. He termed this attempt at establishing a foundation 'transcendental'. Accordingly, all knowledge is gained via the senses, but this knowledge must be anchored in a perception of spatial and temporal continuity (intuition) and in categories of perception that must be present a priori (i.e. in advance). These are the prerequisites for experience itself. They themselves, however, cannot be derived based on experience itself. This view had an enormous impact on the subsequent history of intellectual endeavor despite being unable to validate this specific a priori.
(c) Biologists find it difficult to accept that a clear prerequisite for engaging with the world cannot be substantiated or validated. Ernst Haeckel foresaw the solution,

and Konrad Lorenz rose to the challenge when he was appointed head of the Department for Comparative Psychology in Königsberg, in the 'after-shadow' of Kant. In biology, doesn't the phrase 'given to life-forms in advance' mean 'innate'? Accordingly, innate forms of perception must fit the world for the same reason that a fish's fin fits the water even before it slips out of the egg (Lorenz 1941). The a priori in ontogeny can be *a posteriori* learning products of phylogeny—a product of adaptation. Evolutionary epistemology was born and began to spread with the books of Lorenz (1973a), Vollmer (1979) and Riedl (1980).

The adaptational explanation for our innate forms of perception and our categories proved to be the prerequisite for the solution, albeit still insufficient. It soon became supplemented by a constructivist element (Riedl 1995a) in the sense that the history of every biological system sets limits to its own adaptational possibilities. This perspective shed lights on deficits in our own adaptations.

Philosophers (Engels 1989; Pöltner 1993 and others) were skeptical about validating cognition based on the evolution of organisms (overviews in Riedl and Wuketits 1987). Many researchers, however, derived clear benefits from this approach (Riedl and Delpos 1996b). As has been demonstrated time and time again over the course of history, philosophical problems can be resolved scientifically.

2.1.3 Forms of Perception Versus Communication

Beyond dealing with the conditions of cognition, we need to examine the conditions behind our language or, more specifically, behind the thought processes underlying language. This aspect remarkably influences how we believe we need to see the world and how we engage with it.

For an easier orientation, the issues of (a) adaptation (b) its limits and (c) the forms and the development of selection criteria are discussed separately.

(a) It proves to be relatively easy to demonstrate that human perception has arisen *adaptively*. Among all conceivable programs that could have arisen in our cognitive apparatus, those that promoted survival under the conditions early humans faced in everyday life have gained foothold. This pertains both to the perception of spatial and temporal continuity (intuition) in the sense of Kant and to the categories (cognitive processes) that we in our terminology refer to as forms of perception.

In our daily lives, it has proven reasonable to reckon time as being one-dimensional and, independently thereof, to view space as being three-dimensional. This notion remains operational even though, in mega-cosmic dimensions, it has been disproved by the discovery of a generally valid (i.e. also meso-cosmic) space-time continuum. Our physiological clock, however, ticks in one dimension only, and our own bodies are built based on three spatial axes that we perceive independently of time. The same proves to hold true for cognitive processes as well. I have described these processes in the form of four hypotheses, namely the hypothesis of the 'apparently true' (German: *anscheinend Wahren*), of the 'comparable' (German: *Ver-gleichbaren*), of the 'fundamentally causal' (German: *Ur-Sachen*) and of the 'purposeful' (German: *Zweckvollen*) (Riedl 1980). Their concerted action forms the innate, 'ratiomorphic' cognitive apparatus, an apparatus that we are largely unaware of (Lorenz 1973a). Ratiomorphic because it operates in a manner that resembles reason but is actually entirely unrelated to it. This insight is a cornerstone for the remaining chapters.

This is also valid for our adaptations to social life, whether it be greeting behavior, displays of dominance and submissive behavior, killing inhibition, etc. (Lorenz 1974a; Eibl-Eibesfeldt 1984). These help structure human groups and thus play a role in the survival of the species. The focus here, however, will be on cognitive processes.

(b) The *limitations* imposed upon these adaptations are somewhat more difficult to recognize. They are therefore treated separately and in the context of the above four innate hypotheses (Chap. 3; Sect. 5.1).

(c) A third issue is to differentiate *the development* of the forms of perception from those of communication, language and logic. This calls for examining the forms of selection and the constraints—the limits of adaptation based on the system conditions in the organism.

The best approach is to separately present (c1) forms of perception, (c2) communication and (c3) the consequence of the differences between perception and language.

- (c1) The *forms of perception* help us to react correctly to extra-subjective reality. This is an adaptive process driven by the need to attain 'correspondence' with the environment. Then, secondarily, the conditions of internal organization—the functional interactions or 'coherences' within every organism—decide what specific adaptations can be realized and, if so, in what manner.

  This can best be illustrated based on an anatomical example. Being able to fly offers great advantages. Many insects and birds have successfully responded to that selection pressure. Parts of their organization therefore adhere to aerodynamic principles (Fig. 2.1). Nonetheless, the coherences within various body plans, for example those of a dragonfly and a swallow, have led to entirely different solutions.

  This underlines that our own problem-solving processes, much like the dragonfly's solution to flying, merely represent one solution out of many. The same type of constraints also governed the development of our sensory system. We can never develop the 360 degree, all-round vision attained by the dragonfly. A third, parietal or pineal eye has also failed to prevail (probably due to 'wiring' problems): the rearview mirror provides an ersatz. Storing two series of numbers simultaneously would require us to have two memory units. Such limits to higher performance are discussed later.

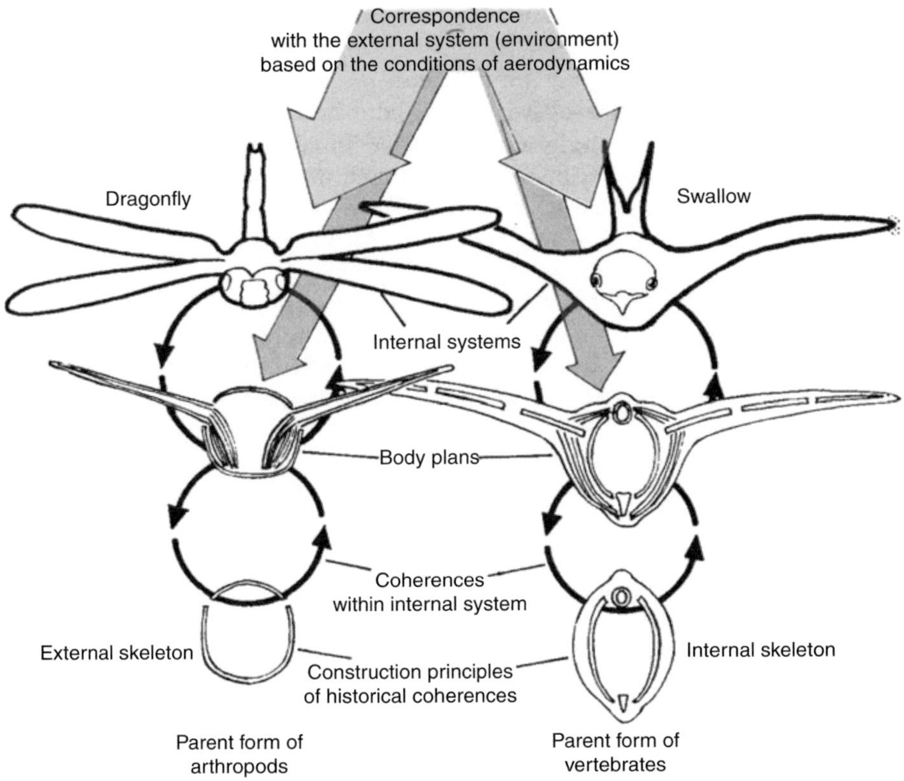

Fig. 2.1 Correspondence and coherence. The interplay between the external and internal conditions based on analogous solutions to flying. Note that the conditions of correspondence, namely of adhering to aerodynamics, are almost identical for the dragonfly and swallow, but that the conditions of coherence of the body plan and of the construction principle lead to different solutions (after Riedl 1992a, amended)

- (c2) In the development of *communication*, the priorities of the selection criteria are reversed. Communication involves the interaction between two individuals. It begins with sexually reproducing, single-celled organisms and has remained preserved—chemically coded—in the mutual recognition of sperm and egg cells in humans. The selection pressure lies in the coherences, in achieving reliable recognition and reliable understanding (Fig. 2.2).

  The employed codices are fully detached from the environment (other than being transported within the respective medium). Originally, nothing was communicated about the environment itself. This changed only much later with the development of the remote senses, ears and eyes, body language and ultimately acoustic signals. This enabled some birds and mammals to communicate messages about the environment itself, for example 'caution, predator approaching!' or 'quick, food!'

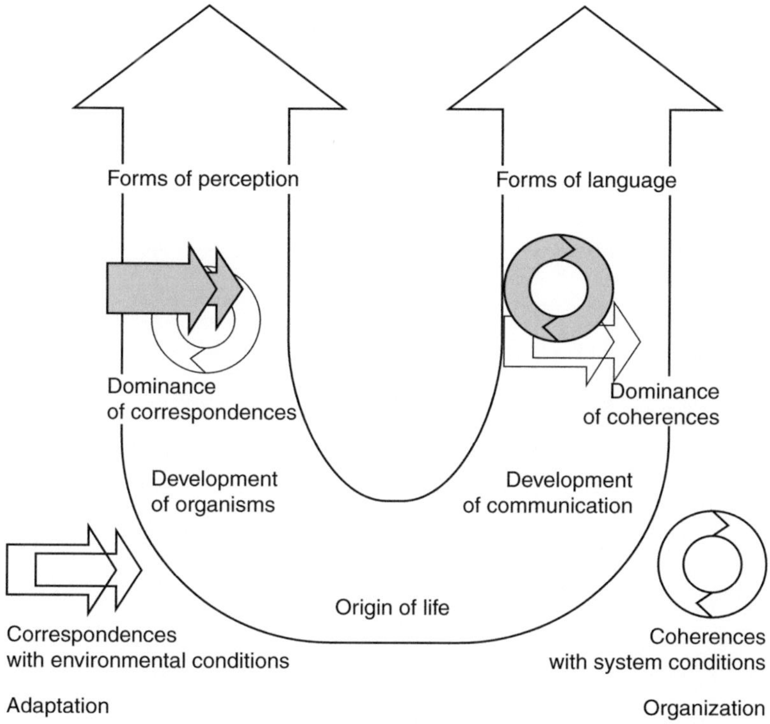

Fig. 2.2 Dominance of the correspondence and coherence conditions in the development of organization versus in the communication within the internal system of the group (sketch of underlying principle; the biological example in Fig. 2.1). The illustration of the developmental pathways in Fig. 2.3

Initially, body language also merely expressed the individual organism's condition, which can be received as a message by the observer. And the signals later used to send messages about the surroundings bore no similarity to the approaching enemy itself or to the food source. This very same symbolism has been retained in human language as well. The few onomatopoetic words (such as 'bump' or 'sizzle') play a subordinate role.

This explains why one and the same message, for example 'a threat has been sighted', is coded very differently by crows, by my dog, or even in English versus French. Understanding the correspondence with the environment requires knowing the codices that have developed from the coherence conditions. Figure 2.3 provides an overview of the developmental trajectories behind the forms of perception and language. The details are discussed later, but the necessary terms and their correlational framework are already introduced here.

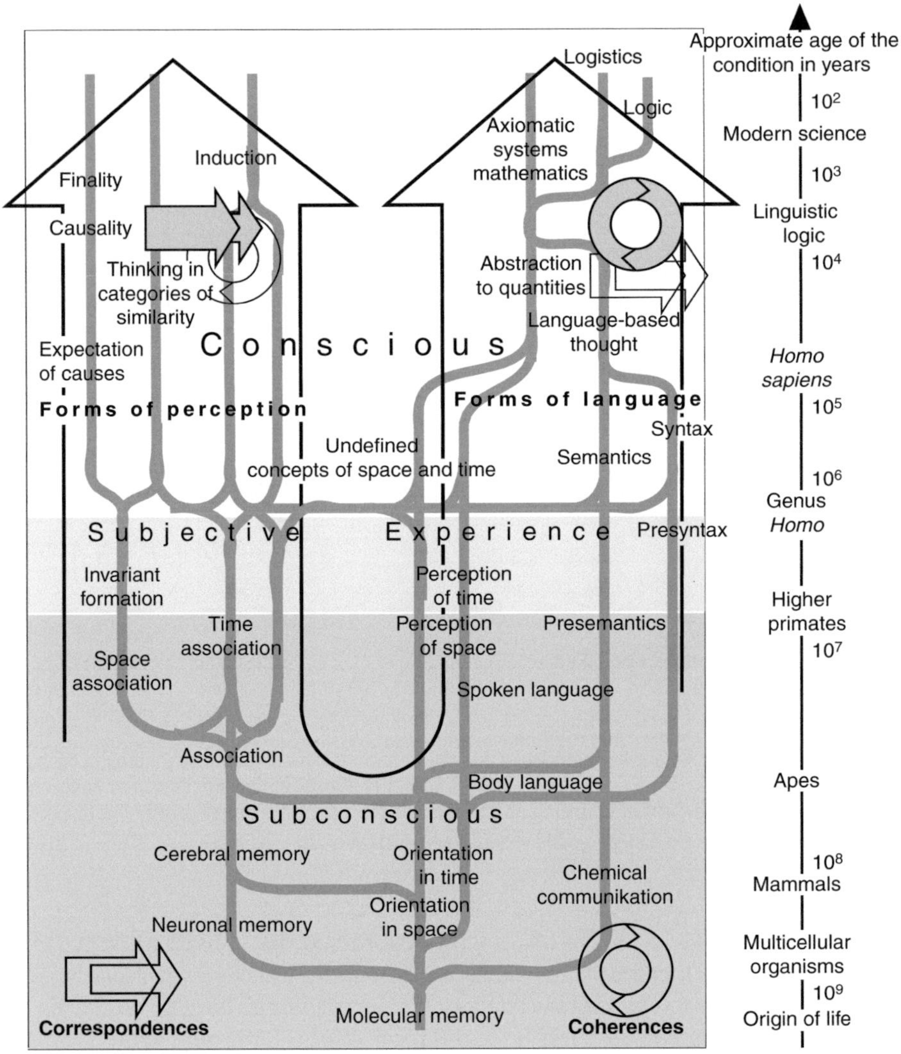

Fig. 2.3 The development of the forms of perception and language based on stages in history, the key developmental pathways and their links. Note the time-scale and the dominance of the correspondence and coherence conditions

- (c3) This differentiation of the two *evolutionary processes* is crucial in addressing the cognitive issues. It turns out that each key epistemological question has an alternative solution (Riedl 1994).

  This is equally valid for (i) the origin of the knowledge, (ii) the 'primary' reality, (iii) the primary sources of truth, (iv) the causes behind things, (v) the primacy of induction and deduction, as well as for (vi) adaptation or organization. Accordingly, we can place our trust (i, ii) either in perception or in reason and ideas in their different guises; (iii) in the correspondence of perception with reality or in the contradiction-free nature of logic, (iv) in attempting to understand the world the world from the perspective of drives or of purposes, or (v) in theories versus evidence, or (vi) in adaptation versus a conceptual framework.

Any further elaboration would require a lengthier interpretation of the history of philosophy than necessary in the present context. Nonetheless, the interplay of both alternative positions is evidently essential for making progress. I therefore provide (Fig. 2.4) only a single example from the 'sources of truth' to illustrate the general problem. This shows that accepting one or the other of these alternatives leads to conceptual burdens or historical loadings (German: *Bürden*) (B) in further developing each perspective. This lead to constrictions or constraints (C) and further to a determination/disposition (D)—to what I call BCD series—in one of the alternative epistemological positions.

Summarizing the BCD series of the six fundamental epistemological questions according to their alternatives (Fig. 2.5) yields two major pathways—an empirical and a rationalistic one—in our understanding of knowledge. Throughout cultural history, they intersect only rarely (for details, see Riedl 1985).

This type of representation is unusual in the history of philosophy (for an orientation, see Eisler 1927–1930; Ritter as of 1971; Mittelstrass 1980–1984; Sandkühler 1990 or Vorländer 1990). The traditional emphasis is on the internal coherence and originality of the individual philosophical systems. My approach, in contrast, attaches greater importance to the continuity of the underlying paradigms and, above all, to capturing the relevant terminology. It goes without saying that empiricists recognize the existence of reason and that even the most extreme rationalists do not ignore experience. Nonetheless, each of the two positions typically prioritizes one of these mindsets.

Logically, empiricism defines reliability as being 'derived prognoses that can be seamlessly confirmed by experience'. This is a cybernetic process involving incremental optimization (this paradigm of 'empiricism' is addressed more specifically below).

Our equally innate forms of communication raise entirely different expectations. Here, what we consider to be logical and reliable is that which is contradiction-free within a particular conceptual framework. This determines our expectations in an intellectual current termed 'rationalism'. It has its roots in classical logic, namely in the desire to use rational means to remove the contradictions arising from our ratiomorphically underpinned language. Major philosophical systems as well as mathematical theories are also born of this desire, often with the irredeemable ambition to prove the internal consistency of the system (compare e.g. Gödel 1931).

Too little consideration has been given to differentiating between those criteria selecting predominantly for correspondence versus those for coherence. This also helps explain why the empiricist and rationalistic interpretations have historically remained cemented in such conflicting positions in their pursuit of truth (Riedl 1992a). The following account is primarily rooted in empirical experience but questions the veracity of logic when required.

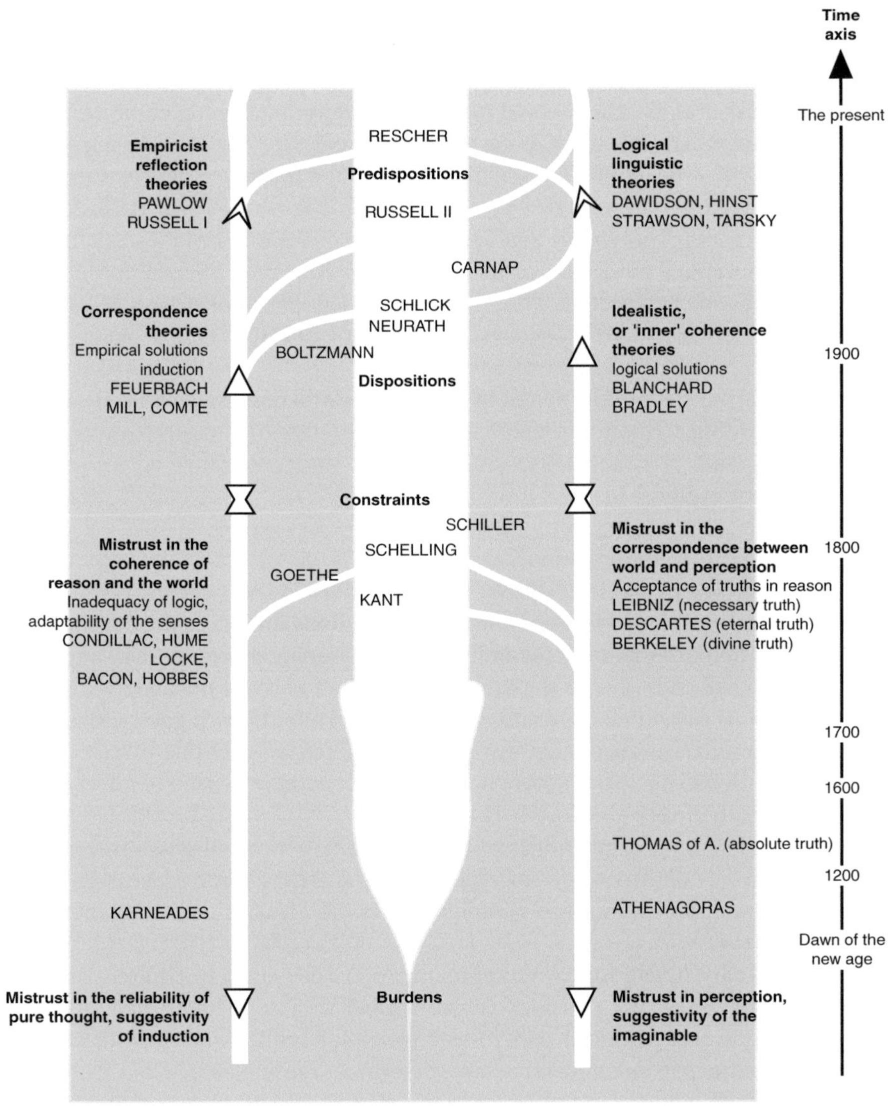

Fig. 2.4 Series consisting of burdens, constraints and dispositions (BCD series) in the history of philosophy based on the example of 'sources of truth'. Note, beyond the few transitions, the different burdens, the approaches based either on the senses or on reason, the development of two major flows with their one-sided constraints, as well as the resulting dispositions and predispositions, which continue to pursue their separate routes (after Riedl 1994). Overall framework and symbol explanation in Fig. 2.5

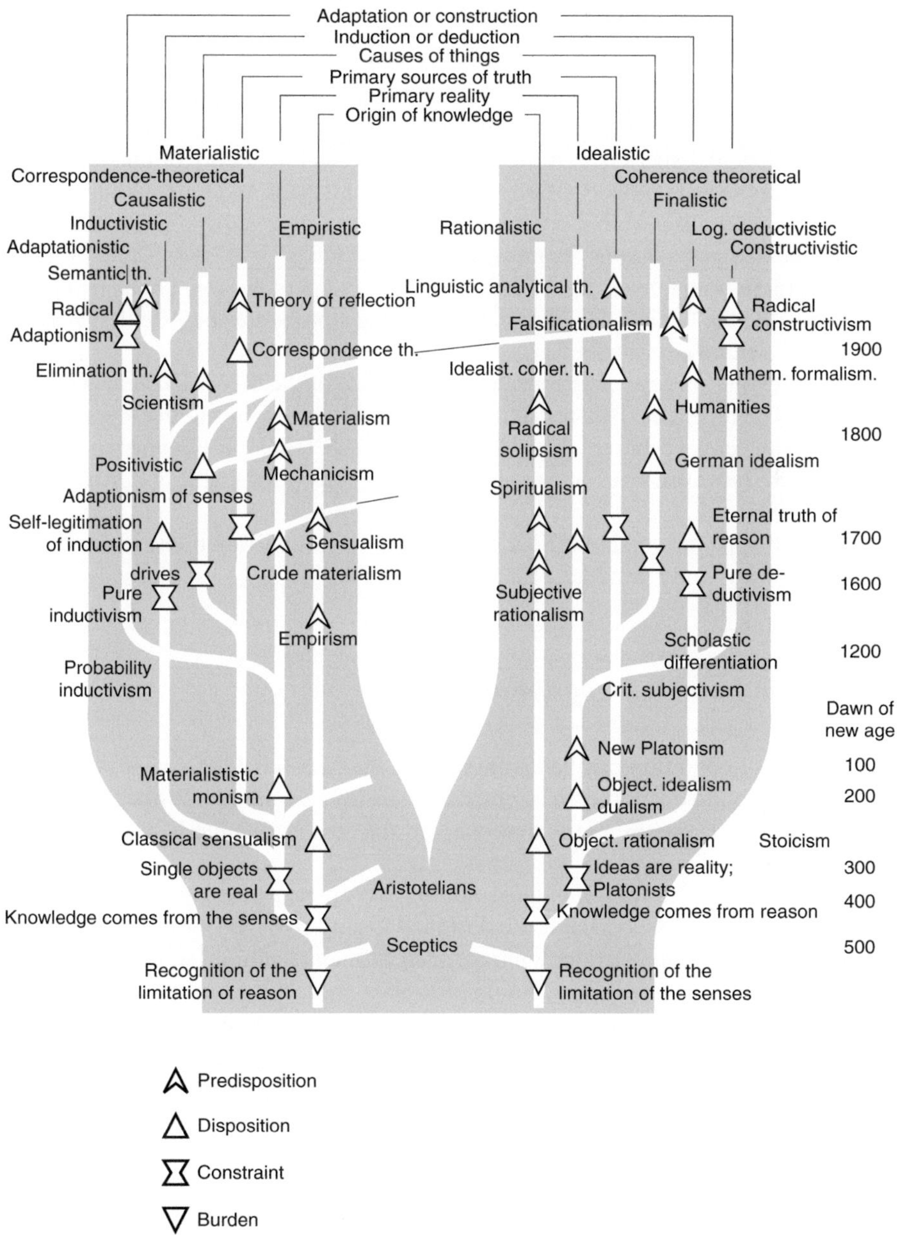

Fig. 2.5 Development of empiricist versus rationalistic axes in cultural history according to how they developed based on being governed by innate forms of perspective versus by communication and language. Represented by the pathways of alternative solution attempts. The few transitions are now indicated (compare Fig. 2.4; after Riedl 1994)

2.2 How Knowledge Is Gained

EE distinguishes itself from the transcendent and transcendental position on knowledge gain by its empirical approach. No statements should be made that cannot be confirmed through experience. Admittedly, EE is a satellite theory of that on evolution: it presupposes the theory of evolution and both stands and falls with it. Today, however, evolution is a well-supported phenomenon—despite certain shortcomings and despite the reservations expressed by 'creationists' and once even by Karl Popper (a mere 'metaphysical research concept'). Popper later renounced that position. I consider the theory of evolution to be the most probable one (Riedl 1996) and as being sufficiently established.

Clearly, evolution cannot be repeated. In fact, one of the theory's tenets is that its path cannot be repeated. This restricts experimental approaches to the level of species and other close relatives. This book's approach relies on circumstantial evidence which, however, consolidates itself into degrees of probability bordering on certainty.

The next steps are to show (Sect. 2.2.1) how cognition functions, (Sect. 2.2.2) what effect it has and (Sect. 2.2.3) what promotes it.

2.2.1 The Levels of Cognition

Three levels of cognition can be distinguished: the genetic, the associative and the cultural. They build on and serve as preconditions for one another. While this is self-evident, the interesting aspect is that one and the same principle is at work on all three levels.

This raises the first terminological hurdle: we operate based on a longitudinal theory, but our linguistic concepts are designed to cut across the individual layers of complexity. We tend to consider a phrase such as 'gene learning' to be a metaphor. In fact, however, all knowledge-gaining processes prove to be highly congruent.

Correctly describing the interrelationships requires expanding terms such as 'knowledge gain' and even 'learning' and 'learning success' to include the levels of adaptation and organization in genetic memory. Conversely, it requires accepting that terms such as 'conditions for survival', 'selection' or 'elimination' are functionally applicable to higher levels such as culture as well.

The following discussion introduces the features common to all three levels. These are (a) the structural framework of (b) reciprocal, (c) iterative, (d) spiral processes (Riedl 1980; compare Fig. 2.6).

(a) The *structural framework*: gene learning ranges from adaptively producing the structures and functions of our bodies to developing successful reflexes. As has been demonstrated for organisms from snails to humans, these reflexes can be conditioned, i.e. linked with additional perceptions.

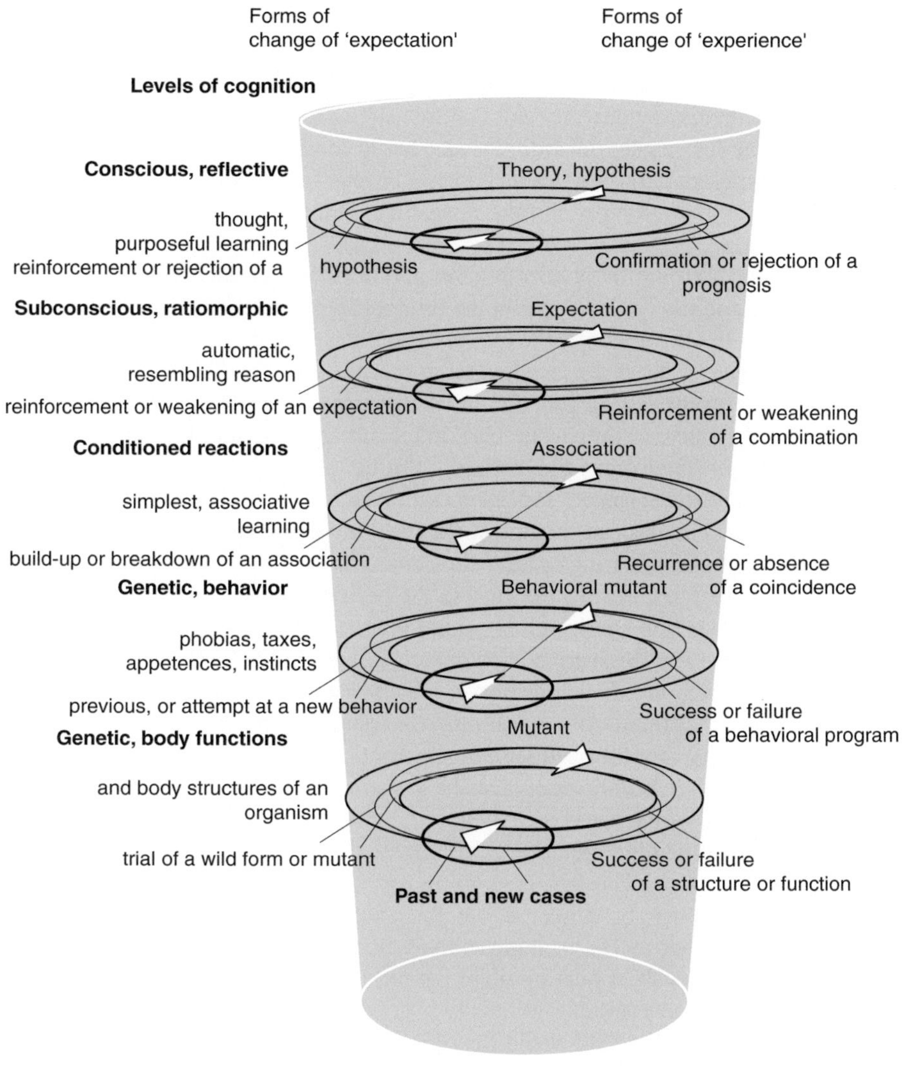

Fig. 2.6 The levels in the spiral process of cognition and retention of the basic structure of a cycle between experiment (expectation) and experience (confirmation vs. rejection). Note the changing terminology for corresponding processes in each of the five levels

A simple example is our blink reflex, which can already be triggered by a whiff of air across the cornea. This can be linked with a second (conditioned) reflex via simple neural pathways. If, in such an experimental set-up, a bell is regularly rung before activating the air flow, then, after several repetitions, our eyelids already close when the bell rings. Such linkages of repeated coincidences are termed ‘conditioned reactions’ or ‘conditioning’. They form the basis for all associative learning, up to and including unreflected experimentation.

In evaluating such processes consciously, we come to expect that comparable events involve a generally valid relationship. This can be described as invariant generation—a generalization, a trial-based or heuristic operation (Fig. 2.6).

(b) The *reciprocity* in all these levels consists of alternating between an organism’s problem-solving attempts and the respective confirmation or refutation by the environment.

Genome learning involves mutation and selection. A change is imposed on the genes, and the environment in the broadest sense then decides about failure or success. In one outcome, the experiment along with its carrier are eliminated straight-away or at least prevented from reproducing genetically, i.e. excluded from further transmitting genetic information. In a second outcome, the experiment can be selected for (positively), the result adopted and then spread—subject to further ‘environmental testing’—through the population.

In associative learning, repeated coincidences are linked into predictable correlations. This process begins at the level of neural pathways, to the extent that the necessary ‘wiring’ establishes contact, i.e. the architecture enables a linkage. This is not always the case in the peripheral nerve system. For example, the patellar (knee jerk) reflex cannot be conditioned acoustically: the pathways in the ear and thigh are not in contact. Unacceptable food can be addressed solely by nausea via the vagus nerve, which makes a lot of sense biologically. In the brain, in turn, almost everything can apparently be interlinked (including nonsense). If the conditioned stimulus is discontinued, i.e. the expected coincidence is no longer confirmed, then the association is deleted (Fig. 2.7), which also makes a lot of sense.

The same type of association is also involved in cultural learning, yet with a two-fold extension. The first is learning by watching others. This strategy initially arose, haltingly, higher up in the animal kingdom (see references in Bugnyar & Huber 1997), peaking in the schooling of our children. The second involves consciously reflecting upon the process (this is formulated cautiously because reflection embodies all levels of consciousness).

Evidently, one and the same task leads to mutually exclusive solutions—depending on whether we place our trust in forms of perception or in logical reflection. This dichotomy mirrors that presented in Figs. 2.2, 2.3, 2.4, and 2.5.

This was demonstrated in an experiment which tasked test persons with indicating the degree to which they thought a process to be random or systematic. It involved judging the development of a series of events. Specifically, a regular switch from black to white to white was presented every 10 s—with one deviation at the eleventh step. The subjects were called upon to make a decision at the eleventh step (black). The series is designed such that both possibilities are equally (im)probable. Quite reasonably, almost all test persons were initially uncertain, but increasingly tended to expect lawful order, until ultimately deciding at the eleventh step either for lawfulness or for randomness (see Fig. 2.8 for an example).

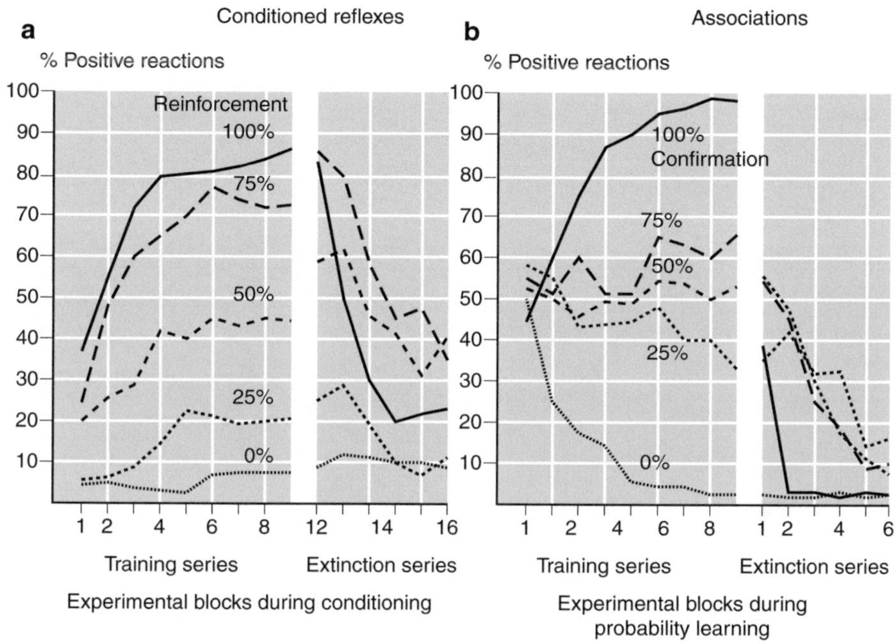

Fig. 2.7 Learning and extinction of reflexes and associations. (**a**) The reaction to a relationship between the two offered stimuli (positive reactions) gradually corresponds to the frequency with which these stimuli coincide. The more regular the reinforcement, the quicker the reaction, but also the quicker the extinction when the confirmation fails to materialize (after Grant and Schipper 1952). (**b**) The expectation that an event will occur (positive reactions) gradually corresponds to the frequency of the reinforcements. The more regular the reinforcements, the quicker the learning, but also the quicker the recognition that the confirmation has failed to materialize (after Grant et al. 1951; juxtaposition from Riedl 1992a)

The course of this problem-solving process showed a dependence on knowledge of and trust in logical-deductive, mathematical operations (Riedl et al. 1991; Wagner et al. 1992; Riedl 1992a). Schoolchildren, at the end of the above series, continue to operate entirely ratiomorphically and cybernetically based on their innate forms of perception. Trained mathematicians, in contrast, operate rationally, based on probability theory (Fig. 2.9). All persons, however, continuously sought confirmations versus disappointments in the same set of information by back-and-forth calculations, albeit involving associatively very different expected relationships.

Framing this process in a more formal conceptual context yields what might be called an inductive-heuristic half-circle (compare Fig. 2.6) in which an expectation, a hypothesis or a theory develops based on known cases. This stands in opposition to a deductive-logical half-circle in which the derived expectations are tested on new cases. The term induction is initially unproblematic—it leads from the specific case to general applicability and rightly establishes the so-called inductive sciences. Deduction, in contrast, harbors two

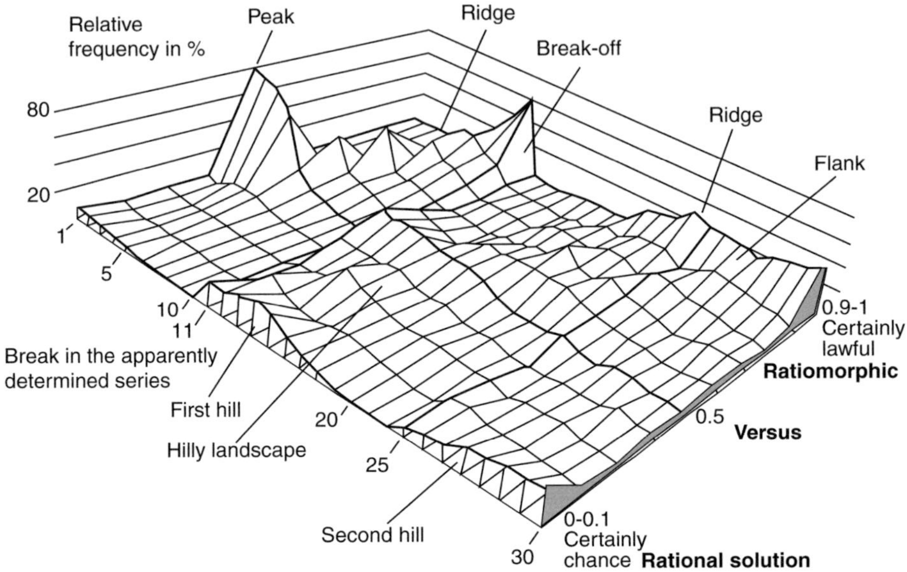

Fig. 2.8 The development of ratiomorphic and rational solutions, illustrated based on the behavior of first-year students of mathematics. At the early phase of regularity (see task definition in text), the majority opinion swayed toward the expectation of lawfulness. The occurrence of the irregularity (event 11) caused a split into two diametrically opposed solutions (after Riedl et al. 1991)

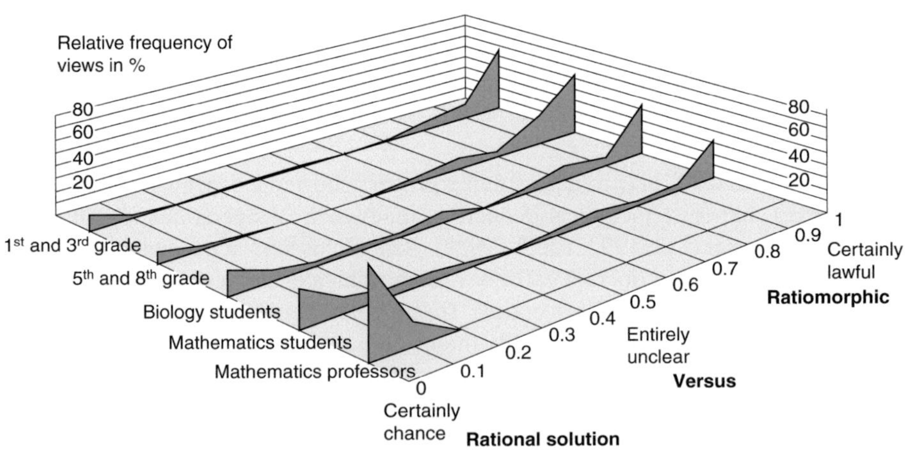

Fig. 2.9 Frequencies of ultimate solutions arrived at ratiomorphically and rationally in relation with the level of rational schooling. The same series of events are considered to be lawful ratiomorphically but to be chance rationally (the case of the mathematics students is detailed in Fig. 2.8; note the different darkly shaded final frequencies; from Riedl et al. 1991)

options: the process can be interpreted as being cybernetic in the case of a ratiomorphic approach, or rational when logical reasoning is applied. This will be dealt with in greater depth later.

(c) *Iteration* is a mathematical term and commonly describes a simple algorithm that helps optimize a solution by recurrent application, as occurs in division. The learning processes on all three levels are iterative. In most cases many—in the realm of genetics very many—cycles are necessary until an adaptive success prevails and establishes itself in a species. Associative learning follows this path more loosely. Reflective learning, in turn, can spawn entirely fantastic ideas, whereby imagination can replace testing the idea against reality. Believing in pure nonsense (Lorenz) therefore remains the privilege of humans.

Note here that biologists are less interested in the deductive process, philosophers of science less in the inductive process. Popper's 'falsificationism' (1973b) even dismissed induction. Nonetheless, everything that is falsified in nature and perishes is unquantifiable and of little interest (a pair of snails produces thousands of progeny: a stable population means that, ultimately, all but two must perish in one way or another). The key question is how nature continuously creates new, successful entities. How the great thinkers arrive at their theories by induction remains difficult to fathom but, again, the key question is how their theories stand up when tested. That is the point on which Lorenz and Popper were unable to reach agreement (Kreuzer 1981, 1982).

(d) The decisive point is that these revolving processes represent *spiral processes* rather than circular logic. In such spirals, the gradient gained per revolution reflects the altered, corrected or gained experienced. There should be no mistake about that. Even the repetition of a perceived coincidence changes the expectational stance. At all levels, these learning processes optimize something. At the genetic level an adaptation or internal organization is optimized, at the associative level a reaction, expectation or theory, at the cultural level a worldview, whether it be wisdom or folly.

2.2.2 What All This Can Tell Us About the World

From an epistemological perspective, merely recognizing this learning process already yields valuable insights into extra-subjective reality. Let's assume we currently know nothing about this world (and, in fact, we do not know all the details). Even then, sound knowledge about the above levels of learning mechanisms would alone suffice to draw three conclusions:

Firstly, the world must be highly redundant, i.e. its manifestations must repeat themselves often, on many cases incredibly often. Otherwise, iterative learning processes could never have arisen or been successful. This conclusion is abundantly supported. Consider the number of leaves we have seen to justify talking about the term 'leaf'. Or the waves that create our image of the sea, or the grains along the seashore that give meaning to the word 'sand'. And there are apparently 10^{80} quanta in the cosmos.

Secondly, things in the cosmos do not repeat themselves in an identical manner. The more complex these things are, the more valid the conclusion. Over the course of our lives we may see 10^6 people, a forest ranger may see 10^8 pine trees. How many leaves has each one of us seen? How many pine needles has the ranger seen? (A solitary pine tree can have around 10^8 needles.) Nonetheless, no two people or pine trees are identical (and, upon closer examination, the same holds true for pine needles as well). This explains why a photographic memory is insufficient and why the ability to generalize is essential.

And thirdly: The world must adhere to the laws that it itself has generated—otherwise, any comparison, prognosis or even memory function would be invalid. Clearly, such laws can emerge and wane, but compared with our lifespans and even with our cultures, they can be taken as constants. Within our lifetimes, 'mother' always means the same thing, going back to the time when it was once called 'mater'. 'Humankind' has been valid for 10^6, 'life' for 3.5×10^9, 'Earth' for 5×10^9, and gravity for probably 1.2×10^{10} years.

The peculiarities of memories are also reflected universally. This begins with the super-long-term memory in the chains of molecules making up genomes, followed by the associatively acquired short- and long-term memory within lifespans. Ultimately, it includes the social and cultural memory held in libraries, which is no longer restricted to human lifespans but defines a constancy that dissolves generations into fleeting moments.

2.2.3 The Purpose Served by Such Knowledge

Learning and memory always help make correct prognoses: correct prognoses are life-sustaining, incorrect prognoses life-threatening. Less dramatically and with an eye towards modern convenience, correct prognoses promote the quality of life, incorrect ones reduce it. Note that this is also valid for all three of the above-mentioned levels.

The term 'prognosis' may seem somewhat overstated considering the 'blindness' of the mutation process and when applied to the cultural paradigms of a society (see below). This again illustrates that our language often lacks adequate terms when longitudinal theories are involved, when lawful order cuts through several levels.

At one end of the scale, the mutation process itself is clearly designed for innovation. How else can we explain that the crucial task of securing successive generations is entrusted to fragile molecular threads. Giant chromosomes bundling 200–300 DNA chains are, in fact, known from somatic cells. Here, every mutation in one chain would be corrected by a hundred others. They would be immune against disturbance, but at the same time not very inventive. Evolution, however, is based on competition among innovations. Those organisms that first resolve and conform to the conditions of the next, greener meadow with a new make-up will be its first occupants.

The same holds true at the other end of the scale of the term 'prognosis'—for the paradigms of a culture. A case in point is the Age of Enlightenment. The underlying prognosis of that concept is that a larger number of people will improve their personal circumstances by applying human resourcefulness. Like biological mutations, most cultural 'mutations' rarely prevail. And because ideologies tend to immunize themselves against their refutation, they often fail to die off in a timely manner. The principle, however, remains the same.

2.3 The Nature of Our Knowledge

It is perfectly dark and rather quiet inside our brains, yet our reality is flooded with light and color, sounds and tones. Were there no eyes to see the cosmos, we would have to conclude that it is perfectly dark. Do we merely dream up colors and tones, do we construct them, and what then is reality?

Donald Campbell (1974) suggested the term 'hypothetical realism' for the perspective taken by evolutionary epistemology (EE), and this term was soon incorporated in its teachings. Only the philosophers seem to be concerned (for example Engels 1989) about seeing one of their central topics wane.

The term 'hypothetical realism' describes an expectational stance. It does not cast doubt on the reality of the outside world, but posits that nothing can be said about the 'true nature' of things (how these really are). Between these two extremes, however, we can assume a similarity between our interpretation and extra-subjective 'real-ity'. How else could we explain why our actions in this world are so regularly successful?

Otherwise it would be difficult to explain why we usually find our way back home, why all of our predecessors, going back to the amoeba, successfully survived, and why contradiction-free prognoses can be made about the entire mesocosmic world.

Of course, our worldview is a reconstruction, but one that is reconstructed with the means and tools that stem from that very world. Colors and tone pitches are proxies for electromagnetic and material wavelengths, etc. Consider also that the impulse clicks in the optic and auditory nerves are identical: the address in the brain where those impulses arrive is what establishes the respective interpretation. All perception is a symbol for expectable reality.

This confronts us with the limits of human cognition. They reflect the handicaps of (Sect. 2.3.1) consciousness and (Sect. 2.3.2) language and the (Sect. 2.3.3) discrepancy between our problem-solving capabilities.

2.3.1 Construction and Reality

We generally take enlightened consciousness to be the faculty that both defines humanness and enables us to probe deeply into the nature of humankind. This is reason enough to begin with this topic, especially because it reveals the tricks that our consciousness plays on us.

At issue is (a) our cognitive constructs and the way we envision (b) the reliability of our insights and (c) the laws underlying those insights.

(a) All the symbols developed to interpret reality are concocted by evolution—'constructs' as it were. And it would be senseless to maintain that our interpretational approach is the only possible or even the best one. Recall the dragonfly and the swallow example (Fig. 2.1). Conversely, implying that this interpretation means our worldview has nothing to do with reality is also incorrect. Nonetheless, this is the very position held by constructivists (Förster, Glasersfeld, Maturana; subsumed under 'radical constructivism' by Schmidt 1987).

In this sense, human organization, like that of any other organism, is clearly full of such constructions. Importantly, these cannot be understood as adaptations to modern conditions. Rather, they reflect phylogenetic constraints, detours and compromises. Our ancestors were constructed in torpedo form, later added four legs in a bridge-like construction, whereby these 'torpedo bridges' were ultimately uprighted to a bipedal tower—all steps typical for pervasive 'evolutionary bungling'. The same holds true for our senses and our brains. Our intellectual flights may well be as clumsy as the flight of a crane fly. Keep in mind, however, that crane flies do indeed become airborne, guaranteeing the survival of the species.

The constructivists, in turn, hold that each one of us in fact imagines his or her own 'airspace' and flies solely within that construct. The fact that each of us can do that if we so choose is not the issue. Where, then, lies reality?

Let's assume that three people are asked to wander through a forest alone. When queried afterwards about their impressions, each provides a unique account: the poacher hears hunters that aren't necessarily there, the love-struck poet hears sylphs, and the diviner hears the grass growing. That is the subjective reality (world B). Yet none of them stumbled over a tree trunk or injured their eye on a branch. In that sense the world must have appeared the same to all three. This represents collective and, in the present context, objective reality (world A). The difference lies in selection. In world B, one can imagine almost anything and survive (at worst in a hospitalized state). In world A, mistakes are lethal.

Radical constructivism falls into the trap of 'solipsism' (Stirner 1845), which assumes that the only reality is that of individual consciousness, for example that of you yourself as the reader of this book. Accordingly, the whole world exists merely in your mind (but why then do you read?). Radical constructivism is refuted by life itself.

(b) Is there any *reliability* in prognoses about reality? Apparently so! To begin with, (correct) prognoses play a vital role for all living beings: Life survives because it makes correct prognoses about existential matters. And for all of us, the reliability of such prognoses depends on two conditions:

The first is the degree of order behind extra-subjective reality, whereby order is defined here as 'law times application' (Riedl 1975). A whole cosmos of matter adheres to the law of gravity, which is simple to formulate. In contrast, a complex piece of regional cultural legislation can soon become invalid and forgotten, creating no predictable order. The second condition is that, roughly spoken, reliability depends on how many of our prognoses are regularly confirmed by experience.

We describe the first of the above two parameters of order as natural laws. These, however, as held in libraries and archives—Popper's world 3 (see Pooper and Eccles 1977)—already represent symbols of symbols. They are linguistic or formal symbols for those symbols our sensory apparatus uses to reproduce interrelationships in the real world. This clearly distances us from nature, at least from its complex features. What counts, however, is the degree of probability with which our prognoses are confirmed by experience.

Importantly, the issue here revolves around probabilities. The macroscopic world can be understood causalistically, i.e. as adhering to laws. Nonetheless, microphysical chance all too often exerts its influence into the macro-world. Short chains of cause and effect suffice. A case in point is mathematically ideal billiard balls positioned one meter apart: the fact that they are composed of material means that the seventh ball will no longer necessarily hit the eighth. This is because the inherent uncertainty in the position of the surface molecules, multiplied seven times by itself, is already larger than a billiard ball (Sexl 1982).

(c) *The world in the background* is, simply put, not entirely deterministic. 'Laplace's spirit' has been defined as knowing the position and movement of all particles in the cosmos. That spirit would be able to predict every event in this world, for example that my next sentence will begin with a capital 'I'. If we were able to convince him of the validity of Heisenberg's uncertainty principle, then he would have to calculate one alternative for each of 10^{80} quanta in the universe multiplied by the 10^{31} picoseconds of its existence: two quanta either meet each other or do not. Assuming that he calculates these 10^{111} alternatives for the course of events in the cosmos, we would, in the present context, merely be told: 'almost anything is possible'.

This underlines the contribution of true physical chance in all complex systems. Evolutionary processes—and in fact all creative processes—thrive on chance. Premises themselves do not yet harbor creativity. Were that the case, then we could create every desirable invention here and now because we already master most of the technical premises.

We cannot necessarily improve the reliability of predictions by formalizing the underlying lawful order. This is especially important for biologists. For example, we all tend to rely blindly on the law of gravity. When we open our fingers, our pen is sure to drop. However, if by chance just once the dancing molecules all darted in the same direction, then our pen would for a brief moment cool down to a temperature approaching absolute zero and would hit the ceiling at relativistic speed. This is very clearly a rare event—perhaps one in 10^{100} cases—considering the number of molecules in a pen. In contrast, Haeckel's Law holds that ontogeny repeats elements of phylogeny (namely the so-called 'palingenetic' characters). For example, every vertebrate animal develops the primordium of the chorda dorsalis. This pathway has never been short-circuited—in none of the 10^5 species represented on average by 10^8 individuals and 10^9 generations, equivalent to 10^{22} cases.

Life turns out to be a balancing act, remaining in a dynamic equilibrium (Bertalanffy 1968) far removed from a physical steady state. According to the law of entropy, all closed systems must end in thermodynamic chaos, namely in a complete uniformity of material and temperature gradients. The order that builds up living beings—as open systems through which energy and matter flow—circumvents this law. The trick: discharging more disorder into the environment than the organism needs to structure and internally maintain its own body. Basically, it feeds on order and creates 'negative entropy' (Schrödinger[1] 1957) (compare Fig. 4.3).

This insight raised the challenge to find a measure for order or negative entropy, no doubt closely related to information or even instruction (Riedl 1975). This quest has remained elusive because physicists, who typically deal with entropy in a purely quantitative manner, failed to follow Schrödinger. Explanations have been forwarded for this omission (Riedl 1991). Order encompasses several dimensions. It is organized hierarchically, and 'values' can be attached to it depending on its content of redundancy (see also Chap. 4).

2.3.2 Emergence, Notions and Language

The development of enlightened consciousness—and language—contributed to the rise of humankind. In all philosophy (to the extent that it is not influenced by EE), the investigation of cognitive processes has always been anchored in linguistic expression.

This is a curious situation for the present discussion because we now know the level of cognitive gain that was necessary to spawn spoken language. Comparative linguistics, as represented by Chmosky (1970), Lennenberg (1972), Mayerthaler (1981) and many others, has also contributed substantially to our understanding of the prehistory of language.

This calls for examining those phenomena—for example (a) emergence and (b) phase transitions—that have failed most to help explain (c) language and, subsequently, (d) logic.

(a) The term *emergence* describes the appearance of new, mostly higher or more complex features. The current opinion is that matter arose from the first elementary particles, the 'energy clumps', of the early cosmos (hadrons and leptons). This subsequently gave rise to life and, ultimately, also to consciousness and culture. This raises three questions. Did the process necessarily have to develop in this manner, was higher development itself necessary, and were the higher forms already contained in the lower ones?

[1] Interestingly, in his book "Proving Darwin: Making Biology Mathematical", published in 2012, the mathematician Gregory Chaitin undertook this very step.

In fact, nothing points to the necessity of culture developing in the cosmos. The situation is somewhat different regarding the increase in complexity as such. A general principle involving certain maintenance or preservational conditions (German: *Erhaltungsbedingungen*) may be at work here (Riedl 1976).

The term used in biology is 'anagenesis'. This refers to the phenomenon that, once all ecological niches are occupied by representatives of a particular organizational level, a transition to a higher form of organization creates a new niche and new maintenance opportunities.

In the inorganic realm this is described by conditions of stability, in the cultural realm by standards. For explanations of this important circumstance see Chap. 6, Sects. 6.1.1 and 6.1.3. The point here is that the cosmos, beyond being subject to entropy, also features the possibility of negentropy.

An equally decisive and controversial question for our worldview is whether the features of complex systems were already contained in their constituent components. If yes, as maintained by the reductionists (e.g. Medawar & Medawar 1986), then emergence would be a non-issue. Today, however, the concept of emergence is once again becoming mainstream (compare Mahner and Bunge 1997).

To date, no such features, even in traces, have been found in the above-mentioned constituent components. In my opinion, they cannot already be contained therein. Even predicting them based on any combination of the properties of the component elements seems fruitless: the number of possibilities is simply much too high. Could any early human—well acquainted with horse hair, wood and chicken gut—have guessed that their combination would once give rise to the violin? The alternatives taken when new systems emerge must be accepted as being singular historical processes. Complex systems are not entirely reducible using a causal approach; their structural organization is unrepeatable and in that sense also 'irreparable'. Our duty is to respect and protect them.

When still new, this topic was known as emergentism, a term coined by English emergence philosophy of the 1920s; its chief proponent is usually taken to be Lloyd Morgan (based on a book published in 1923). The underlying insight was that, to paraphrase philosophers, 'new levels of being show new qualities that cannot be traced back'. Why this is so will be dealt with later.

(b) *Phase transitions* characterize the systems dealt with by physics, chemistry, biology, psychology, sociology and the cultural sciences. This is reflected in the necessity of using entirely different terminologies to describe the characteristic features in the respective sciences. We ourselves are ill-prepared to deal with such transitions. This helps explain why the sciences have carved the world up based on levels of complexity, why they have a wide range of 'cross-sectional' theories regarding their specific levels but find it difficult to then combine these levels via 'longitudinal theories' such as a general theory of evolution.

We have difficulty accepting that new features apparently arise spontaneously. All of our terms such as 'creation' or 'emergence' are based on the underlying notion that novel properties can arise only from the void or—adhering to the concept of 'evolution'—merely need to be unwrapped.

Lorenz (1973a) used the term 'fulguration', which suggests an 'innovate spark', best illustrated based on single events. Phase transitions, in contrast, are lengthier and typically involve multiply-linked processes acting in concert. They define many biological and cultural processes.

(c) The *effect of language* on our thought processes should not be underestimated. Although thought is not based solely on linguistic elements, language-based thinking leads us to suspect a connection between expression and reality. This may often be the case, but can mask cognitive shortcomings (Riedl 1987a).

This addresses the issue of 'linguistic universalities' and also touches upon 'linguistic relativism'. The nouns that every language relies on to describe things arise through gestalt perception (Mayerthaler 1982). This gives them a static-typological character, allowing no transitive transformations into new features. The phrases a 'large group of trees" and a 'small grove', for example, still retain the distinction between a tree and a forest. The definitions of reptiles and mammals are unequivocal, giving rise to the expectation that the first mammal could very well have hatched from a reptile egg. The moment we attempt to bridge the gap transitively using phase transitions, language turns out to be an obstacle.

(d) The same also holds true for *classical logic*. Developed above all by Aristotle, it strove to exclude the antinomies inherent in language, such as in the sentence 'I am a liar'. Moreover, all vagueness was to be eliminated. This required the postulate *tertium non datur*—the hope that no third option between 'true' and 'false' needed to be introduced.

History shows that it took centuries to realize that this approach is irreconcilable with empirical knowledge gain. The 'fuzzy logic' developed in recent decades has finally tackled and helped gradually reduce our uncertainties (for an orientation, see Kreiser et al. 1988; McNeill and Frieberger 1994).

Nonetheless, we are educated to believe or to expect that logical conclusions offer a degree certainty going beyond the underlying premises. Such conclusions are, of course, suggestive: 'All humans are mortal, Socrates is a human, ergo Socrates is mortal.' Yet they remain certain only insofar as their premises are certain, thus also trivializing the certainty arrived at. Primitive peoples and our own children avoid logical conclusions (Luria 1976; Scribner 1977; Piaget 1978), something we tend to consider uneducated.

Our daily lives are guided by the deceptive hope for 'truth-enhancing conclusions'. The widespread, rationalistic positions taken by philosophy and the theory of science also lull us into behaving as if logic is a solid foundation for evaluating cognitive processes. Recall, however, the insight that the chain involving communication, language and logic took a different developmental path than our perception of extra-subjective reality. Importantly, the 'reality' of the former has nothing to do with empirical experience: it is content in hoping to achieve 'internal consistency' within the respective conceptual system. This calls for avoiding the rationalistic trap of confusing one 'truth' with the other.

2.3.3 Perceiving (Cognition) and Explaining

David Hume (1711–1776) determined that, rather than extracting causality from experience, we can only ascribe causality to it. We merely observe the post hoc, the succession of events; the *propter hoc*, the 'because', must be added reflectively in a trial-and-error process. This insight made an impression on Kant, inspiring his critical essays and, as mentioned earlier, his a priori.

Based on EE, a priori are interpreted as being adaptations to the world. At the same time, a wealth of adaptations clearly shaped humans and their senses long before we were able to anticipate the 'because' issue.

The initial step involves organizing the 'things' in this world based on incoming sensory data. The second step is to link them causally. This process is primed by two innate hypotheses, namely that of 'the apparently true' and of 'the comparable'. Both essentially suggest that—in the event of repetitive coincidences (i.e. the simultaneous occurrence of features)—we assume an obligatory relationship.

This is the appropriate time to juxtapose Hume's term *propter hoc* with *simul hoc* ('because simultaneous'). The two form the basis for association, for 'gestalt perception' and for 'recognizing fields of similarity'. While this will be treated in greater detail below, the key point is that *simul hoc* involves an automatic, ratiomorphic, pre-conscious rather than a reflective process; *propter hoc*, in contrast, describes a rational or reflective process, even if it be supported ratiomorphically.

The final step is to return again to the cognitive process (see paragraph 2, Sect. 3.2.1) and differentiate it (illustrated in Fig. 2.10) into the automatically operating process of cognition and the consciously supported process of explanation.

This calls for dealing with (a) terminological issues and examining wherein the two processes (b) correspond and wherein they (c) differ.

(a) *The terminology* creates two problems. Although these are primarily of a linguistic nature, adequately defined terms are a cornerstone of any effort. This necessitates examining the dictates of language-based thought.

The issue involves (a1) expanding or, more decisively, (a2) sorting familiar terms.

- (a1) The first problem was already touched upon above: the difference between the demands of a longitudinal theory and the *confines of familiar terms* rooted in the cross-sectional perspective. The task is to extend the use of the terms 'expectation', 'prognosis' and 'experience' (in their functional sense) to also cover pre-conscious processes. Conversely, 'internal' and 'external system' need to be applied to the structure and environment of a culture.
- (a2) The second problem is weighty and inconvenient because it forces cleanly distinguishing often casually used and colloquially interchangeable terms (Fig. 2.10 provides the necessary orientation).

The problem is that—if coining new words is to be avoided—traditional, *context-specific terms* must be applied to juxtapose the inductive/deductive

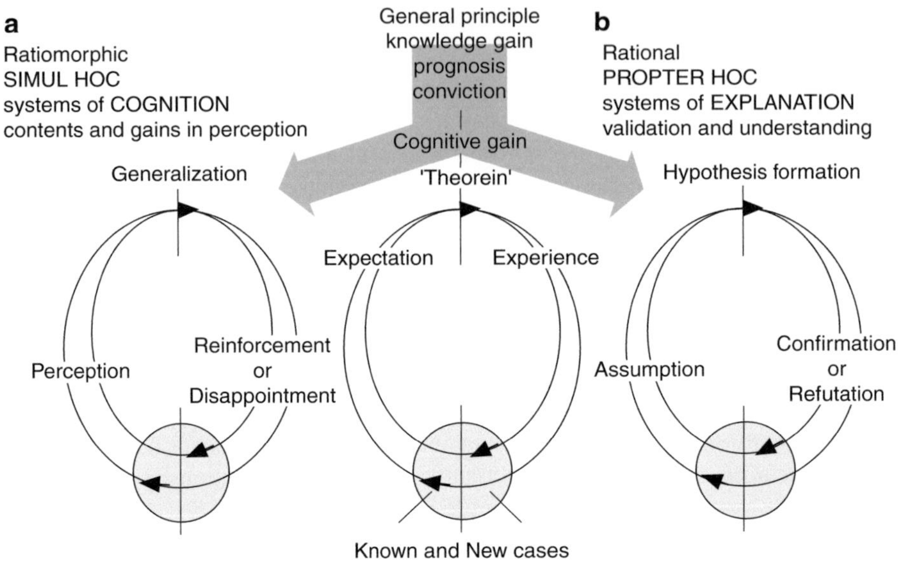

Fig. 2.10 Terminology of the knowledge-gaining processes in general form (center) and in the juxtaposition of cognition and explanation (details in text). These terms are typically poorly differentiated in both colloquial and scientific languages. This figure provides an orientation for much of the remaining text

components (refer back to Fig. 2.6) of two processes. These are, roughly speaking, the ratiomorphic and the rational processes. It seems opportune to borrow the terms for the ratiomorphic pathway of cognition from physiology and ethology, the terms for the rational pathway from psychology and the theory of science. This approach puts us into unaccustomed territory.

Terminologically, three levels (i–iii) need to be distinguished:

(i) In the ratiomorphic algorithm, I replace the terms expectation and experience in the general model with '*perception*' juxtaposed with '*reinforcement*' or '*disappointment*'; in the rational algorithm, '*assumption*' is juxtaposed to '*confirmation*' or '*refutation*'. Note, however, that 'rational' processes still contain considerable subconscious input.

Two outputs of both processes should be distinguished: the individual output of each cycle and the overall outputs of the algorithms.

(ii) A broad spectrum of words is available to describe the individual output of each cycle in the ratiomorphic versus the dominantly rational process. They range from 'invariant formation', 'object constancy' or 'generalization' to 'concept' and 'term', then to 'hypothesis' and 'theory'—reflecting a successively increasing degree of rationality. I juxtapose '*generalization*' with '*hypothetization*', which closely approximates my concept.

I refer to '*theories*' when a common concept for both individual outputs is required—theory in the sense of the Greek '*theorein*', i.e. a 'composite picture of things.'

(iii) No gradient is distinguishable in the terms describing the overall products. Moreover, the available terms are poorly defined and in some cases even permutable. In the ratiomorphic algorithm I will refer to '*gain of knowledge*', in the rational to '*gain of validation* (German: *Begründung*) *and understanding*'. This in the sense that a dog or a hapless day-dreamer know their way home without any need for validation. Both would do well to solidify their recognition of home, albeit without necessarily needing to forward an explanation. Dilthey (1883), in his efforts to delimit the humanities from the natural sciences, attributed an 'understanding' method to the former, an 'explanatory' method to the latter (Riedl 1985). The usage introduced above requires viewing the act of explaining as part of the understanding process.

To define both overall products using the same terminology, I refer to gains in conviction, foresight or knowledge.

(b) Both processes have *many commonalities*. Both presuppose a lawful, highly redundant world, yet one in which nothing is repeated identically. And both require a memory, whether it lie in the peripheral nerve system, in a brain or in a culture.

This calls for distinguishing (b1) the underlying structures, their (b2) differentiation and their (b3) symmetries.

- (b1) *The underlying structure* of both processes is identical. This is unsurprising considering that certain basic structures extend across the full range from genetic to cultural learning (Chap. 2, Sect. 2.2.1, Fig. 2.6). The matter becomes more interesting when demonstrating the methodological correspondence of cognition and explanation in both the natural sciences and the humanities.

  Recall that iterative cycles involving alternating inductive-deductive processes are detectable in all three layers of learning. They join to form a spiral process, whereby the cycles never revert down the same path. Rather, their slope corresponds to the gain in knowledge, to improved prognoses, and clearly to increased conviction.

  The method itself, if viewed in light of practical research, goes by different names. In the 'exact' natural sciences its differentiation is related to the 'subsumption scheme', in the humanities to a methodologically honed 'hermeneutics'. This relationship has remained unrecognized and will therefore be argued in detail in the following sections. The two methodological approaches prove to be intimately related both to the complex structures of the world and to the structure of human thought, which developed in dialog with nature.

  Neglecting this relationship means continuing to endure the perceived contradictions between the causalistic and hermeneutic approaches.
- (b2) The *differentiation of the methods* is related to the recognition that individual cycles of experience—taken separately—gain can easily lead one astray. This is similar to the legendary Russell's chicken: the chicken holds the person feeding it to be its benefactor, a connection strengthened with each feeding. Little does it know that it is being fed to land on its patron's dinner table.

The mistake lies in impermissible extrapolation. Nonetheless, prognoses by definition always contain extrapolations. At what point do these become impermissible? A useful answer is to avoid placing trust in a simple chain of confirmations, in a singular expectation or in an isolated theory. The truth involves testing within a larger context. Nature itself provides one such context. Its matrix of conditions and tiered structure underline that every complex system or subsystem must itself adhere to the conditions both of its constituent components and of its broader environment.

Technically speaking, this requires a hierarchic system of theories, whereby the theories subsumed at one level of complexity and based on observable cases are themselves expected to serve as the case studies for the theory at the next higher level. In principle, each is a precondition for the other: cognitively, higher-level theories develop from subtheories and recursively test these. Figure 2.11 depicts a simplified example for the cognitive process. It describes the process behind the insight that places the genus *Homo* in between its species and the hominids in general.

The second example involves the relationship of explanations (Fig. 2.12). Again, it shows in simplified form the observed cases are connected, from Galileo's theory of terrestrial mechanics and Kepler's theory of celestial mechanics to Newton's theory of gravity.

The relationship between the cognitive pyramid and the explanatory pyramid on the one hand and their relation to hermeneutics and subsumption on the other is a cornerstone of this book.

This raises the problem of 'empirical truth'. As noted above, little trust should be placed in simple chains of confirmation, singular expectations or isolated theories. The appropriate hierarchic framework, however, can help approximate empirical truth. The task is to attain conformity by making that network as dense and encompassing as possible. Three goals need to be set. First, the meshes must be dense enough so that no facts can slip through. Second, the framework must encompass all disciplines open to empirical knowledge. Third, the theories in the network should support rather than contradict each other, and all the possible prognoses they generate should be confirmable by experience. This is a challenging, ambitious target that fosters humility.

Neglecting these conditions reduces 'empirical truth' to a mere figure of speech.

- (b3) Regarding *the symmetries*, additional agreement between the cognitively based and explanatory views need to be highlighted. No clear differentiation of this type has ever been attempted. This represents another cornerstone in my argumentation and proves to be fundamental for a deeper understanding of complex systems.

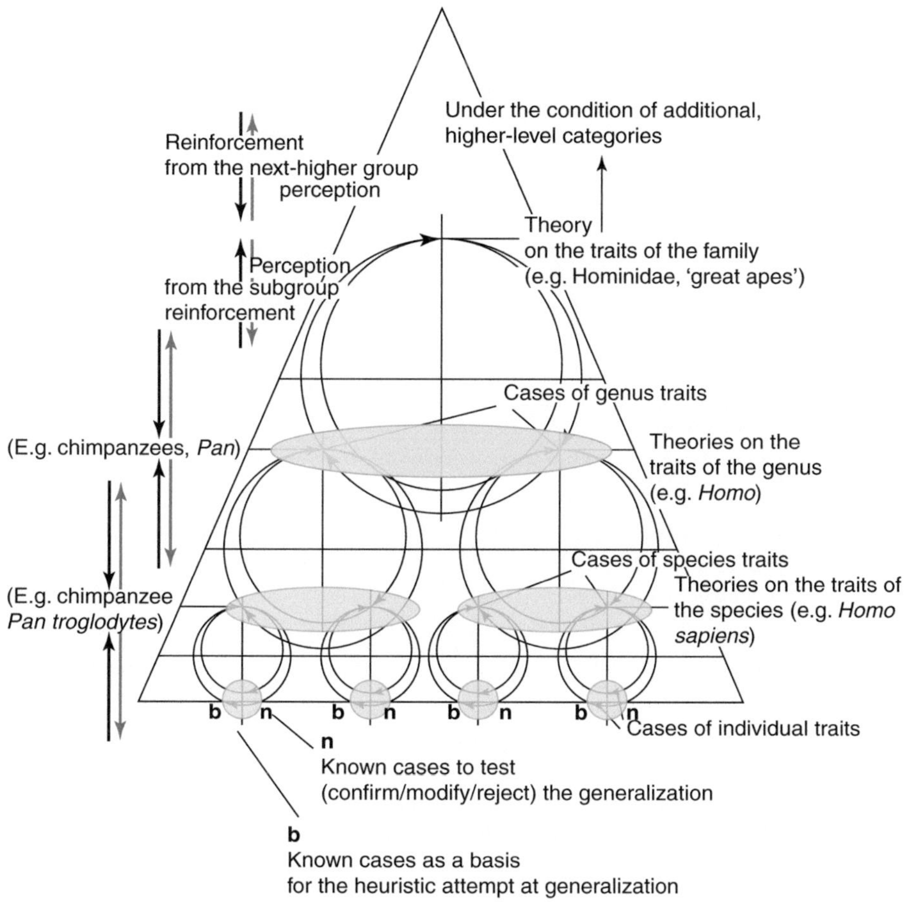

Fig. 2.11 The hierarchic system of generalization, illustrated based on integrating the cases of characters of chimpanzees and humans into the family of 'great apes' (symbols after Fig. 2.10; Type A, 'system of cognition'). Note that the generalizations gained and tested based on the cases will themselves become cases for the next-higher generalization. Moreover, the interplay between perception and confirmation remains preserved at each respective level. For simplification, only two of the respective subsystems are identified

The processes of cognition and explanation correspond to each other in three forms of symmetry: in the (i) arrangement of structural hierarchies, in (ii) their juxtaposition with hierarchies of classes and (iii) in connection with three cognitive pathways.

(i) All complex systems reveal two halves in their *structural hierarchy*. Double pyramids best represent this configuration (Fig. 2.13), one pyramid encompassing 'unique, individual' components, the other the 'standard building blocks'.

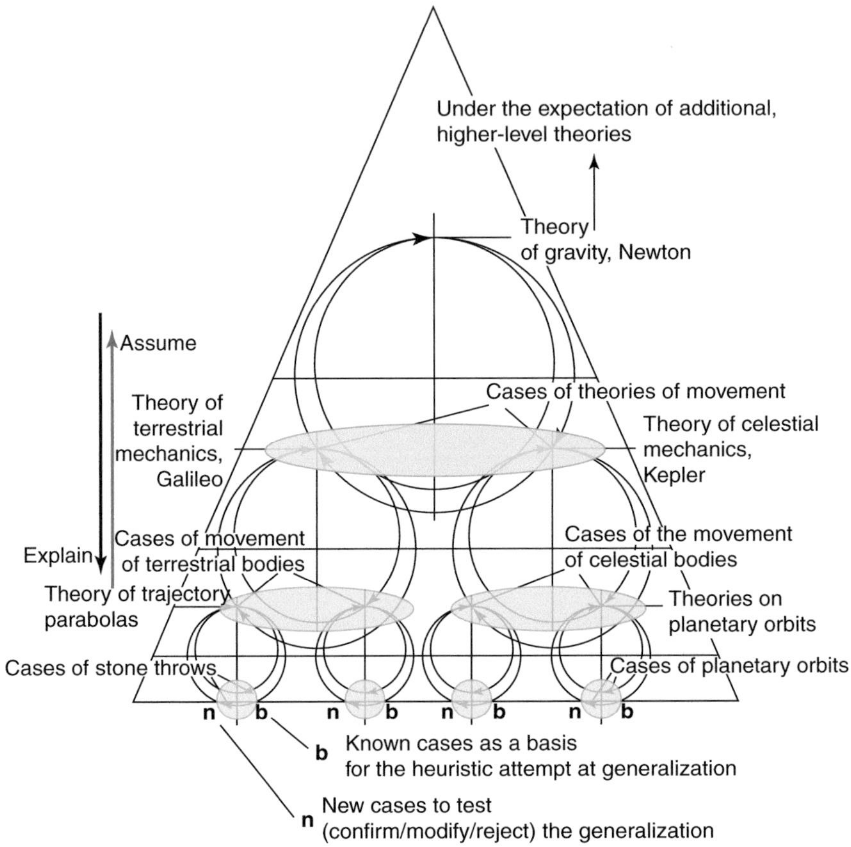

Fig. 2.12 The hierarchic systems of hypothesis formation, illustrated based on integrating the cases of trajectory parabolas and planetary orbits into the theory of gravity (symbols after Fig. 2.10; type B, 'systems of explanation'). Note that the theories built and tested based on the cases will themselves become cases for the next-higher theories, but that the interplay between assumption and explanation extends through the entire system in the form of intention (simplifications as in Fig. 2.11; both examples represent class hierarchies)

Both pyramids differ phenomenologically, conceptually and in their genesis. The content of one develops over a series of unique, unrepeatable, historical occurrences and is therefore composed entirely of non-interchangeable individualities, whether it be the solar system, a mountain range, the morphological type of an organism group, or a period style. The other structural hierarchy, in contrast, consists of largely interchangeable *standard or mass building blocks*, whether these involve quanta, atoms or molecules, a compound, a cell or organ, the individuals of a species, or even the type of font in a text or an automobile from a particular model series.

The pyramid bases of such individual and standard hierarchies adjoin one another. This usually represents the cross-section of greatest differentiation and diversity of a complex structure. It also corresponds to the horizon of our immediate mesocosmic perception and, accordingly, is also the place where research efforts tend to be initiated.

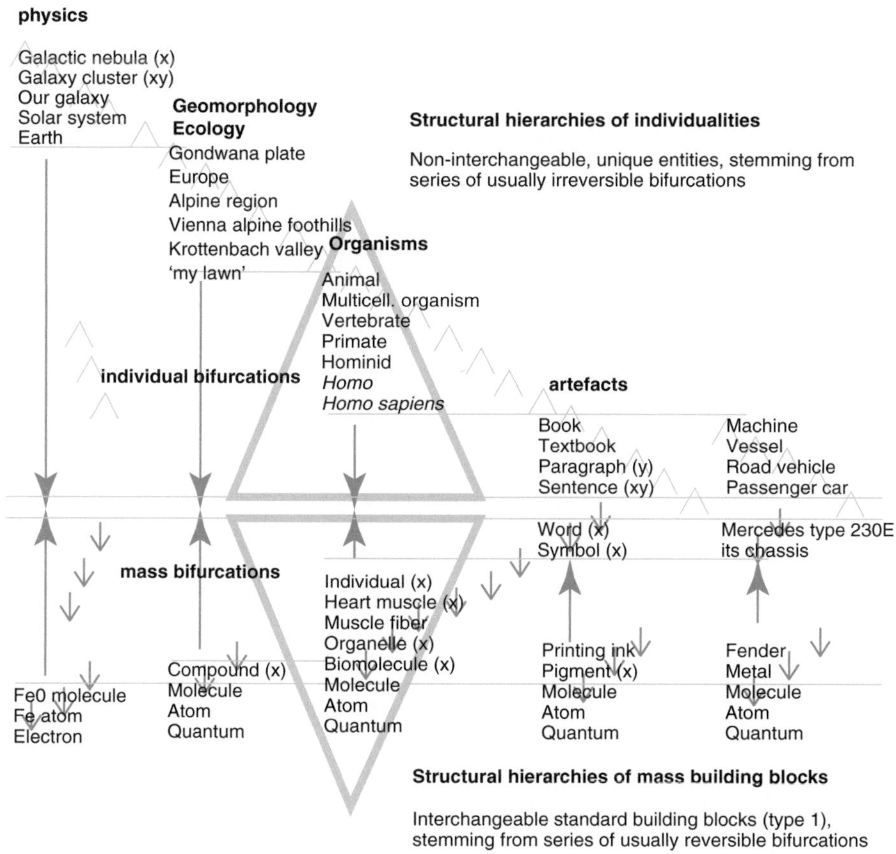

Fig. 2.13 Individual and mass building blocks in structural hierarchies, arranged based on key degrees of differentiation and selected scientific disciplines (compare Fig. 1.1; the cases of chemistry or geology as well as psychology or sociology are omitted for space reasons but are easy to associate). 'type 1' means that a second type ('homonomies') remains to be discussed. The gray angles mark sites of further differentiation, the gray triangles the position of both hierarchies (compare Fig. 2.14)

Neglecting this differentiation lumps the interchangeable together and with the unique, masking a key difference in the historicity of the components of complex systems.

(ii) The double pyramid of structural hierarchies is mirrored by double pyramids of *class hierarchies*. More specifically, they pass through one another. This can be visualized conceptually as the two sets of pyramids positioned at right angles to each other (Fig. 2.14).

The structural hierarchies comprise the specific objects composing a complex system, whereas the class hierarchies comprise the conceptual entities to which the respective objects belong. These, in turn, are separated based on the classes of the unique components or of the standard building blocks. This is

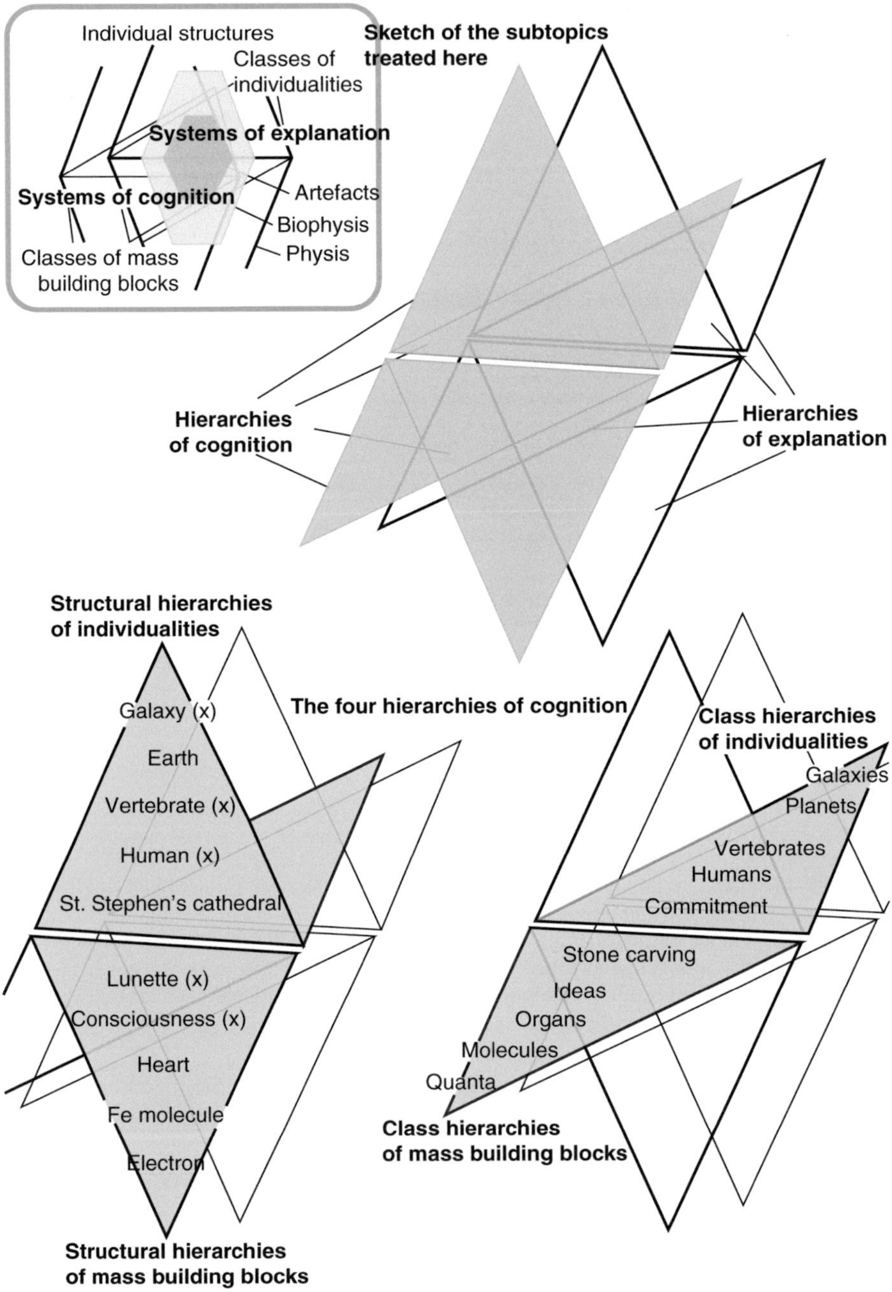

Fig. 2.14 The two double hierarchies, namely the class-structure double pyramid of the cognition processes (as a precondition) before the class-structure double pyramid of the processes of explanation and understanding. Both end (compare Fig. 2.13) at the limits of our cognitive ability. Most scientific questions represent a section in the above. Top left: the scheme of the conceptual relationship along with its contents

equally valid for the classes of unique structures (such as the above-cited solar system, mountain range, taxonomic units or period styles) as it is for terms describing building blocks (quanta, atoms, molecules, compounds, cells, organelles, individuals, fonts or cars of a particular model).

The relationship between both double pyramids is obligatory. Every class in a complex system must be structurally organized. Conversely, every complex structure must have its place, its affiliation or classification into the categories describing its genesis. In the case of genealogical processes, this would be the respective family tree (Fig. 2.14).

Neglecting the difference between hierarchies of structures and of classes precludes any validation of structural terms or of class concepts.

(iii) A preview of the *connection between the three cognitive pathways* is necessary here. The point of departure for perceiving objects, as noted above, is usually at an intermediate level, i.e. at the interface between the two hierarchies. The paths of our perception of and assumptions about (inter-)relationships (see Fig. 2.10) then move outward toward the respective ends of what the 'state-of-the-art' allows us to comprehend. Specifically, the paths go through the successive structural and class hierarchies, ultimately ending at the laws governing the macro- as well as the microcosmos. The reinforcements and confirmations, in turn, take the opposite course, away from the most overarching insights and laws inward to the diversity of the objects themselves. Importantly, this path proves to repeat the course of the objects' ontogenies. In this sense, the development of our perception of the world recapitulates its creation (see further Chap. 5, Sect. 5.3.3).

Neglecting this relationship relegates our understanding of lawful order to the level of a cognitive idiosyncrasy or randomness.

(c) Beyond the commonalities listed above, the processes of perception and explanation also show *fundamental differences*. Whereas the commonalities are based on the relationship between our knowledge-gaining mechanisms (the categories of understanding) and the basic structures of extra-subjective reality, the differences are rooted in cognitive behaviors. Three manifestations are evident:

Differences are evident in (c1) the method, (c2) in the result and (c3) in how the processes are assessed scientifically.

- (c1) *Methodologically*, cognition and explanation differ in the same way that our innate, subconsciously operating, problem-solving mechanisms differ from the approach involving superimposed consciousness, language and culture. Although explanation builds on cognition (and needs insights into its workings), cognition operates based on *simul hoc*, explanation on *propter hoc*. The first relies on observation and the systems of gestalt perception, the second adds a speculative element.

  Again, discussing "the world and cognition as a problem" first requires dealing with the respective terms and concepts. Some simply need to be defined from the onset (see 'a' in Table 2.1). The concept of 'interdependence' or 'reciprocity' (German: *Wechselbezüglichkeit*), however, requires immediate attention. As noted above (Chap. 2, Sect. 2.3.3 (b2)), no 'theorin' can stand on its own. Generalizations as well as hypothesis-building lead us to expect a network or at least a hierarchy of interrelated theories (Figs. 2.11 and 2.12).

Interdependencies link these theories together. They remain operatively active in the cognitive process, albeit often subconsciously. All hierarchic levels are interlinked, although this cohesion appears to dissipate during the explanatory process because understanding and explaining seek something akin to 'ultimate causes' (but more on this later) (Table 2.1).

Neglecting the above differences yields the two mutually exclusive understandings of the world outlined below.

- (c2) Both approaches differ equally strongly *in their result*—much like recognizing regularity in complex patterns differs from the analyzed functions. Or like the reliability based on synoptically examining variability differs from the reliability based on excluding as many variables as possible. Or like the richness of phenomena differs from their mechanistic skeleton. This is all the more true because the natural sciences have constricted the concept of causality. This is reflected in the attempt to understand complexity based on solely one of the four forms of causality identified by Aristotle: the *causa efficiens*, translated as 'power', force' or 'energy' (Table 2.2).

  Neglecting this circumstance condemns us to misunderstand the foundation of all scientific endeavor.
- (c3) Based on such differences, cognition and explanation are conferred *different status*. The scientistic perspective held, and often continues to hold, that eternal

Table 2.1

Cognition *Knowledge gain*	*Explanation* *Gain in understanding*
As a precondition	As a subsequent operation
Primarily innate	Culture and language dependent
Largely subconscious	Largely conscious
Ratiomorphic	Rational
Through gestalt perception	Through logical operations
On observable objects	By adding speculative elements
Cybernetic	Probabilistic
In a recursive process	In 'if-then' argumentation
Hermeneutic[a], morphological[a]	Scientistic[a], causalistic
Interdependent at all levels	Apparently 'holistic'
Largely synthetic	Largely analytical

[a]'Hermeneutic' refers to a method involving reciprocal enlightenment; 'morphological' means the same thing applied to comparative anatomy and systematics. The two following chapters examine this in greater detail. 'Scientistic' refers to methods modeled after those of the inorganic sciences

Table 2.2

Cognition *Cognitive contents*	*Explanation* *Contents of understanding*
Patterns of 'natural order'[a]	Patterns of natural laws[a]
Constraints in structural conditions	In functional conditions
Lawful simultaneity	Lawful successiveness
Simul hoc	*Propter hoc*
Four *causae* incorporated	Only *causa efficiens*
Historical products	Supposedly perpetual products
Largely irreversible	Largely reversible
Dominantly qualitative	Dominantly quantitative
Rarely formalizable	Formalizable
Difficult to rationalize	Rational
Usually unrepeatable and undoable	Usually repeatable and doable

[a]Most feel they know what natural laws are, but not necessarily natural order, particularly because it can also include forms of order in the cultural sciences. Both terms are detailed in the following text

laws are involved. Cosmologically, however, even the laws of physics are merely 'older' (Thirring and Stöltzner 1994). Accordingly, explanation, as a conscious process, need not be made intelligible from the onset because it is fundamentally plausible. The opposite holds true for cognition.

Several other things have gone unrecognized as well: (1) lawful 'simultaneity' can reach the same level of reliability as lawful 'successiveness'; (2) the explanations for what we purportedly perceive must be changed if and when our knowledge framework changes; (3) in contrast, changing explanations hardly have any influence on the recognized frameworks themselves (see the many iterations of the theory of evolution). We tend to overlook the fact that no explanation can be more reliable than our knowledge about the object itself. Moreover, things that go unrecognized can, by definition, never be explained (Table 2.3).

It may be disconcerting to see the foundation of all scientific endeavor relegated to the pre-scientific level in the juxtapositions of the above tables. The two complexes introduced below shed light on this situation: they are necessary even if some may consider them naive.

Firstly, recall that the process of knowledge gain is often removed from the realm of conscious observation: its methodology has remained obscured and difficult to convey. This makes it easy to overlook how highly adapted the ratiomorphic method is. As long as life has existed, it has been tried and tested to fit the 'realities' all our ancestors faced.

Secondly, the very method we consider to be especially scientific is based on a specific condition and opens a decisive opportunity. That condition involves successively reducing the complexity of phenomena to a level where the remaining parts can be imitated. The opportunity, in turn, is the ability to intervene in nature, something typically referred to as 'subjugating nature'. This ultimately

Table 2.3

Rank of knowledge and cognition	*of understanding and explanation*
Precondition, prerequisite	Subsequent possibility
Dependent on the explanation	Dependent on the recognized
Does not change with changing explanation	Changes with the mode of the recognized
Invites explanation	Appears to replace cognition
Is hardly open to experimentation	Is based on experiments
Gains certainty through many comparisons	Gains certainty through numerous repetitions
Extends into high complexity	Reduces complexity
Holistic	Reductionistic
Is characterized as being descriptive	Is characterized as being explanatory
Is characterized as being pre-scientific	Is characterized as being scientific

translates into gaining influence, into asserting the power of sponsors and of their scientific teams over competitors and over society in general.

A final word on the structure of reductionism. The 'formal reduction' of phenomena down to their lawful order addresses our legitimate need for precision. There is also little need to disparage 'pragmatic reductionism': it reduces things into manageable units. The matter becomes more dubious when accompanied by 'ontological reductionism', with its pretension that the level at which we can intervene represents the full story. This brings us to the concern, raised by Lord Snow back in 1959, that the natural sciences rather heedlessly change the world before they have ever fully understood it.

The differences in the processes of cognition and of explanation have now been outlined. And because cognition always precedes explanation, the system context of the former needs to be presented first. This calls for taking a two-pronged approach, namely first addressing our phylogenetic dispositions and then describing in detail the specifics of their application in practice.

Chapter 3
The Systems of Cognition

Perceiving elements of the world is as old as life itself. The life-sustaining reactions of bacteria or simple protozoans demonstrate that these responses involve molecules. These molecules react highly specifically: among the profusion of molecules encountered, they recognize particular ones. Cognition therefore begins at the molecular level. Perhaps the most familiar case involves the duplication of the molecular chains making up genetic material along with their transcription and translation into matrices of amino acids.

This is equally valid for internal organization as it is for the organism's innate reactions to the environment. And it echoes Aristotle's postulate that we can perceive the world because the elements making up our surroundings resemble those of our senses. That's the whole story! 'Cognition', however, goes beyond mere depiction to encompass successful reactions. Such reactions are life sustaining and can therefore be taken as being highly reasonable. The topic here is human cognition, and it is not easy to concede that our senses splendidly reconstruct rather than depict the world to enable correct reactions. This calls for illustrative examples.

The task (Sect. 3.1) begins with the conditions of perception, then Sects. 3.2 and 3.3 presents two fundamental, phylogenetically based solution strategies, and, finally, Sect. 3.4 describes the worldview arising from such phylogenetic blueprints.

3.1 Conditions of Perception

The initial step is to recall the linguistic hurdles facing longitudinal theories such as the one here. Concepts such as 'perceiving' or 'recognition' and 'cognition' may appear somewhat overstretched when applied to the reactions of bacteria or protozoans. Nonetheless, the identity of the mechanisms operating here requires acknowledging the gradual transitions between reaction, perception and cognition.

R. Riedl, *Structures of Complexity*,
https://doi.org/10.1007/978-3-030-13064-0_3

The present approach involves three steps: Sect. 3.1.1 discussing the genetically acquired foundations, followed by Sect. 3.1.2 introducing the associative processes and then Sect. 3.1.3 demonstrating how these are manifested in our cognitive processes.

3.1.1 Perception Means Problem Solving

Correct perception equates with vital reactions to incoming data. This does not require believing that the world is depicted within ourselves, as proponents of the image theory would have us believe. Rather, a simple hypothesis suffices. Namely that the evolved, symbolic representations of the outside world in which we operate successfully guide us—as if within an image—through that world. The same holds true.

A paramecium gliding through a water drop cannot afford to become stuck in a barrier. Accordingly, encountering a solid object with its front end triggers the release of molecules that reverse the direction of the flickering cilia. The animal backs up. The reverse beat soon ceases, but initially one-sidedly. The result: the paramecium turns and goes on its way again (Hinrichsen and Schultz 1988). This correctly interprets everything implied in a barrier: it will be impenetrable, have a limited size and will not follow the change in course. This underlines the validity of interpreting perception and cognition as ‘correct problem solving’.

As noted earlier, it is entirely dark and very quiet inside our brain. Nonetheless, it mediates a world replete with tones and bright colors. Our eyes, for example, are unable to detect electromagnetic waves. Nonetheless, coding such waves into impulse frequencies via specific nerve tracts conveys differentiated brightness and colors. This navigates us quite well though our reality, i.e. the world in which we need to operate successfully in order to survive. This reveals two frameworks: expansions and restrictions. The first can be illustrated in our ability to perceive the brightness of the stars—a supplementary realm in which we are clearly not actors. The second involves the constraints in mammal evolution that have barred seeing UV-light and minimized infra-red perception.

The same holds true for our built-in, unconditional reflexes. The cornea of our eyes itself fails to perceive the possibility of an oncoming grain of sand when hit by a windblast. In fact, the situation is quite the opposite: of all the types of eyes that our bodyplan might have brought forth, the fittest—and therefore the one that survived—is the one in which a reflex circuit automatically shuts our eyelids when strafed by wind.

3.1.2 Fundamentals of Association and Conditioning

The above also holds true for associative learning, which is based on conditioned reflexes (Chap. 2, Sect. 2.2.1 (a)).

Such reflexes know nothing about the world itself. Rather, they recognize coincidences, i.e. simultaneous occurrences, of biologically relevant sensory data whenever this is anatomically possible. This is remarkable from two standpoints involving reasons and causes. Section 3.2.2 addresses the question why such coincidences are linked, i.e. become associated, with one another. The underlying cause is the following:

Coincidences are already linked among nerves themselves, reflecting a peculiarity, actually an inefficiency, of neural pathways. When repeatedly conducting a signal, they alter their 'resistance'—even if it might seem better to report the same stimuli in the same manner regardless of their frequency. The system uses this anomaly for an entirely new performance. If two such pathways are repeatedly and almost simultaneously taken, then a formerly latent connection between them can be switched on. This represents additional 'wiring'. Once such a simultaneity ceases, the connection is deactivated. This can be referred to as 'neuronal memory'. Associations, like many other phenomena, can therefore be attributed to 'evolutive jerry-rigging'.

One consequence of this peculiarity is that only simultaneous or temporally very closely spaced pairings of sensory impressions are automatically associated. In contrast, more widely separated events require a different type of memory in order to be linked, namely 'cerebral memory' or memory in the narrower sense. This involves retrieving pertinent contents or conducting a mental operation. Again, *simul hoc*—the simultaneity of perceptive elements—can function entirely subconsciously, whereas *propter hoc* requires active reflection.

3.1.3 The Transition to Cognitive Processes

The first of two of innate reactions discussed here is 'innate hypotheses' (Riedl 1980). Hypotheses themselves are not innate to us. Rather, innate here refers to reactions in the form of expectational stances. Their outputs—if we lift them into the conscious realm—closely resemble reasoned hypotheses.

Importantly, these ratiomorphic reactions rely on a phylogenetically acquired, built-in foundation. They operate on associations. Although they extend into a realm resembling reason, they fall far short of involving reflective reasoning. This makes them a stepping stone in the development of conscious thought and its mechanisms.

The transition from conditional reflexes to ratiomorphic operations and further to conscious processes is gradual. Accordingly, every association already involves a ratiomorphic achievement, even if it be weak. And the ratiomorphic operation, in turn, is a step in the direction of (and a foundation for) consciousness.

The next, peculiar task is to examine how the suggested expectations about the world around us manifest themselves consciously. This involves the 'hypothesis of the apparently true' (H1) and the 'hypothesis of the comparable' (H2). The former leads us to associate consecutive, the latter simultaneous coincidences. This insight into the ratiomorphic mode of operation yields two additional prognoses about the structure of extra-subjective reality.

3.2 Processing Consecutive Coincidences

The program behind 'hypothesis of the apparently true' (H1) leads to certain expectations. Specifically, prognoses that are confirmed by experience should increase the reliability of subsequent prognoses.

The following discussion describes (Sect. 3.2.1) the components of the algorithm, (Sect. 3.2.2) why this type of processing is successful, (Sect. 3.2.3) the types of mistakes they induce and (Sect. 3.2.4) how these can be avoided.

3.2.1 The Composition of the Algorithm

It is immediately apparent that this algorithm cannot involve a logical operation. Logic alone, for example, is insufficient to conclude that the sun will rise tomorrow—even with any number of experience-based confirmations. This drives philosophers to despair. Nonetheless, even in academia we behave as if the sun will rise.

If an outcome (even a surprising one) of an observation or an experiment repeats itself regularly, we soon believe to have discovered an obligatory connection rather than a series of flukes. Repetitions of an experience make us anticipate the next experience, and the ongoing confirmation of such events leads to a level of certainty interpreted as 'empirical truth'.

This takes us back to a competence already exhibited by children (Chap. 2)—one that is still operational in mathematics students and only incrementally replaced by an alternative solution strategy.

In fact, a mechanism very similar to the conditioned reflex is at work here. In the above-mentioned conditioned eyelid reflex example, sounding a bell before each experimental windblast soon leads test persons to anticipate the next jet of air after each bell. Advance warning of an upcoming disturbance via conditioning makes sense biologically. And it is equally sensible to delete this association if the disturbance subsequently fails to be confirmed (compare Fig. 2.7).

The certainty gained in scientific investigations dissipates in much the same way if initial experience cannot be confirmed at a later date. Similarly, in conditioning, an occasional lack of confirmation becomes relativized as long as verified prognoses dominate (recall the problem-solving behavior of math students; Fig. 2.8). After all, no observation or experiment is entirely error-free: 'nobody is perfect'. Finally, memory does have its limitations and can play tricks on us.

The ratiomorphic apparatus operates cybernetically. It operates on the premise that, initially, nothing can be known. Every gain in predictability relies on weighting the sequence of confirmed versus dashed prognoses, whereby imperfect perception coupled with a limited memory capacity play a role. Recent confirmations tend to mask past inconsistencies. Finally, our short memories have the disadvantage that certain types of lawfulness or regularity—namely those that repeat themselves only after long series of events—will go unrecognized (Riedl et al. 1991, 1992; Riedl 1992a).

The rational problem-solving strategy operates based on entirely different premises (see Chap. 5). In short, logical operations involve the theory of probability, memory is considered to be unlimited, and mistakes are not permitted (mathematics tests are a case in point). The drawback is that complex lawful order will go unrecognized if simple regularity is expected (Riedl et al. 1991; Wagner et al. 1992; Riedl 1992a).

3.2.2 Wherein the Seeds of Success Lie

Any algorithm incorporated into our expectational stance must have increased the fitness of our ancestors over the long term. What is the formula for success?

In recalling the bell/wind gust example, sounds are clearly not perforce related to blasts of wind. The connection merely involves the expectation of a potential connection—if the coincidence repeats itself. It may seem disconcerting to expect lawfulness in every coincidence that is simultaneous or in close succession. After all, don't random coincidences by far dominate in our daily lives? Certainly, but not when such coincidences repeat themselves, such as thunder and lightning, or a warning call and some imminent danger. Experience clearly confirms that randomness becomes increasingly less likely the more often and invariably such coincidences repeat themselves.

This behavior of the learning mechanism provides additional insight into the structure of extra-subjective reality. The world need not be entirely unchanging and redundant: successive events can also harbor potential relationships.

3.2.3 Wherein the Deficiencies Lie

The deficiencies of such an algorithm lie in its simplicity. More specifically, it was developed and tested to help solve problems in an environment that was far simpler than what we are up against today in our science- and industry-based societies.

How long does it take to incorporate and modify such an algorithm or hypothesis in phylogenetic history? We can only make an intelligent guess. For physical traits at the species level, the time span is one to several million years (at the complexity level of mammals). The principle, the mechanism behind the hypothesis, however, is as old as conditioned reflexes, i.e. as old as the molluscs for example (ca. 500 million years). Assuming that adapting the algorithm takes only a million years, and considering that the complexities of society have evolved over a few centuries or less, leads to the clear conclusion that society's progress has overwhelmed the algorithm's adaptive capacity.

The algorithm is too simple for our world. The shortcoming of the hypothesis is that it suggests we can indefinitely extrapolate from single series of confirmed prognoses. Revising faulty prognoses without suffering harm might have been easier in earlier times, but often proves lethal in the complex challenges we face today.

Interestingly, the deficiencies in a hypothesis no longer subject individual persons to selective pressure. In fact, the most successful and reproductively active of our fellow citizens often appear to be most easily led astray. At the same time, we also hope to have attained a level of humanity that does not step in to regulate reproduction in potentially mentally compromised members of society.

Impermissible extrapolation is the fundamental error the hypothesis insinuates in empiricism (recall Bertrand Russel's chicken). The paradox: the extrapolation always collapses precisely at the peak of presumed certainty.

This type of extrapolation harbors a well-known error, namely the expectation that something better will come from providing more of the good. In fact, the misguidance is somewhat more blurred. Two cases in point:

Physicists once thought they would be solving a problem of humanity by harnessing nuclear energy. Nonetheless, the social and political framework in which all human problems play out was simply not their playing field. The envisioned blessing was an impermissible extrapolation. Today, we can only hope that the specialists in nuclear fission remain firmly put in splendid isolation (such as once in Russia) and do not provide politically dubious states with nuclear know-how. The next impermissible extrapolation? Interventions into cell nuclei sound like a good candidate.

3.2.4 How to Overcome the Deficiencies

Avoiding impermissible extrapolation may appear to be very simple, but this is by no means the case. We are programed for extrapolation, as underlined by our insuppressible development of successive expectational stances. This is simply a life-sustaining function. At the same time, no one has ever forwarded a genetically anchored plan for defining permissible extrapolations. Just the opposite: human thought thrives on analogies and metaphors. Our language is full of 'table legs', 'sea stars', 'river arms' and 'sickle moons' and the like. The temptation is built in. It is an indispensable drive yet a poor guide.

Some extrapolations are unavoidable: 'Tomorrow I will be fed again' (thinks the chicken…) or 'more nuclear-power-based electricity will be sold'. If the extrapolation framework is exceeded, then its veracity can only be tested outside that framework. This may appear trivial. Less trivial, however, is the fact that everything lying outside such a frame is based on a series of further experiences, which, as discussed in Chap. 2, show a hierarchic relationship.

The chicken may very well correctly predict one of the chicken farmer's actions: he or she will again provide feed. But the bird cannot be aware of the farmer's other activities as purchaser, seller and consumer, or what 'wise' animal husbandry entails. Physicists can correctly predict one component of humanity's problems: energy will be used. But they are not necessarily fully informed about or may underestimate the issues involved in the concentration of political power, in stabilizing social and political systems, and even in the final storage of the fuel rods or global warming.

This again raises the respective structures of generalizing and hypothesizing (Figs. 2.11 and 2.12), which reveal a hierarchic constellation of *theorin* (see Fig. 2.10 for terminology). The cases that give rise to a high-level theory are themselves composed of theories based on confirmed expectations from the next lower layer, which in turn are controlled by the higher-level theory.

The process of cognitive generalization (Chap. 4, Sect. 4.2.1) will be examined in more detail under the terms 'reciprocal enlightenment' or 'hermeneutics'; the process of explanatory hypothesizing (Chap. 5, Sect. 5.3.1) will be referred to as the 'subsumption scheme'.

The chicken farmer's role as benefactor can be understood only based on theories governing all of his or her actions. Equally, the solution to the problems facing humanity can be grasped theoretically solely based on all those theories that describe the individual problems that humans suffer.

The 'hypothesis of the apparently true' cannot stand on its own. It involves confirming the expectation of consecutive coincidences. This presupposes that such repeated coincidences involve comparable perceptions. In the conditioned eyelid reflex example this seems unproblematic: the sensory channels that report the air stream and the bell tone are unambiguous, enabling renewed recognition. The situation soon changes in more complex perceptions, prompting the second hypothesis.

3.3 Processing Simultaneous Coincidences

More closely examining the program behind the 'hypothesis of the comparable' (H2) reveals an expectation: in comparing things we tend to negate the dissimilar and embrace presumed equivalencies.

This calls for going into more detail because the phenomenon is less well known and because it is fundamental for understanding the ratiomorphic operations behind cognitive gain.

To simplify comparison with the above discussion on consecutive coincidences, the same pattern of section headings is used here as well: (Sect.3.3.1) describing the algorithm, establishing its (Sect. 3.3.2) success and its (Sect. 3.3.3) deficiencies, and (Sect. 3.3.4) demonstrating how the latter can be overcome.

3.3.1 The Composition of the Algorithm

As noted above, both hypotheses presuppose one another. The reaction to consecutive events presupposes their (renewed) recognition, just as the reaction to simultaneous events presupposes their repetition. Accordingly, recognition is the more complex process or, more precisely, it needs to be successful far into the realm of complexity (for details and references, see Riedl 1987a).

The discussion is structured accordingly. The sequence is (a) invariant formation, (b) the processes of gestalt perception and (c) our perception of structural hierarchies.

(a) The increasing ability of organisms to form *invariants*—to arrange objects into classes and to distinguish between such classes—is a biologically interesting phenomenon. Although the term invariant is traditionally applied at the level of human psychology, the continuity of the relationships is the more interesting aspect here.

Lower organisms distinguish only three or four classes of phenomena in their environment: basic/acidic, warm/cold, wet/dry and, later, also dark/light. Desiccation, for example, can trigger encystment or mean immediate death.

Even in the insects and higher mammals, however, filters have been built into the sensory apparatus to ensure that only a tiny, select set of sensory data determine the key reactions ensuring species survival (Lorenz 1978). The term applied here is fixed action patterns (FAPs). Many of these remain inflexible and are independent of experience, others can be modified by experience.

Infant behavior illustrates the latter case. Newborns initially react to a nodding balloon with a grinning mouth the same as to the mother's nearby face. Only later does the balloon face trigger anxiety; at an even later stage, toddlers become wary of true strangers.

The same holds true for words used during early childhood. Seeing a train from the balcony and naming it 'choo-choo' means that everything spotted from the balcony that moves will be named 'choo-choo', be it an automobile, baby carriage or marching column (Piaget 1978). The fact that baby talk separates individual and collective terms at a fairly late stage further supports this relationship.

The way we develop terms resembles a sandglass. The concepts are initially very broad, then attain a specific narrowness in naming individuals, and once again expand to include notions such as 'the good' or 'the beautiful' and the concepts of entropy and evolution. Invariant formation, which arose at the pre-verbal stage, has clearly programed our language-based thought: definitional communication with a signal-like character (rather than transitive communication) stood model for the development of language. Accordingly, human language was distilled down to classical logic in its definitional, invariant form. That logic, in turn, retroacts on how we envision 'correct thinking' in everyday life.

It pays to go into more detail because a wealth of information is already available. This information indicates what we can still discover about the differentiation of our genetic predispositions in areas where little information is available.

(b) Our knowledge on *gestalt perception* has boomed in recent decades. Early gestalt theory has joined forces with ethology and bridged the gap to sensory physiology and neurophysiology, topping it off with cognitive psychology. This combination helps demonstrate the tiered structure of our competence in converting flimmering light into images of the structures making up our world.

Two key points: The first is that all the faculties outlined here represent learning and adaptational outcomes of our phylogenetic history. They operate fully automatically and subconsciously. The second is that these faculties all probably require testing against the environment in order to develop fully. This can be illustrated based on the important concept of the 'innate master teacher' (German: *angeborener Lehrmeister*) (Lorenz 1973a; more recently in Heschl 1998). This concept holds that while the 'masters' of all these faculties are innate, each master must pit itself against the environment in order to finalize the specific 'teachings' in the overall system.

These faculties are presented in the sequence (b1) the retina, (b2) the linkage, (b3) the first syntheses, (b4) the so-called abstraction and (b6) the composition of the forms.

- (b1) At the first level, a competence known as the 'constancy phenomenon' is already established *within the retina* itself. The dominant color in the field of vision is removed from the color impression. The result is that a face in a shaded forest or in candlelight does not appear greenish or reddish. The biological advantage is evident.

  This phenomenon can be convincingly tested by facing the full sun (or standing directly in front of a strong lamp) and shading your eyelids with your hand until the remaining color sensations disappear. This can take a few minutes. Lifting your hand away yields a bright orange-red surface through the flooded eyelids. Then, within one or two minutes, that color pales and gives way to a neutral (grayish) tonality.

  The retina is also the site where 'lateral inhibition' circuitry yields 'contour intensification'. The result is that all brightness and color borders are exaggerated: the edges of dark areas become darker, the light ones lighter and the color lines more intensive. In nature, most color and light borders correspond to object outlines: it is biologically very advantageous for these borders to be distinct as possible, even under shaded or twilight conditions.

  Experiments underline the significance of the above 'master teacher'. Raising kittens in surroundings featuring solely vertical light-dark contours causes them to hold their heads at an angle when confronted with horizontal contours later in life (details in Riedl 1987a).
- (b2) The *'linkage problem'* refers to the attempt to understand how a connection is established between two different sensory inputs that arrive simultaneously but at different places. A simple example is the concurrent perception of an outline and a color.

  Here, a suggestion by Hebb (1949) has been elevated to Hebb's principle. It postulates a cooperative interaction of neurons, similar to conditioning and the conditioned reflex. The concept holds that the neurons distributed in the brain, by synchronizing their discharges, can join to form units (Malsburg and Schneider 1986). This leads us squarely to the question of how complex perception is represented in the brain (Engel et al. 1993; Singer 1995).

The jury is still out on this issue. Nonetheless, very solid cerebro-physiological hypotheses describe the simultaneous processing of perceptions. Such perceptions typically involve the lawful or regular simultaneities characterizing the features of complex systems. This process holds equally true for birds, mammals and humans and underlines the considerable age of the *simul-hoc* hypothesis so crucial in the present context.

- (b3) The interplay of the retina and cortex yields first *synthetic outputs*. Namely, rather than perceiving the position of light-dark boundaries point by point, we perceive them in an already interlinked form. Even more high-level pathways convey the twists and turns of that outline.

  This is the framework that helps us form the so-called 'good gestalt'. The boundaries we register in our field of view do not remain 'on hold' in a jumble of data. Rather, we seek something akin to a solution. This requires two synthetic tasks. The first involves seeking closed contours in the tangle of lines. This reflects the fact that most objects, be they pebbles or lakes, leaves or trees, have closed contours.

  If the created image offers competing and undecidable solutions (as in Fig. 3.1), then the algorithm fails to arrive at a conclusion. The image appears to vibrate: the system refuses to accept defeat and continues its search.

  A simple home experiment with a cardboard box—split down the middle and with two viewing holes—illustrates this situation. Offering one eye a horizontal, the other a vertical pattern of stripes does not yield a checkerboard pattern. Rather, the viewer experiences an endless search for a solution involving moving spots of horizontal and vertical patterns.

  Secondly, the algorithm attempts to select the best (simplest, clearest) interpretation among the stabilized solutions (compare Fig. 3.2). This is an exceptional achievement that once again offers adaptive advantages. At the same time, it also provides new, astounding insights into the relationship between the world and our senses. The simplest solution proves to be the best because the number of competing solutions increases with increasing complexity, reducing the chances of finding the correct one.
- (b4) All further faculties involve the cerebral cortex. The initial step is the '*abstraction*' of the pre-formed gestalt from the background. This is simple in principle: whatever moves together, such as two points of light in the night sky, belongs together. This interpretation clearly reflects the real world situation.

  Movement, however, is a relative parameter. Everything can move with and against each another: objects, eyes, head, the observer as a whole. We tend to pick the simplest solution: the small object—shifting relative to the large one or often relative to the entire field of view—is de facto the one that is moving. This strategy can be confirmed based on its errors. Examples include the 'bridge effect', in which the bridge piles appear to be plowing through a standing waterbody, or a church tower that seems to be making headway against the clouds.

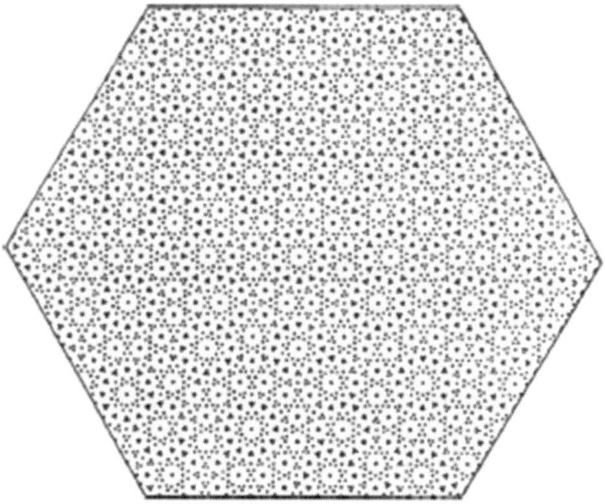

Fig. 3.1 Autonomous structuring processes of perception illustrated based on a pattern with competing solutions. The picture seems to be 'simmering' because the intimated circular figures suggest their perception while at the same time dissolving each other (from MARR 1982)

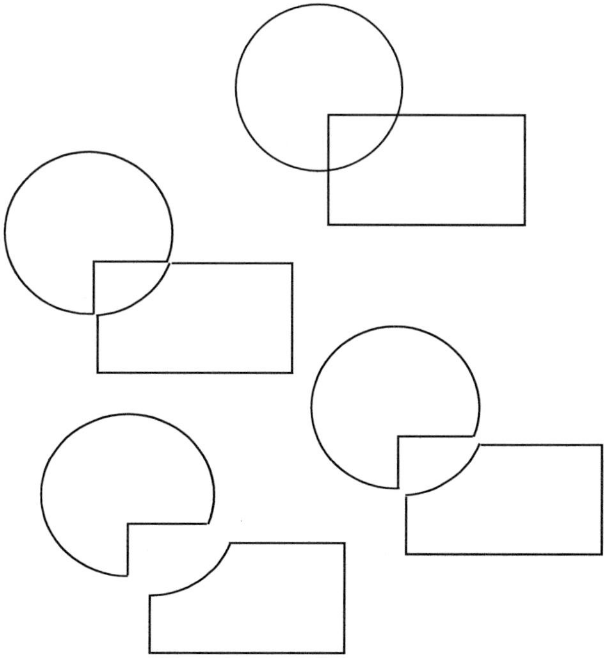

Fig. 3.2 The 'good gestalt' as a structuring principle. This graphic suggests finding a simplified solution: forming a combined circle and rectangle when the figures approach one another. In general, this process is 'inescapable'

Reactions to spatial depth, i.e. perspective, along with responses to 'collision courses' mark the transition to the subsequent, experience-supported processes. Again, experiments demonstrate that both are already genetically fixed in visually orienting organisms.

Kittens, even those that have never experienced a vertical drop, still cross a glass-covered gap with hesitation and caution. Infants viewing a film depicting a ball heading toward them show defensive reactions. These behaviors anticipate a characteristic feature of real-world perception. Whenever the size of an object rapidly increases, it will almost invariably be approaching us rather than inflating itself. The biological significance is indisputable.

- (b5) The above-described faculties are supported by innate (historically gained) experience in the environment. Some faculties are also supported by a more direct adaptation involving *individual experience* in the environment.

  These include the development of 'object constancy' in young animals and small children. If an object disappears, for example a ball behind the sofa, it also disappears from the mental picture. Additional experience (and motivation) is required to retain the object in the imagination, to expect its return and to search for it. This process is initially undirected, later more targeted. Once again, the program turns out to correspond to the environment: only rarely do lost objects truly disappear, for example a sugar cube in the depths of a coffee cup.

  This leads to a faculty that Piaget (1975) refers to as an 'evolution of reality forms'. In a first phase, small children apparently attribute true reality only to those objects they can actually get their hands on. This is followed by something akin to 'figurative reality', for example when a child recognizably expects to be able to pull a ball toward itself with a stick, even if that stick is lying on the ground and not yet physically connected to the ball. The subsequent, third phase develops via object-related experience—a type of naive realism that resembles our unreflected expectational stance. This might also be referred to as an expansion of reality because it involves an expansion of perceived space: the subject can complete a task in that space and, ultimately, supported by experience, 'act' in imagined space.

  Our assessment of perspective is strongly promoted by experience. It rapidly develops into a vital function by helping us gauge our own fast movements and resolve the position of friends and enemies. This is so rigidly anchored that we are easily tricked by all manner of 'illusory perspectives'. This is further substantiated by the fact that children's drawings, as well as paintings and graphic arts in early art history, only gradually introduce perspective.

  The next step beyond object constancy is the capacity to 'complete' concealed parts of objects. This combines experience with the respective objects with a mechanism to extract the memory of the full object stored in our brain based on the visible part. The development of this faculty is outlined in more detail in Sect. 3.4.1. Biologically, it is vital to mentally assemble the full image of a lion or tiger even if only its long tail is visible.

- (b6) Finally, the '*composition of forms*' is instructive. The term is deliberately borrowed from the fine arts because the processes are nearly identical. This can be simply demonstrated by imagining a figure just discovered in the distance. The composition proceeds hierarchically. The initial perception focuses on its main axis: the object is either standing or lying. This is followed by the axes of the appendages, the directional axes from the head down to the fingers (see Wainwright 1988 for the significance of axes in the organismic realm). This parallels the creation of a figural composition (Fig. 3.3).

  These preconditions will at some point enable us to understand the greatest performance of subconsciously operating perception, namely the 'recognition from all perspectives'. We recognize a cat based solely on its silhouette, whether the cat be rolled up, outstretched in the midst of a jump, or arching its back with bristled fur. No computer has yet equaled this performance.

  We know less about the degree to which 'seeing' fields of similarity is genetically programed, although it probably is (see Sect. 3.4.2 for more details).

(c) Our sensory apparatus has apparently also prepared us for the *hierarchic structure* of the world. We recognize that we are composed of parts, and at the same time recognize ourselves as being part of a broader framework. This is peculiar in the sense that all these levels involve entirely different manifestations. Thus, organs, individuals and hordes clearly differ from one another in appearance

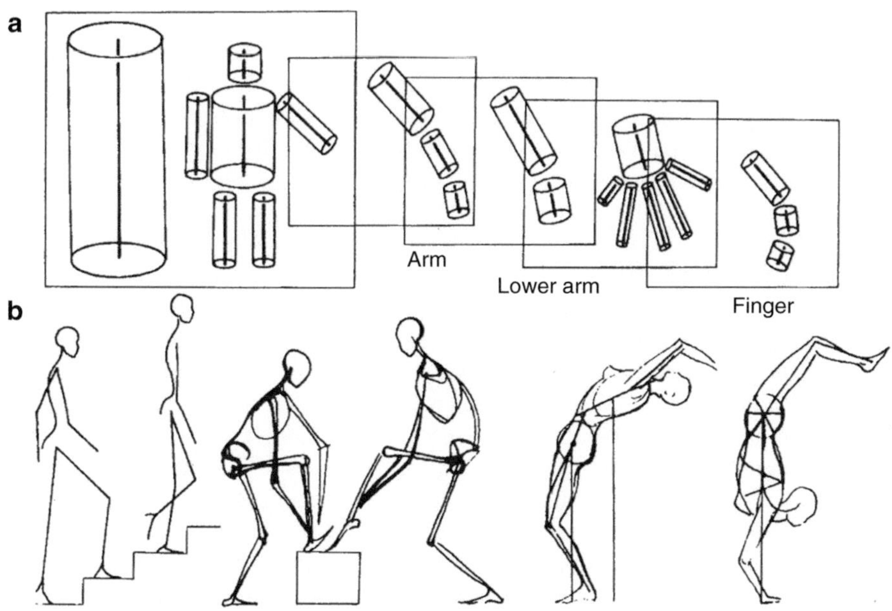

Fig. 3.3 Hierarchy in deciphering and composing forms. The decomposition according to axes with mounting perception (**a**), corresponds to an artist's structuring of composition (**b**). (Compiled after several authors; from Riedl 1987a)

and behavior. Moreover, a hierarchic structure is no simple construction. Chap. 5 discusses this construct in relation to our understanding of causality.

Structural hierarchy is present in all complex systems: from quanta, atoms and molecules to cell organelles, cells, tissues, organs, organisms, up to speech development involving sounds, words, sentences and context, and finally encompassing the group structures characterizing businesses both large and small. It is no coincidence that these hierarchies extend into our artefacts—from furniture, rooms and houses to cities.

The universality and significance of this approach to ordering the world no doubt served as the 'master teacher' introduced above. Importantly, our ability to operate based on hierarchical patterns offers clear advantages that explain why we continue to build on this understanding.

3.3.2 The Reasons for Success

Recall that incorporating and preserving genetic programs makes sense only if they yield tangible successes. I presented a range of such successes in describing the algorithm above. A summary suffices here.

This also provides an opportunity to differentiate two things: (a) our innate foresights into the environment and (b) how we attain such foresight.

(a) *a priori expectations*—based solely on our knowledge about this learning algorithm—provide insight into the fundamental structures of the world. We expect to encounter recognizable configurations, which typically manifest themselves as closed contours and exhibit constancy (even if they sometimes partially or temporarily elude perception). We experience that their parts move together, that their apparent size changes with distance, and understand that this explanation is more likely than assuming they are rapidly inflating or shrinking.

Chance plays no role at all here. We exclude it even though it surrounds us in vast profusion. The reason for rejecting it? The repetition of shapes and forms—in combination with the 'hypothesis of the apparently true'. This all presupposes *simul hoc*, the regularly repeated, combined appearance of features. Importantly, the terms 'feature', 'character' or 'trait' underline that it is valuable to differentiate and remember specific things. Simply put, the issue is the non-arbitrary combinability of the features in complex objects.

Even minimally loosening the fixed combinations of features that our memory relies on leads to the fantastic world of Hieronymus Bosch—a bizarre combination of individually recognizable parts. Dissolving the cohesiveness of features leaves us with what we describe as being indescribable (Fig. 3.4).

(b) This algorithm reaches its *goal* of helping us to interpret reality and react appropriately by three means:

- First: the amount of regularity (lawfulness) we can always expect from our environment is also present in the form of an expectational stance, i.e. it is genetically anchored and reliably available.

Fig. 3.4 The dissolution of a gestalt based on a figure in Hieronymus Bosch's painting 'Hay wagon' (Prado Madrid). Proceeding toward the top right, the still describable parts are further decomposed into the 'indescribable' (from Riedl 1994)

- Second: wherever experience can intensify, expand upon or improve some element of these programs, it will. This conforms to the innate 'master teacher' metaphor. In the deepest layers of the mechanism, for example contour intensification or object constancy, this is apparently no longer necessary.
- Third: the hermeneutic principle prevents our search for a solution from being abandoned, even in artificially constructed, 'undecidable' cases. Again, we are not conscious of this hermeneutic or 'reciprocal illumination', making it difficult to make convey the process (see Chap. 4, Sect. 4.2.1 for more details regarding its application). At this point we have addressed only the 'back-and-forth' between the synthetic and analytical processes of the program. The synthetic ones are represented by all the processes from the 'wiring' in the retina up to establishing the 'good gestalt'. The analytical ones range from the 'good gestalt' to recognizing objects (based on calculated perspectives) and their composition.

3.3.3 The Deficiencies of the Program

The program was created for a specific environment and was selected in that environment through rigorous elimination. Accordingly, it has perfected our faculties and activities (to the extent that life can be perfect). Its deficiencies first crop up where, in modern society, we have arrogated to ourselves powers we lack. Specifically, this refers to our increasing attempts to intervene in nature to a degree for which we are cognitively ill equipped.

This goes back to the context discussed under the 'hypothesis of the apparently true': our faculties have simply been overwhelmed by the complexity of our modern environment. Again, the manifested deficiencies are not subject to selection pressure. In fact, the situation is nearly reversed: those who most blatantly plunder the world's resources tend to reap advantages over more considerate neighbors.

The result is a drift into collective nonsense. The (a) causes and (b) effects are treated separately here.

(a) First the *causes*: The innate hypothesis that prompts us to compare, to equate and to complete appears to be quite reasonable and justifies the moniker 'ratiomorphic'. What better way could our sensory data have enabled a structured conception of the complex world and supported such a high level of predictability? The other side of the coin? We need to recognize that the remaining deficiencies reflect an insufficient acceptance of emergence as a phenomenon and, ultimately, also of transitivity.

Both are obscured by an apparent contradiction between our make-up and cultural history: our (a1) reluctance to accept emergence as a natural phenomenon as opposed to our (a2) desire to set limits to transitive processes. This particularism has consequences for (a3) our language-based thought.

- (a1) The phenomenon of the *emergence* of new qualities, still viewed as forming an entity in the emergence philosophy of the 1920s (Morgan 1923), raises new questions today. The cognitive aspect is addressed here.

  Higher development or 'anagenesis' was introduced above as a seesaw process in which organisms—as well as cultures and their artefacts (which themselves are part of the selecting environment)—are incrementally elevated into a higher level of differentiation. This astounding, world-changing phenomenon is not the issue here.

  Rather, the pertinent aspect involves our earlier question about whether the new features are predictable. Raising this question may be justified from a theoretical standpoint, but is apparently not possible in practice. The first reason lies in the enormous number of possible combinations inherent in even a modest number of constituents. The second is the complex and unique combinatorial aspect: this is a historical process with all the random decisions along its course, making it irreproducible. Correctly guessing the outcome is highly improbable.

  This needs to be acknowledged. We always do so grudgingly because we simply lack the sensory apparatus for the physical randomness at play here: our cognitive apparatus operates under the premise of excluding phenomena governed by chance. De facto, all major languages also lack a term for a process that creates an entirely new entity.

True 'emergences' have never been directly observed. In fact, they can best be described as historical, irreversible phase transitions that escape human observation. Examples include the transition from inanimate to animate nature, from reptiles to mammals, or simply from apes to humans. Our expectation is that the new qualities should develop harmoniously from the old. Alternatively, everything is already contained in the old, even if well concealed.

The early 'preformation theories' touchingly illustrated this concept: a tiny person was thought to already be discernible in human sperm. Accordingly, human generations followed one another like Russian nesting dolls. In principle, this naiveté is foisted upon us by our make-up, abetted by the continued hope for a tangible mechanistic explanation of the world.

Either everything is already encapsulated in an initial, ancestral condition, as also postulated by fractal theory, or the phase transitions unavoidably arise from the constituent components, as put forward by synergetics. Although we cannot change our predisposition for simplification, we can override it through introspective insight.

- (a2) As a counterpoint to the above, we like to set boundaries—even where none are actually present. We tend to feel uneasy when confronted with the limitless. This can be convincingly illustrated by experiments involving transformations of simple geometric figures. Test persons set boundaries when presented with such shifts. These were typically defined by the spatial axes of the test page or other prescribed geometries (Fig. 3.5).

  This is peculiar indeed. For example, we can all follow Escher's illustrations, in which almost everything can transform itself into something else. We ourselves can come up with all manner of similar transformations. What, then, underlies our particularistic stance, our terminological straitjacket? When, in Escher's artwork, a fish transitions into a bird, there is an intermediate phase in which a tiny something (or almost nothing) of each is present and therefore escapes being pinpointed by name.

- (a3) *Language* plays a role here. Nonetheless, it would be incorrect to assume that language alone determines our thought processes. The thought processes built into our make-up must have yielded at least the 'linguistic universalities', i.e. those features common to all—even the most exotic—human languages. A case in point is noun–verb separation, whereby the nouns are more interesting in the present context.

  Clarity was clearly crucial even back at the beginnings of communication. It probably started with conveying information about one's own state and then extended to messages about the environment, for example signals for enemy, friend and food. Despite all subsequent differentiation, this signal character, in the sense of being unambiguous, has retained its dominant function in semantics.

  Language does ultimately act back on thought. When talking about something, there is no way around language-based thinking with its linearity and its intransitory, definitional features. The act of designating classes also suggests that we should increasingly sharply define everything. This is a mistake.

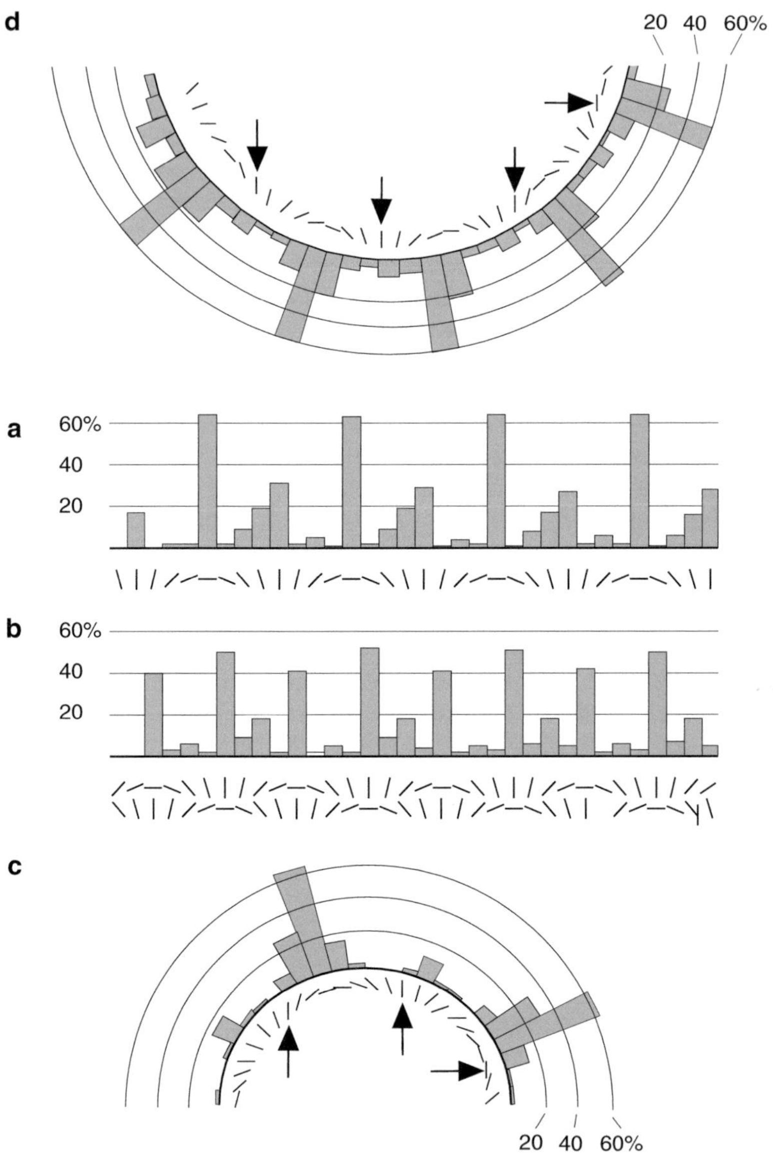

Fig. 3.5 Drawing borders in a continuum. Test persons were tasked with defining borders in the transformation in the position of a line (**a**). No commentary or objections were provided. The columns indicate the frequencies. Even when the task is made more complicated (**b**) the frequencies remain very similar. The relationship with the respective geometry is shown in (**c, d**): the frequencies correlate with the arc; the geometrically equivalent positions are indicated by arrows (after Riedl 1987a)

(b) The *consequences* of such predispositions are as peculiar as they are contradictory. Transformed elements are set in stone, nested in hierarchic systems. When emergence is involved, we postulate that the underlying transformation can be reconstructed. Recognizing this predisposition helps resolve the contradiction and explain the odd and unlucky fragmentation of our sciences.

This calls for examining (b1) the nested systems followed by (b2) the limits of our sensory apparatus and (b3) the effect on the fragmentation of the sciences.

- (b1) European, language-based thinking shows that we break objects down into a *nested system* of class concepts. While this does yield hierarchic order, it tends to deter transitions between the classes of a particular hierarchic level or between the hierarchic levels themselves.

  This transition problem is evident in many concepts such as 'mountain chain', 'mountain' and 'peak'; or 'trunk', 'branch' and 'twig'; as well as 'castle', 'house' and 'hut'.

  Even purely quantitative changes complicate phase transitions for us. A classical question is: how many sand grains does it take to make a sand pile? Providing a specific number is clearly absurd, although we recognize that grains roll whereas a pile flows.
- (b2) We have an *insufficient sensory apparatus* for exponential developments, as exemplified by the rice and chessboard example. The desire to put one kernel on the first square and successively double the amount on each of the subsequent 62 squares is unrealizable. The final volume of rice grains would cover every continent on the planet arm deep.

  The 'water lily example' demonstrates our lack of feeling for phase transitions leading to catastrophe. The scenario is that water lilies cover one-thousandth of a lake's surface, that their coverage doubles every year, and that the lake will collapse when entirely covered. Most of us would need pencil and paper (or, better, a calculator) to recognize that the catastrophe would be precipitated within a mere 10 years.

  This faulty predisposition leads us into a real cognitive trap when we attempt to codify our definitional conceptualization using the classical logic behind mathematics. The operation 'multiplied times 2' suggests identical effects regardless of the scale on which it is applied. And little thought is given to the fact that such identity, if material things are involved, must ultimately collapse. Consider eleven orders of magnitude. Today, major companies already deal with billions of such real things, namely units of payment, hard dollars. If revenues double, we still tend to consider it as being identical with a doubling of, let's say, a child's allowance from ten to twenty dollars.

  A next step is to imagine this difference on a scale involving orders of magnitude. Our bodies, enlarged by eleven decimal places, would attain the diameter of Earth's orbit and be equivalent to so many times the mass of the sun that gravitational forces would cause that material body to collapse. We would ignite and reach the sun's temperature, shrink into nothingness and remain as a black hole in the cosmos. Equally, we fail to anticipate the effect that simple doubling a single company's revenues has on the world around us. The consequences triggered by mega-economic events simply transcend our sensory competence. The consequences for civilization are becoming readily apparent.

Researchers have examined human choices involving alternative, deceptively simplified and misleading solution strategies. The clear conclusion: individual experience cannot provide guidance (Brehmer 1980). Rather, innate hypotheses (Riedl 1992a) control the process (Fig. 3.6).

- (b3) Definitional, language-based thought is also responsible for fragmenting *the sciences*. This reflects the levels of complexity we ourselves are programed to distinguish. Physics, chemistry, the biosciences (from molecular biology to ecology and ethology), psychology, sociology and the humanities (compare Fig. 1.1) are all steeped in their own particular terminology. Their boundaries lie at phase transitions. As noted earlier, we need 'longitudinal theories' to deliver an expanded, more encompassing set of terms.

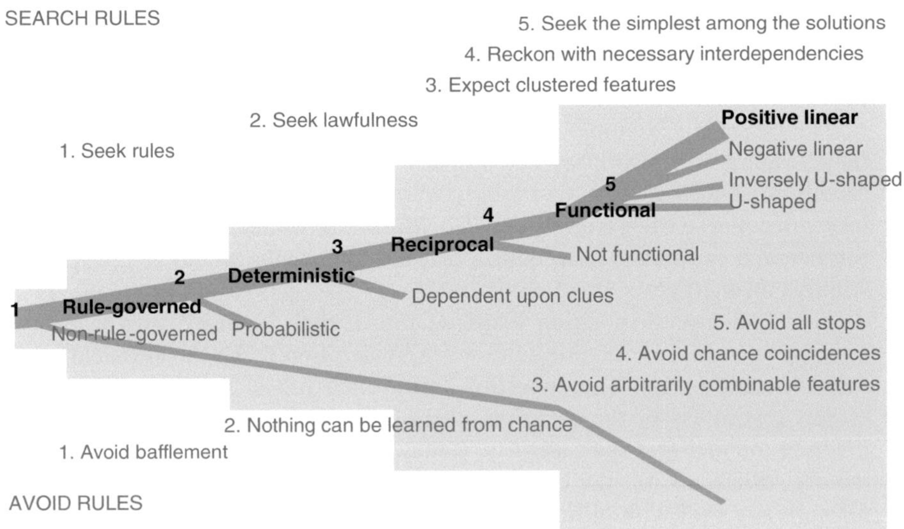

Fig. 3.6 The behavior when confronted with alternative solution strategies. The more readily accessible alternative always proves to be the preferred one (analysis by Brehmer 1980). I have added the search and avoid rules according to the program behind the innate hypotheses (after Riedl 1992a)

The increasing fragmentation of science is correlated with an ever larger number of disciplines. This has reduced our institutions of higher learning into training centers. The fields are treated as separate entities: the interrelationships fall by the wayside. Nonetheless, we behave as if knowledge in an individual discipline suffices to make generally valid prognoses about the world. Again, the flaw of empiricism lies in impermissible extrapolation. What we have learned regarding successive coincidences (Sect. 3.2.3) now proves to be valid for the simultaneous ones as well.

3.3.4 The Path to Overcoming These Deficiencies

The 'hypothesis of the comparable' proves to be richer in features than the 'hypothesis of the apparently true'. This holds true for the operational process, the grounds for its success and for its deficiencies, and also for overcoming those deficiencies. The wealth of qualitative characters is decisive.

This again calls for differentiating (a) the deficiencies of the ratiomorphic program and (b) the limitations that our language imposes due to these shortcomings.

(a) Overcoming *ratiomorphic misjudgements* in successive coincidences (Sect. 3.2.4) once again involves correcting misguided extrapolations. Two aspects are involved. First, we underestimate exponential developments, the complications of emergent transformations as well as the effects of complexities and of orders of magnitude. Second, we severely overestimate the applicability of insights from one discipline to other disciplines.

    Neither of the two pitfalls is easily avoided. We will continue to be misled (in a reversal of the equally unavoidable 'deceptive perspectives'). Moreover, extrapolation will never be eliminated. On the contrary: it is an indispensable element in cognitive gain. We can no longer alter our sensory apparatus, but we can supersede it based on experience.

(b) The definitional character of the *semantics of human language* presents a special problem. It is immutable. Our innate program hinders transitive semantics in colloquial language. We have little choice but to accept this. At the same time, we should not lose sight of the restrictions and compromises this language type imposes in dealing with complex surroundings.

Two types (b1, b2) of semantic hurdles must be taken and (b3) a fundamental syntax problem needs examination.

- (b1) One of the two semantic difficulties was broached in Chap. 2, Sect. 2.3.3 (a). It involves pursuing phenomena extending beyond the boundaries of individual disciplines and their respective *terminologies*. Examining developmental phenomena ranging from organisms to the theory of science requires expanding our terms and concepts. For example, 'knowledge gain' should also encompass molecular genetics, and 'adaptation' should be extended into the realm of theory development.

- (b2) A second difficulty is that the individual *class names* for complex objects are unsatisfactorily defined—simply because their boundaries are too strictly delineated. It's too late to change our language in this sense either. One partial remedy is to understand the compromise involved in such language-based thought. We need to accept that the boundaries of all the features of a class never align perfectly. In fact, the descriptive terms in such tiered and nested constructs actually relate to one another like the heights of a mountain chain (Fig. 3.7): some have abrupt ridges and flanks and tightly delimited valleys, others are separated only by flat mountain saddles.

  One way to alleviate such drawbacks is to indicate the relative strength of the delimiting features. Depending on their diagnostic power, they can also be categorized (see Chap. 4, Sect. 4.2). Nonetheless, we can approach but never reach the more transitive semantics characteristic of the Chinese language, in which the determination emanates from the middle of a term (for key references, see Riedl 1985). The concept of the type in biology is an effort in this direction.

- (b3) *The syntactic problem* initially appears to be entirely unrelated to the above, but closer examination reveals that it is. In our 'European' or in the broader sense 'circum-Mediterranean' diction this involves the phenomenon of the linked expressions 'is' (German: *ist*) and 'be' German: *sein*).

  As opposed to the 'circum-Pacific' family of languages, the separation of nouns and verbs is quite distinct (Mayerthaler, pers. comm.). This requires a separate, syntactically highlighted link between the two. Such a special grammatical construction probably arose with the 'Greek declarative sentence' (C.F.v. Weizsäcker 1982): 'Socrates is a human being'. In combination with the concept of classes, this yields the logical conclusion: 'All human beings are mortal, Socrates is a human being, ergo Socrates is mortal'.

  This development, already highlighted in Chap. 2 Sect. 2.3.2 (d), is interesting in the present context because, structurally, it fuels the expectation that classes can be explicitly defined. Complex objects, however, push these boundaries (see above). This manner of drawing logical conclusions is a prerequisite for cultured discourse and therefore retroacts strongly on the definitional mode our language-based thought.

  All these limitations cannot be overcome simply by attempting to eliminate them. From the evolutionary perspective, the task is to know and recognize them (Mayerthaler 1996a) and to acknowledge that they represent mere compromises. Then, we can use experience to help deconstruct such compromises (see Chap. 4).

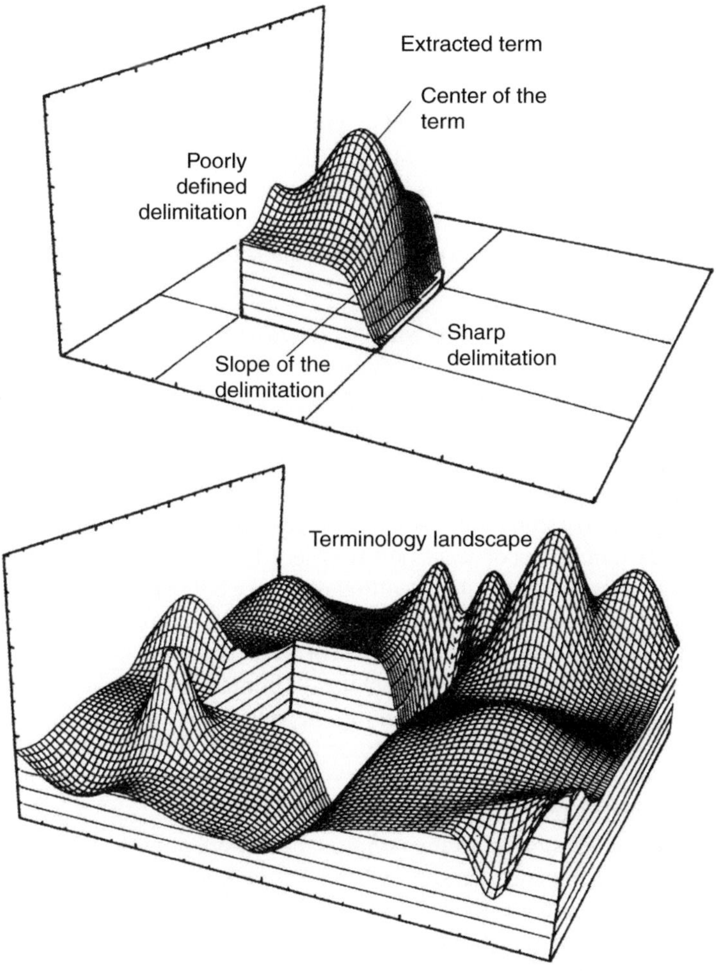

Fig. 3.7 The landscape character of our terms represented as elevations. Note the very differently developed borders delimiting the specified definition of a particular term (from Riedl 1992a)

3.4 On Structures and Classes

Structures and classes first came up in relation to 'recognizing the problem' (Chap. 2, Sect 2.3.3). More detail is warranted here. Recognizing structures is a matter of gestalt perception. Recognizing classes, however, requires expanded powers of memory.

Three levels of this competence can be initially distinguished: (Sect. 3.4.1) cerebral memory. This level is a prerequisite for two decisive achievements of our ratiomorphic-apparatus-based interpretation of the world, namely perception in (Sect. 3.4.2) fields of similarity and thought processes involving (Sect. 3.4.3) hierarchies of classes.

3.4.1 The Evolution of Memory

We are only beginning to understand how memory content is stored. More is known about how it is recorded and recalled. The theory presented in this book helps fill in numerous blanks on how memory evolved. This is a necessary step in understanding the ratiomorphic contribution to cognitive strategies.

The terms imprinting, fading, envisioning and recognition play a role in memory research. The focus here is on (a) imprinting and retrievability, (b) 'intermodality' and (c) renewed recognition.

(a) *Imprinting* is assumed to have been a subconscious process from its inception because it must be older that consciousness. Nonetheless, even in the presence of full consciousness, most of our memories continue to be incorporated unintentionally or even unnoticed. In many cases we have little say about what is being stored.

Fading, a very figurative concept of memory pathways, appears to fall short in helping to understand the mechanisms behind memory retrieval. We need to consider the conditions of accessibility (Riedl 1992b), i.e. the circuits leading to the storage sites. Intentionally recalling olfactory contents, for example, is rare. Conversely, a recurred odor can immediately conjure up full recollection of a school event or an emotional encounter in all its details.

The *accessibility* of memory contents is limited. Deliberate approaches are therefore rarely helpful in determining the full extent of what is stored in our memories. Rather, situations—whether random or willfully induced—trigger hidden memories. This is useful in the further discussion here. But first an evolutionarily instructive phenomenon:

(b) *Intermodality*. This refers to the reciprocal assessment of data inputs from different senses (Stein and Meridith 1993). This faculty interests ethologists more than psychologists because it is best perfected in humans. Tracing its development in the organismic realm, however, reveals the important role memory plays in higher forms of perception.

Intermodality seems to be only minimally developed in organisms even as highly evolved as snakes. In order to swallow their prey, snakes first needs to see it, then feel it—in precisely that sequence. The prey must come to lie such that it can be 'smelled' with the flickering tongue. If the prey changes its position, the snake, although it may well have already 'bagged' its victim, will continue to flicker its tongue in search mode (compare e.g. Sjoelander 1995).

This deficiency may come as a surprise, which only shows how much we ourselves rely on intermodality. This insight is useful in helping to determine what type of competence is missing in our own 'program'. For example the ability to automatically connect temporally or spatially widely separated data. We require protocols and instruments to achieve this.

(c) The next step is to examine the difference between intentionally retrieving or 'envisioning' stored information and the automatic *recognition process*.

Retrieving memory contents probably arose via that automatism, and that process continues to play a major role.

For example, a bird of prey circling high above a mountain chain clearly cannot 'en-vision' the tree and the specific forked branch holding its nest. It nonetheless finds its way back by successively recalling from memory the overall massif, the respective peak, the forested flank, and the tree group—a solution distinct from all other possible alternatives.

Even astute observers will rarely be able to fully reconstruct from memory a path once trodden through an unfamiliar city. Embarking upon the same path again, however, conjures up valuable recognition that helps guide the visitor to his or her destination. Another perhaps a more convincing example: you leave your workplace to get something that is missing, only to forget why you got up as soon as you enter the next room. Returning to the workplace setting immediately 'reveals' what was being sought.

Subconsciously controlled recognition is as crucial in scientific comparisons as it is in everyday life. We can even retrieve connections that we thought had gone unnoticed. A final example:

When hiking through a forest, we step over, or crunch underfoot, many dead branches littering the path after a stormy night. All were at least fleetingly perceived and, we might think, immediately forgotten. Then we encounter a branch that seems to point to a fork in the trail, and we immediately recall having seen the very same situation earlier. This piques our curiosity and we start to pay attention. As soon as this configuration repeats itself once again, we are certain that someone has marked the path.

In every complex system we encounter, every detail is visualized on our retinas and established in some form in our memories. Only very few details can be intentionally called up, but almost all can be brought to light after renewed recognition. This explains why even those things we accurately refer to as traits or characters are produced by ratiomorphic processes. The evolutionary success of programed recognition is self-evident.

3.4.2 Fields of Similarity

Humans have been programed for yet another astounding faculty or competence. Our memory delivers well-ordered rather than jumbled recollections. Perceiving one object immediately conjures up several similar objects. This is particularly evident when confronted with an item that is difficult to categorize, whether it be an unusual tool, a piece of furniture or an animal. A range of more familiar comparative items automatically pops up to support the interpretation.

We still know little about how that happens and what actually triggers it. In any case the selected comparisons clearly reflect the structure of the storage, the structure of the retrieval process, or both. And the phenomenon is apparently ratiomorphic and pre-conscious. Otherwise it would be difficult to explain how pigeons and rats—as shown experimentally (Huber and Lenz 1996)—can evaluate objects based on degrees of similarity.

This mechanism is a valuable tool for experienced biologists, ethnologists or art historians. A paleontologist who extracts the forefoot of a three-toed early horse from a fossil bed will automatically have in mind all the skeletal limbs known to him or her. And these, rather than being jumbled together, will be arranged (Fig. 3.8) according to potential trends in similarity.

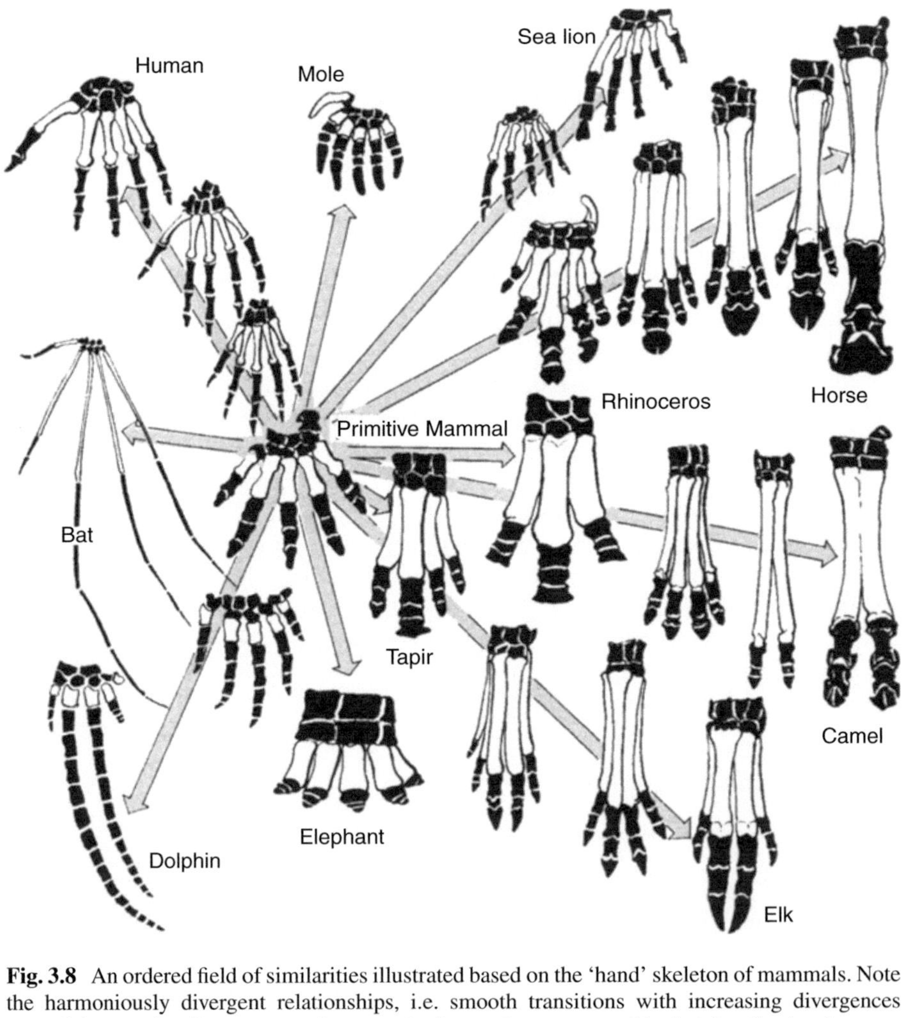

Fig. 3.8 An ordered field of similarities illustrated based on the 'hand' skeleton of mammals. Note the harmoniously divergent relationships, i.e. smooth transitions with increasing divergences toward the margins. In reconstructing genealogies, such patterns enable deriving the developmental pathways and degree of relatedness

Fig. 3.9 A still-to-be-arranged field of similarity illustrated based on possible and imagined tableware. Note that naming them requires varying levels of effort and is difficult for the 'impossible' pieces

Offering test persons a card with an image, such as one from the group presented in Fig. 3.9, reveals two phenomena. If asked to name the respective object, the response time steadily increases as the object tends toward the atypical end of the spectrum. If asked about previously viewed images, the typical ones are specified as having already been seen—even if they were never presented. Our imaginations focus on the respective typical 'central' item.

An entire imagined field of similarity can also shift completely. A case in point: wandering along a sandy beach with all manner of washed-up and half-buried objects. A white, finger-thick crescent projects from the sand. The handle of a pot or bowl? Everything that features such a handle 'comes to mind'. Nudging it, however, unearths the cheekbone of a pig's skull. Every picturized piece of crockery disappears instantaneously, only to be replaced by everything we believe to know about mammal skulls.

Such fields of similarity provide the basis for deriving the key concepts of morphology in both science and everyday life (Chap. 4, Sect. 4.3), namely the forms of homology and of analogy. Experience then helps to eliminate 'false similarities'. Importantly, what is potentially false in the similarities *simul hoc* is often the explanation, the *propter hoc*, that we subsequently append.

3.4.3 On Structural and Class Hierarchies

The above discussion introduced the preconditions that prompt us to recognize structures as well as classes of systems. How are these arranged and interlinked? Structural hierarchies emerge virtually automatically. The *simul hoc* of gestalt perception drives the process. This situation differs in the perception of class hierarchies, which first develop based on fields of similarity.

The approach here is to describe (a) the cognitive process, followed by (b) the patterns in the relationships between class concepts and finally (c) how these relate to the structural concepts.

(a) Objects within a field of similarity are immediately arranged upon their *perception* in a harmoniously divergent manner—around a central or basic, typical representative. In such arrangements, the objects radiating out toward the margins become increasingly dissimilar. Moreover, the individual permutations are harmonious or at least considered to be so. This imparts something akin to a developmental direction characterized by a specific type of incremental transformation.

This process yields groupings based on the features unique to them. Recall the transformation of the mammal forefoot to the even-toed or to the odd-toed representatives (horses versus camels and deer) and other groups (Fig. 3.8). This initiates a conceptual differentiation within the constellation. The result is a class with sub-classes. This is valid beyond the forefoot or 'hand skeleton' example to include all the structural elements of organisms—in the present case breaking the class concept of 'mammal' down into its subdivisions. While this process can be consciously tracked, as has just been done here, the motor once again appears to be ratiomorphic.

Geomorphologists, archeologists, ethnologists and art historians order their fields of interest in the same manner. Space restrictions prohibit going into more detail. The illustrative examples here will continue to be anchored in a 'propedeutics through biology' because this discipline has yielded the greatest volume of viewed and ordered material.

Importantly, the class concept 'mammal' does not stand in isolation. Rather, it is arranged together with fishes, amphibians and others as a class in the 'superclass vertebrates'. The subclasses, in turn, prove to be composed of subordinate classes.

This describes the systematics of organisms extending from species and genera to families, orders, classes(!) and phyla, to the five kingdoms of life. These major ranks are typically further subdivided, yielding a total of somewhere between twelve and eighteen levels. Thus, the five kingdoms with their approximately two million described species encompass half a million class concepts.

This approach is valid for treating all complex systems, including the inorganic realm and artefacts, the only difference being that these are not as minutely differentiated (as in the organismic realm).

What is difficult to recognize and long overlooked in the natural sciences is the fact that, methodologically, a hermeneutic process is involved (briefly mentioned earlier and practically demonstrated later in Chap. 4, Sect. 4.2).

(b) *Grouping* such class concepts again manifests hierarchic order—or sometimes even something resembling a trajectory (as in the system of organisms). Genealogical relationships explain this.

The hierarchic systems of concepts in the realm of artefacts do not necessarily follow this rule. This is because hybridization, in the sense of major interactions between developmental pathways, occurs in the development of cultures but not in the phylogeny of organisms. Thus, early African art clearly influenced European art in the late nineteenth and early twentieth century, but a seastar genome can no longer interact with that of a bird.

This distinction is in pronciple dichotomous, i.e. it exhibits binary alternatives. 'In principle' because more detailed examination reveals that all splits in the history of both organisms and cultures occurred consecutively. The precise succession of such bifurcations often remains unstudied or unrecognized. This gives rise to what might be referred to as 'mass hierarchies' (Fig. 3.10). Such

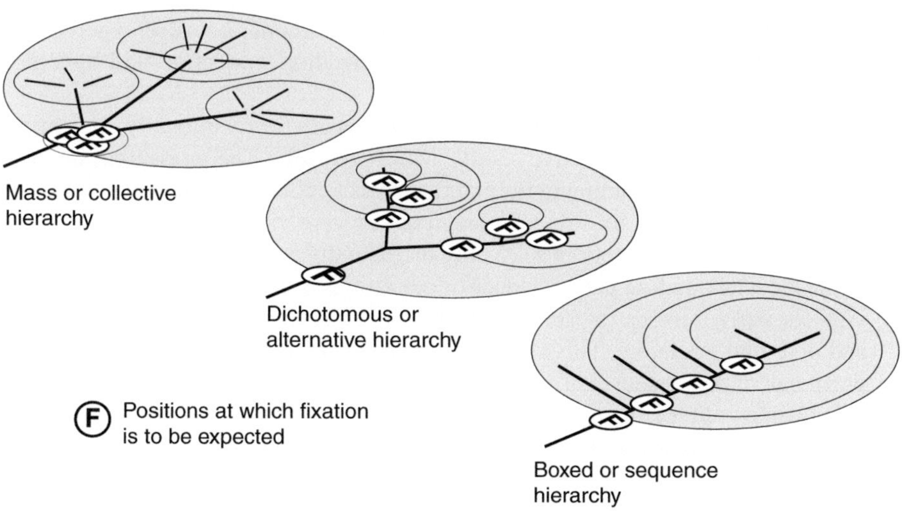

Fig. 3.10 The three forms of hierarchy. 'F' denotes those points at which traits have become fixed (a prerequisite for hierarchic arrangement). Cognitively, human perception assumes mass hierarchies of subordinate system groups whose sequence of branching is yet unknown. Dichotomous hierarchies are typical. These can lead to boxed hierarchies when a particular developmental pathway begins to dominate in large system-groups (e.g. the vertebrates)

constellations arise when young, poorly differentiated branches have not yet formed discrete entities. Examples include species of large genera, dialects of a language or the school of a master.

These can be distinguished from 'nested hierarchies', i.e. those in which the branches largely follow a main axis (Fig. 3.10). This is sometimes the case in older bifurcation series such as those found at the level of classes and orders in the animal kingdom, or in the early divergences within language families or cultures. Such main axes probably arose because one of the branches proved to be more promising than the respective alternative.

(c) *The interrelationship*: all classes of complex things in this world are composed of real objects, and each of these objects is in turn built up by a hierarchy of structures. This explains the very specific relationship between class and structural hierarchies. It also provides crucial insight into how the concept of hierarchies arises in human consciousness—even if this be difficult to convey.

This calls for separately treating (c1) the unifying principle, (c2) the inevitable need for qualitative terms and (c3) the type of linkages between the two hierarchies. The theoretical nature of all these terms is addressed in (c4).

- (c1) The first step is to present the *connecting or unifying principle*. It encompasses the interlinked structural hierarchies making up the system of 'comparative anatomy' along with its class hierarchies, which in turn comprise the system of 'systematics' (more precisely 'comparative systematics', although this term is rarely used). Both systems presuppose a comparative approach. This is interesting for two reasons: First, because neither system could have been developed conceptually without the other. Second, because experienced biologists are only partially aware of applying this connection and have not scrutinized it in detail.

  What connects the two disciplines is the necessity for reciprocally comparing and confirming their respective results. More precisely: The terms used in anatomy first gain their meaning in light of the terms applied in systematics and vice-versa. Terms related to structure and classes give rise to one another. This proceeds automatically and we are only incidentally aware of it, making this one of the most astounding performances of our ratiomorphic program. It was simply vital for early humans to correctly overview the profusion of fruits, prey items and potential predators. This is reflected in the surprisingly accurate systematic grasp native peoples have of plants and wildlife.
- (c2) *Qualities* first crop up as an issue in the methodological concept of 'numerical taxonomy' (Sokal and Sneath 1963; Sneath and Sokal 1973). The notion was to circumvent the qualitative characters of complexity by reducing the task of taxonomists to comparative measurements. This is doomed to failure. The ensuing stalemate between numerical and traditional taxonomists (e.g. Mayr 1965; Riedl 1975) is interesting in the present context.

  The 'numerical' adherents were justified in stating that the 'traditionalists' failed to be cognizant of their own method. The latter could rightly demonstrate that simple measurements alone were insufficient. Both failed to recognize the workings of our innate faculties. Numerical taxonomists had no plans whatsoever to tackle this issue, and the classical crew didn't consider it doable. The crux of the controversy revolves around the 'weighting problem' or how to rate the relative importance of traits. That problem must and can be resolved (see Chap. 4, Sect. 4.4.1 (a) in connection with structure and compare Figs. 4.37 and 4.44).

- (c3) The *configurational* relationship of both hierarchies is simple: they intersect. This can best be envisioned as being aligned at right angles to each other. The borders of the class concepts cross certain borders of the structural terms along the entire hierarchic constellation (Fig. 3.11). Specifically, they intersect at those borders that best delimit the classes. These are the precisely

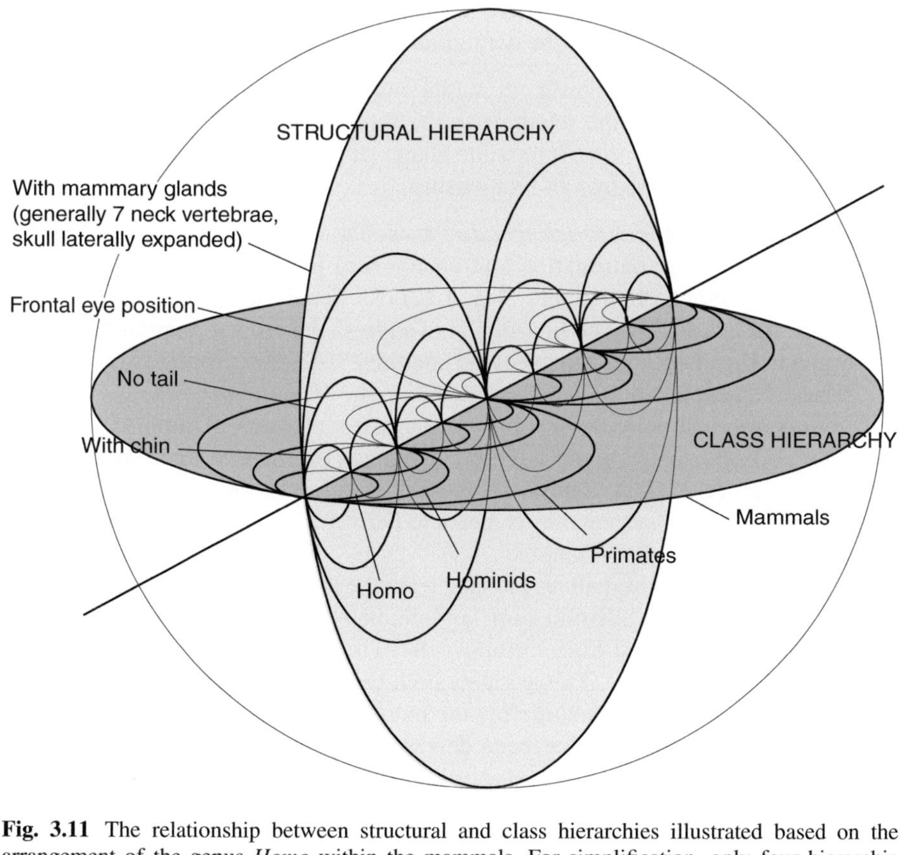

Fig. 3.11 The relationship between structural and class hierarchies illustrated based on the arrangement of the genus *Homo* within the mammals. For simplification, only four hierarchic levels with two cases each are presented, and only one series is labeled: along the structural hierarchy (left) only one diagnostic trait each is inserted, along the class hierarchy (bottom) the systematic unit

defined 'differential-diagnostic characters' (see Chap. 4, Sect. 4.4.1 (b)), i.e. traits that are present in all representatives of one class but absent in all others.

A basic requirement for this two-fold hierarchy is the presence of at least one alternative to each class or structural term. While this may appear trivial, it turns out that all terms require an alternative in order to be grasped conceptually.

The differential-diagnostic structural features that help define the respective classes are hierarchically ranked. These, in turn, yield the hierarchy of the classes themselves. Conversely, the hierarchic ranking of the systematic (e.g. class) groupings yields the ranking of the structural features.

This is not always immediately apparent because the traits best suited to achieve the desired discriminatory power are extracted from different structural hierarchies. Figure 3.11 provides an example. Importantly, the vertebrates as a whole, for example, can be classified based on several trait hierarchies: the musculoskeletal system, the vascular system or other features. The hierarchic principle, however, is never breached because the traits of higher hierarchic levels always provide the basis for the configuration of the next lower ones.

This discussion tests our powers of spatial imagination and calls for three successive steps: presenting (i) the relations to the three 'pathways' and four forms of causality separately from the (ii) possible planes of intersection and (iii) the 'curled over' configuration of the overall relationship.

(i) Note here that the four *forms of causes* as well as the three *'pathways of unfolding'* (perception, confirmation and emergence) insert symmetrically into the class and structure hierarchies (Fig. 3.12) (for details see Chap. 5, Sect. 5.3.3; Chap. 6, Sect. 6.2.3). The first step is to address the crucial relationships illustrated in Fig. 3.11: they represent the cognitive basis that prompted Lamarck to forward an explanation, namely a theory on the origin of species.

It is also instructive to see that the four main methods of comparative anatomy and systematics yield a double symmetry in the two interlinked hierarchies (Fig. 3.13). This is particularly evident in the sector containing the class concepts because systematics is meant to deliver a tangible product and practical definitions are important.

The pattern of alternatives pursued from the kingdom level down to the species is known as an 'identification key'. In the opposite direction, the step-wise arrangement into next-higher groups—from the species up to the kingdom—is termed 'classification'. This pertains both to the overall diversity of forms as well as to identifying and ordering the individual species.

No such terminology has been developed in comparative anatomy. This is because research into a system typically directs the perspective simultaneously in two directions: to the sub-systems (its constituents) and to the next-higher system (defining its affiliation). Thus, evaluating a tissue relies on knowledge both about its cells and about the organ it constitutes. Once again, the herme-

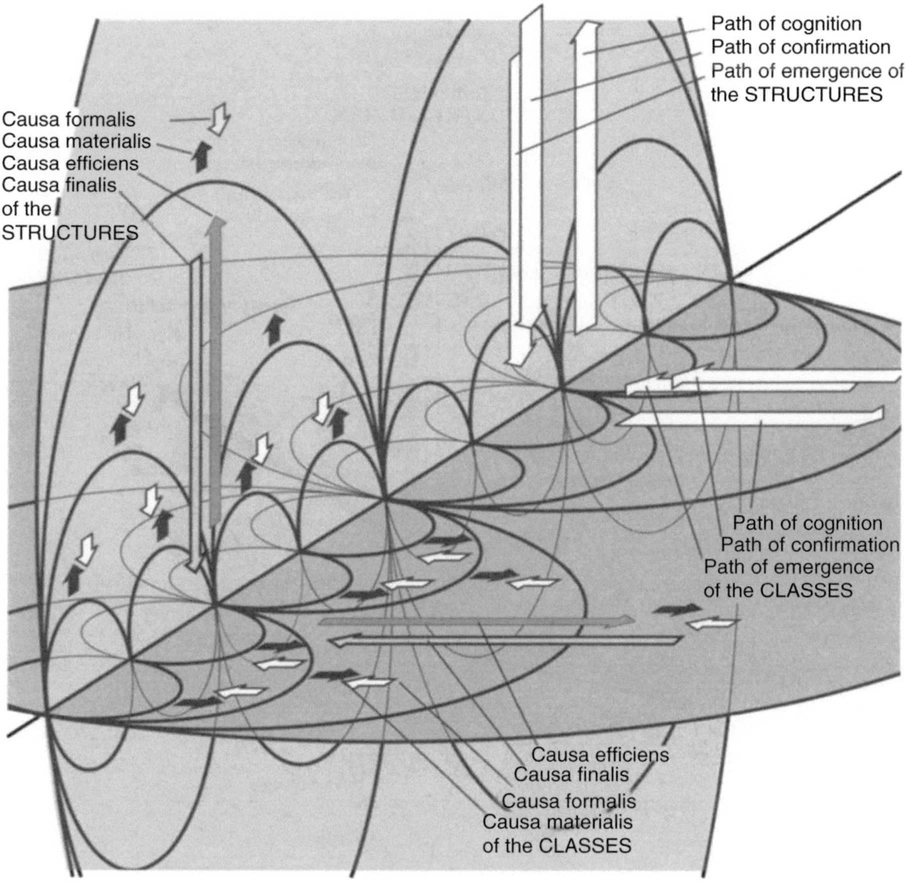

Fig. 3.12 Additional relationships of the structures and classes. Based on Fig. 3.11, the symmetries of the 'three pathways' (that of 'emergence', 'cognition' and 'confirmation') are also entered, as are the 'four forms of causes'. This serves as a background orientation for the topic 'explanation' (see Chap. 5, Sect. 5.3.3; Chap. 6, Sect. 6.2.3). The purpose here is simply to illustrate the connection of structures and classes in the framework of later discussions

neutic principle described above as 'reciprocal enlightenment' plays a role (see Sect. 3.2.4 for practical applications). Importantly, this comparative approach is unreflected, explaining why the process is so difficult to convey.

An additional aspect is allocating the four *causae* and the three pathways to the structural hierarchies (Fig. 3.12). The *causae* turn out to permeate the entire system, from the standard components at its base to the principle underlying the individual system as a whole (Fig. 3.14).

The three pathways, in contrast, mirror one another in relation to the base of the two pyramids, depicting our direct perception of diversity. From here,

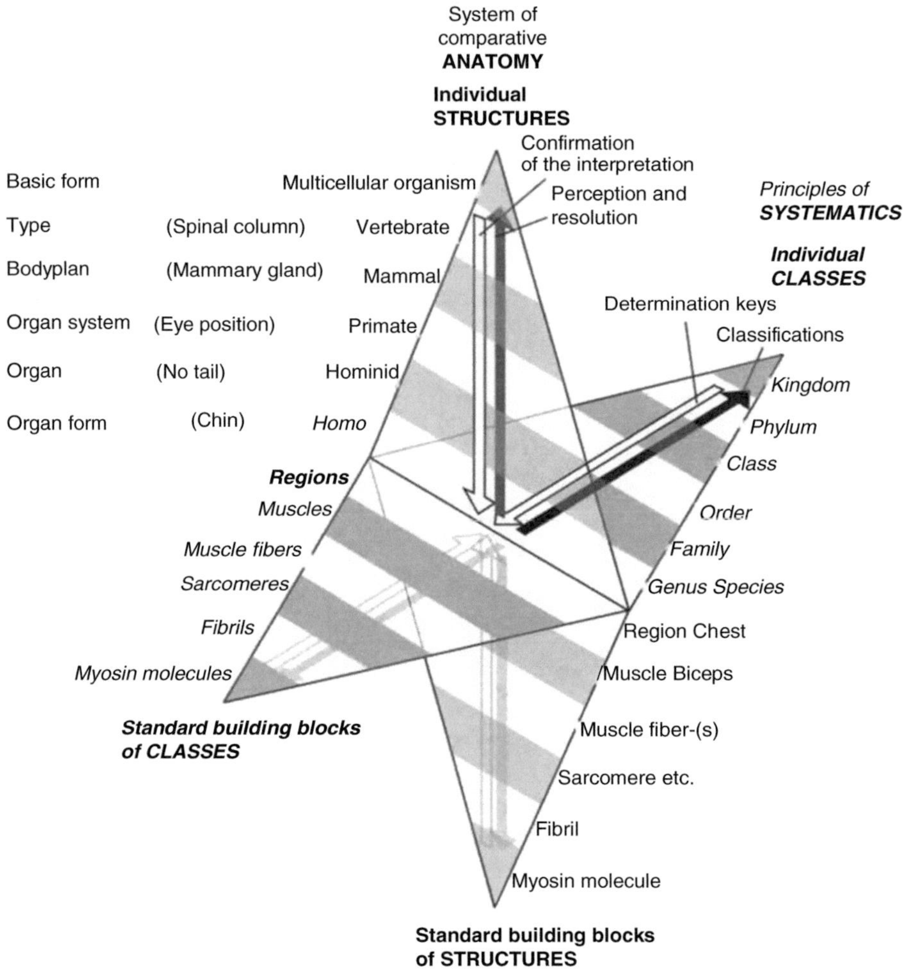

Fig. 3.13 The relationship between comparative anatomy and systematics represented here based on their common foundation in 'individuals' (structures) and 'species' (classes). Note the counter-directional processes of perception and confirmation, as well as of classification and determination. Standard building blocks are not always equally considered in systematics and comparative anatomy (therefore indicated here faintly)

perception progresses across the constituents (the standard building blocks) up to the molecular structures; the same holds true for perception leading to the principles behind the individual system. In contrast, confirmation (or rejection) is generated from those ends back to the base—as is also the case for the conditions giving rise to the complex system (Fig. 3.14).

Recall that the automatic recognition of complex objects (independent of the perspective) is among the greatest achievements of our sensory apparatus. Equally, the automatic nature of the reciprocal insight or enlightenment

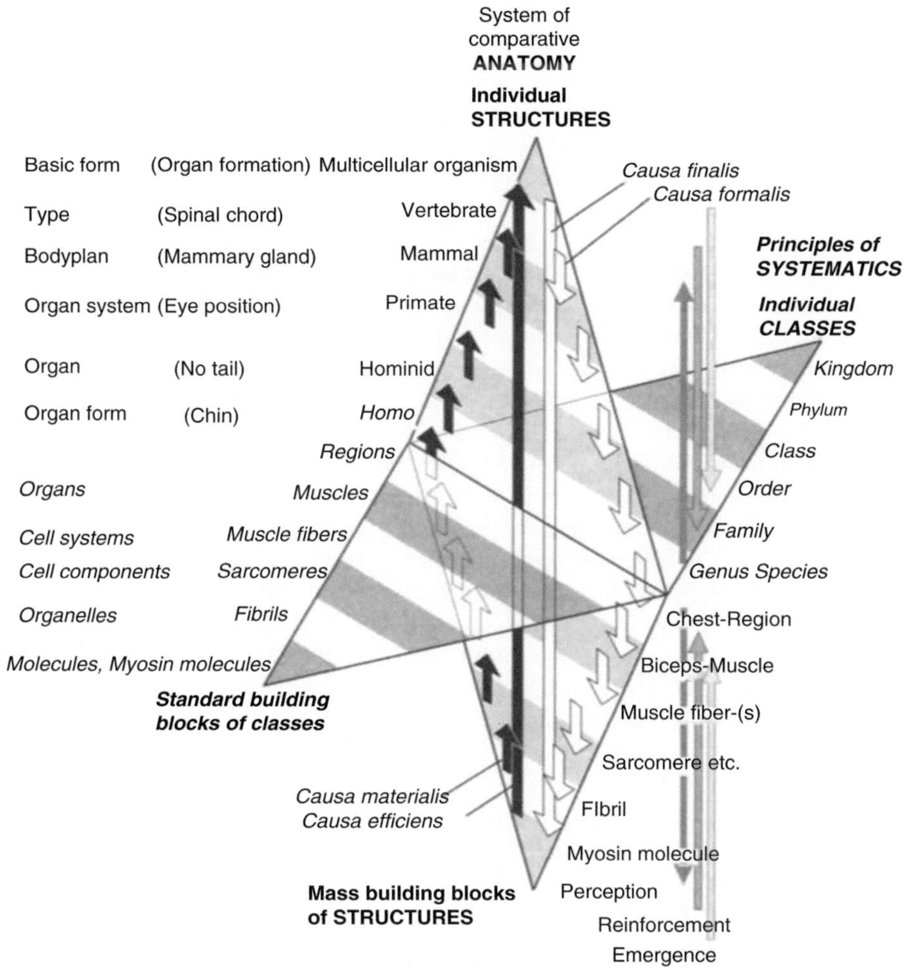

Fig. 3.14 Relative orientations of systematics and comparative anatomy in the further elaboration of Fig. 3.13, with relationships to the standard and individual hierarchies, the four *causae* and the three pathways

outlined above is among the greatest achievements of our innate information-processing apparatus.

This automatism probably explains why the various causal relationships have received little attention when clarifying anatomical features. Closer examination reveals a symmetry here as well (compare Fig. 3.12). The overlying systems embrace the *formal* and *final* causes behind a system, whereas the sub-systems embrace the *material* causes and *driving forces* behind the process (see Chap. 6, Sect. 6.2.3 for details about distinguishing the four *causae*). Note here, however, that the material and formal conditions or causes take on a different ‘appearance’ in each tier and therefore rely on different terminologies. Electron microscopy, cytology, histology, organ research and the study of body plans tend to develop their own idioms. In contrast, the drives and final causes extend unchanged terminologically throughout the structural hierarchies, from the energy gain in cells all the way to the purpose of species survival and reproduction. Overall, the sub-systems answer what a system is made up of (the whereof), the overlying systems answer the why (what for) behind a system.

(ii) Back to the *levels or tiers*: The fact that structural and class hierarchies intersect is an important insight. For the sake of simplification, it suffices to examine this contact at the level of species and individuals (Figs. 3.11, 3.12, 3.13, 3.14, 3.15, Plane A) and to show that the boundaries of the hierarchic terms of both systems meet at this point of intersection. In reality, however, the two hierarchies intersect—depending on the focus of observation—at all levels. Figures 3.15, Plane B and 3.15, Plane C illustrate this.

The principle itself is again simple: the more universal elements of class hierarchies intersect the more universal elements of structural hierarchies. While this may seem trivial, the underlying hermeneutic process is far from it.

(iii) The *overall interrelationship*: From this perspective we can assume that going from the species to kingdom level (i.e. as the number of class hierarchies increases) corresponds with an increase in the structural hierarchies both in the direction of bodyplan composition and of component cellular elements. In other words: the lowest or most variable structures are the tissues and organ types. From this level on, the universality, i.e. conservativism, increases in the direction both of organ types and bodyplans on the one hand, and cell types and ultrastructures on the other.

Accordingly, in the mid-level systemic categories (such as orders) the cell components and organs are the most characteristic features (Fig. 3.16, right). In genera and species, these features are the cell systems and organ types (31, bottom), and in kingdoms and phyla the molecule types and basic bodyplan types (31, top).

Standard building blocks and overall architecture appear to be the two most conservative levels in complex systems (Fig. 3.16). This is also partially true for inorganic systems and certainly so for artefacts, although studies on this aspect are lacking.

This relationship can best be envisioned by ‘curling up’ the two-dimensional ‘double-pyramid’ configurations of both structural hierarchies until their tips (the most fundamental constituents of the standard building blocks and the basic architecture defining a bodyplan’s individuality) meet.

This yields a coherent orientation of the three pathways and enables aligning the system categories according to their scope and to the temporal succession of their constraints. The four *causae* are oriented at right angles to these. This general principle will again be highlighted in relation to patterns of explanation (Chaps. 5 and 6), ranging from the evolution of the cosmos to that of organisms (Figs. 5.3 and 6.19).

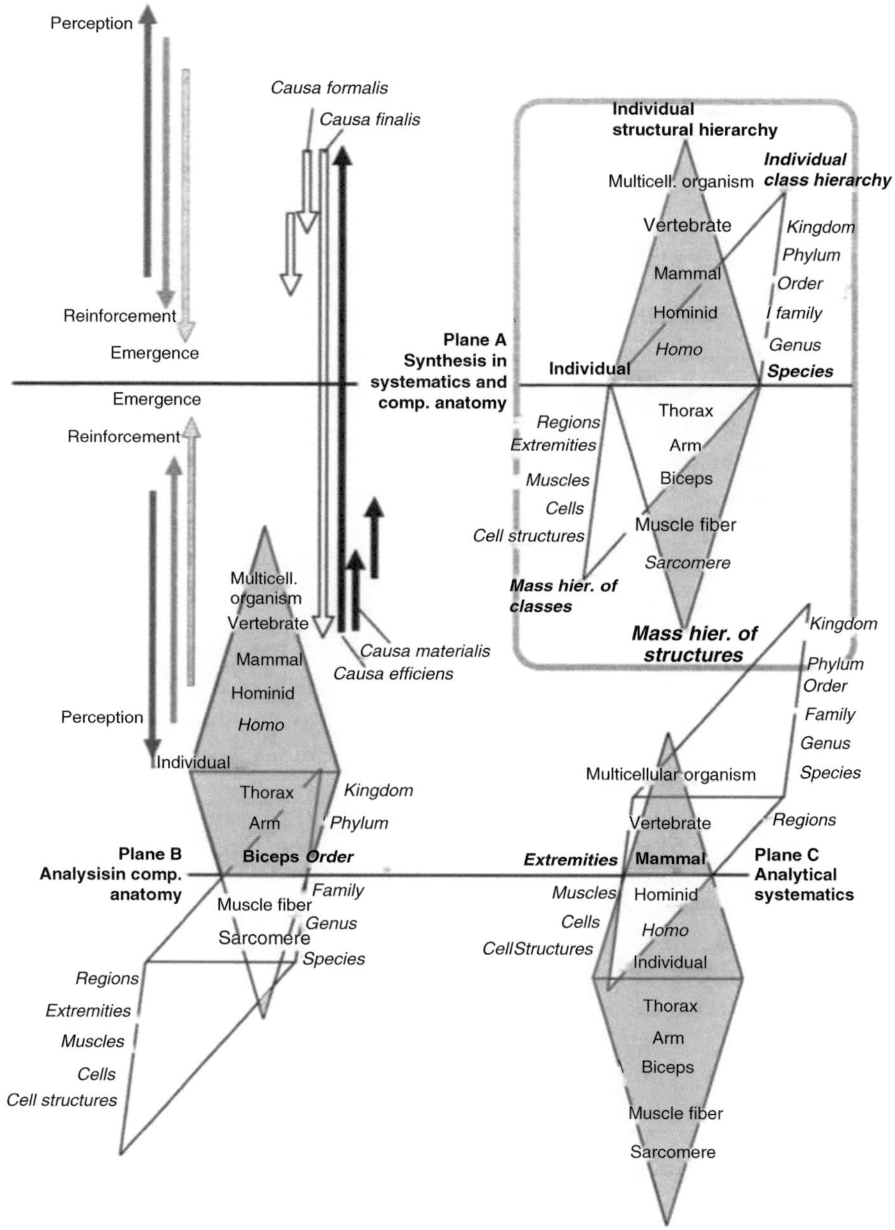

Fig. 3.15 Planes of intersection between structural and class hierarchies; combined with the reference levels between comparative anatomy and systematics. Note that structures and classes can be correlated at all levels and that the paths of determination and classification in the hierarchies run counter-directionally, as do the paths of perception and confirmation (compare Fig. 3.13). They refer in all cases to the plane of observation. The causae, in turn, always run through the entire system

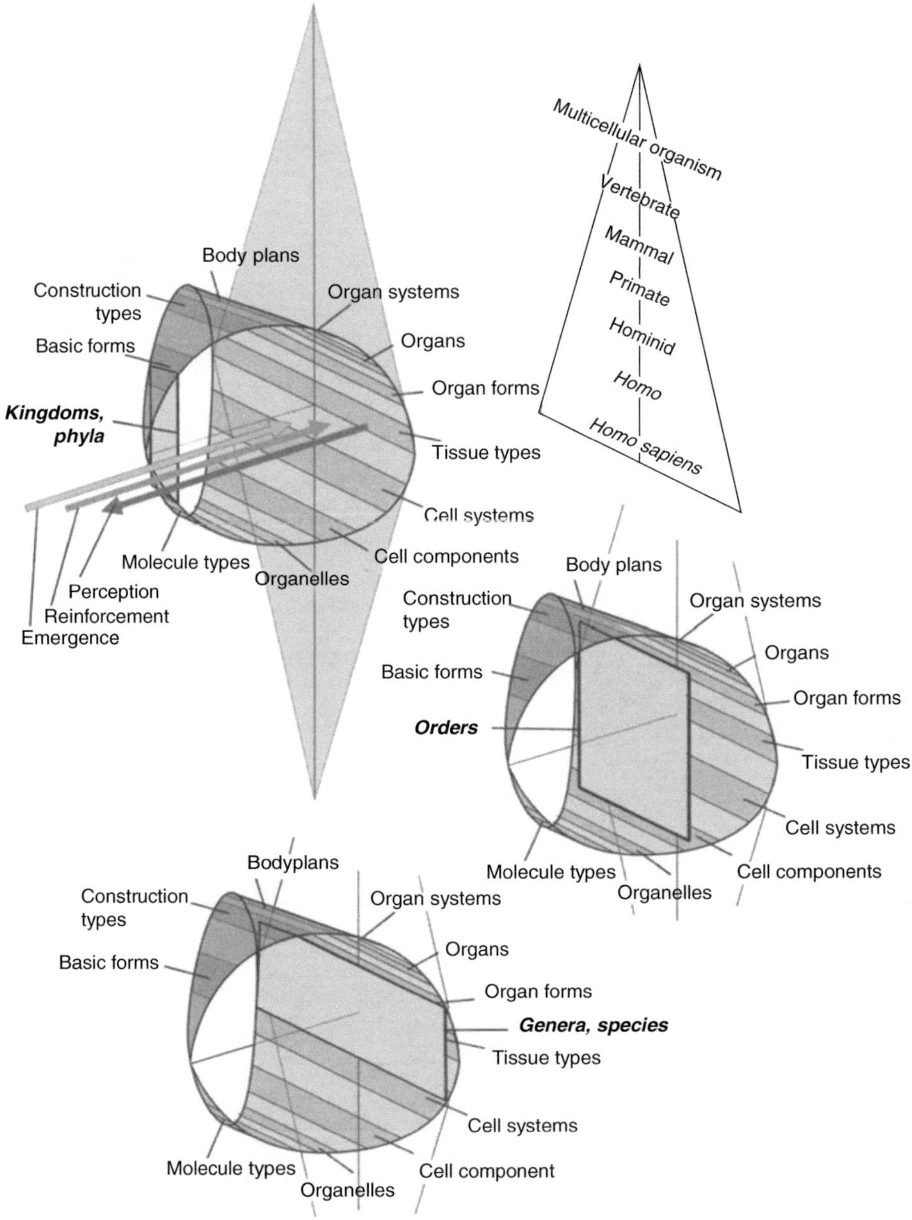

Fig. 3.16 Overall constellation in the system of structures. Representing this requires 'curling over' the 'diamonds' of both structural hierarchies (as presented in Fig. 3.15) such that the basal constituents of the standard building blocks and the basic form of the individual (unique) type touch each other. That, in fact, is how the system arose. This reveals the uniform position of the three pathways and the assignability of the characteristics in the tiers of the system categories

- (c4) *The theoretical character* of all the terms applied here should be kept in mind. Specifically, comparative anatomy, systematics and the links between the two systems represent a constellation of theories. These systems of theories adhere to all three conditions of empirical research. First, that nothing can be known in advance: all applied terms are valid on a trial basis only and are subject to modification and even elimination. Second, only experience can improve the probability that prognoses about the contents of these systems are correct (this process involves testing the applicability of the system categories and structures of known species on newly discovered species). Third, a contradiction-free system—such as the joint configuration of both forms of hierarchy—yields a degree of certainty that can match that of any physical law.

Chapter 4
Structuring the Perceived

The previous chapter was devoted to the general conditions of cognition, to the theory behind our innate faculties. The next step is to apply this information and demonstrate its practical utility, i.e. to determine the structures this yields in extra-subjective reality. Of course, any distinction between theory and practice rings artificial and is misleading at best.

No practical investigation would make sense without its theoretical background. This may seem surprising considering that researchers often find themselves deeply immersed in empirical studies with little reflection on theoretical underpinnings. One explanation is that such backgrounds are anchored in the ratiomorphic realm, proceeding sub-consciously or in a pre-programed manner (see Chap. 3). Clearly, however, investigations that fail to address a question or hypothesis make little sense.

The issue is then the degree to which we are aware of this. Many a study has been driven by simple, unspecified curiosity. Boredom and even playfulness in the lab are known to have yielded valuable results. Chance discoveries, supplied with later, novel interpretations, have also been reported. The history of discovery is replete with stories about wondrous inspiration and associations. 'Good things come to those who wait' (inductive, probably involving the brain's right hemisphere). Such serendipity is an acknowledged phenomenon.

It is equally recognized that critical (deductive) examination operates differently. The probability of reaching some level of empirical certainty depends on three conditions: First, prognoses must be confirmed through experience; second, they should not contradict those made in adjoining systems; third, they must be able to be falsified.

The clearer the expectation underlying an investigation, the clearer it can be refuted or confirmed by experience. Such an expectational stance can either be anchored in the sub- or pre-conscious, reflect some sort of incidental awareness, or involve the formulation of a hypothesis or, optimally, a theory. The rules can be entirely unspecified or very specific. The latter encompass well-thought-out

R. Riedl, *Structures of Complexity*,
https://doi.org/10.1007/978-3-030-13064-0_4

scientific assumptions or models and constructions that clearly transcend mere empirical experience.

The focus below is on practical approaches and real-life scenarios, i.e. on the conditions underlying how we apply our innate faculties. The task is to examine the consciously reflected parts of the method.

The first step is to introduce (Sect. 4.1) a theory of the world, to then demonstrate (Sect. 4.2) how we arrive at insight into the order behind things and further examine this in relation to the principles (Sect. 4.3) of morphology and (Sect. 4.4) of systematics in sciences. The latter are embedded in the structural and class hierarchies discussed above.

4.1 A Theory of the World

We know little about what gave rise to this world and even less about the origins of the underlying principles and fundamental developmental conditions. Nonetheless, we need to make certain assumptions about such matters because everything that we come to think and expect—whether we admit it or not—always harks back to those assumptions. Some philosophers refer to this as 'ontology'. The present discourse (and Konrad Lorenz as well) may well have been influenced by one such position in its dynamicized form, namely by the 'new ontology' as espoused by Nicolai Hartmann (1964).

In brief, the best we can do under these circumstances is to combine the most reliable insights regarding the above origins and differentiations into an expectation that can be tested.

This calls for arriving at an interpretation about (Sect. 4.1.1) the structure of things, (Sect. 4.1.2) their change over time and (Sect. 4.1.3) their general scale, regardless of how much they may be influenced by experience.

4.1.1 The Hierarchic Structure of Things

Hierarchies are a recurring issue here and will now serve as the framework anchoring all remaining phenomena. The issues addressed above as structural and class hierarchies were based on plausibility, but now require critical examination from two perspectives. Our concept of extra-subjective reality must somehow correlate with our powers of conception. After all, the term 'complexity' itself already represents a judgement made by our powers of perception.

The (a) recognized structural forms making up the world and (b) the program behind our thought processes are treated separately here.

(a) *The structural forms*: Theories about the early cosmos hold that matter arose from an initial core of energy. First, heavy quanta were formed, followed by light ones, whereby the latter were captured into orbits of the former. Atoms, composed of nuclei and electron clouds, represented the first hierarchic entity. Unless of course such a hierarchy was already inherent in the organization of quarks, which themselves make up the quanta.

Accordingly, hierarchic organization is an early phenomenon that further propagates in the combination of atoms into molecules, compounds and complex associations of minerals and biomolecules. Whether this differentiation was a necessary development in the known cosmos is a moot point, but a distinct possibility. And it takes temperatures as high as in the sun's core to even partially undo the resulting hierarchic framework.

This structural principle conspicuously and demonstrably continues into the organismic realm. Ultrastructures, cell organelles, cells, tissues, organs, organ systems, metameres and individuals are clearly positioned in a tiered hierarchy. In colony-building forms they can lead to even higher entities, namely groups of individuals: in 'state jellyfish' (siphonophores) this culminates in serial groups of individuals, the so-called cormidia.

The structure of such architecture has two prerequisites: (a1) redundancy, i.e. the use of uniform building blocks and (a2) their assembly into higher entities.

- (a1) The reason for *redundancy* is simple to understand: it represents 'cheap order' in the sense that many standard building blocks can be produced and arranged with minimal instruction. Industry literally builds on this principle. In the organismic realm, all structures—with the exception of cilia (in flagellates)—have apparently arisen based on mass building blocks. This ranges from the genetic molecules to the cell organelles, cells, fibers, eyes (of *Amphioxus*), teeth (of fishes), scales and hair to the extremities and metameres (of arthropods).

  The gradual transformation of such standard building blocks is termed differentiation. Nonetheless, even an architecture as differentiated as that of a human being still retains a high degree of redundancy. Quantifying the amount of information (alternative decisions) necessary to correctly position all the vital molecules yields 10^{28} bits of information for the human body, 10^{11} for a human sperm. Considering that a sperm contains about half of the instructions for building a human, this represents an enormous redundancy (17 orders of magnitude). For comparison, our brain alone contains 10^{11} small grey cells, multiplied times more than 10^3 ribosomes, each of which contains 10^3 same molecules.

  The source of such mass production is evident: reproduction entirely involves autocatalytic processes, i.e. those that produce identical building blocks.
- (a2) *The causes behind the arrangement* of these building blocks into the hierarchy of units evident in all tiers can best be understood using models. Simon's (1965) model is instructive and presented here in a 'translation' of its algorithm by Koestler (1968), namely into a colloquial watch-maker metaphor.

  Two watch-makers make a bet: which one of them can assemble a 100-component timepiece faster? They agree to the condition that anything can be done with the parts as long as it is done by hand and that, when a customer comes into the shop, the work must be set aside. Whatever parts are still loose will fall apart.

The bet stands for competition, the timepiece for complexity, and the customer for stochastic disturbances. Both watch-makers are familiar with the conditions of probability. The first, Technikos, knows that the interval between two customers needs to be long enough to assemble the clock. He consistently manages to put together between 20 and 30 parts. At which point a customer enters and everything falls apart again. The second, Bios, crafts clamps, each of which hold ten functionally dependent parts together. And he regularly manages to assemble ten such units. This means he ultimately needs much less time to put the ten units, each holding ten parts, together into the finished product. He beats Technikos by a wide margin. Resumé: when disturbance and competition are issues, the hierarchic principle is superior.

This highlights a general principle behind economy and dealing with instruction.

(b) *The blueprint behind thought*: As noted earlier, this principle is also valid for human artefacts. It pertains both to mass production and to the subsequent insertion of those products into higher-level systems—the same tiles in equivalent rooms of the same apartments in the same buildings of a housing development. Languages and texts all combine letters into words, words into sentences and sentences into a context (Fig. 4.1). Importantly, they are analyzed in the same hierarchic manner in which they are produced: the syllable is perceived and interpreted based on the word, the sentence used to interpret the word, and the context used to interpret the sentence.

We also clearly think in hierarchic patterns (recall the class hierarchies discussed above). We expect an apple to have a fruit pulp composed of cells and that apples successively belong to the tree fruits, to fruits in general and ultimately to the plant kingdom—all of which do not hold for an Adam's apple, for example.

Which causes which? Two interpretational approaches are available. Either we break the whole world down into hierarchic entities solely because our innate thought patterns dictate it. Or we have learned to interpret hierarchically because the world is actually built in this manner. In the former, the hierarchic structure of the world would be a mere mental tool, a projection of our form of perception. The second interpretation still leaves open the question as to what the specific advantage of such an adaptation might be.

This book tends to the second interpretation. Firstly because it is becoming increasingly unlikely—with improved knowledge about the structure of matter, of organisms, societies and artefacts—that we have been duped into merely projecting our mental perception onto such identical forms of order. Secondly because the advantage of hierarchically disassembling and reassembling the world in our minds has led to tangible results.

The advantage can be illustrated by the 'base two logarithm' example. Finding or correctly determining the correct solution out of 1024 cases would

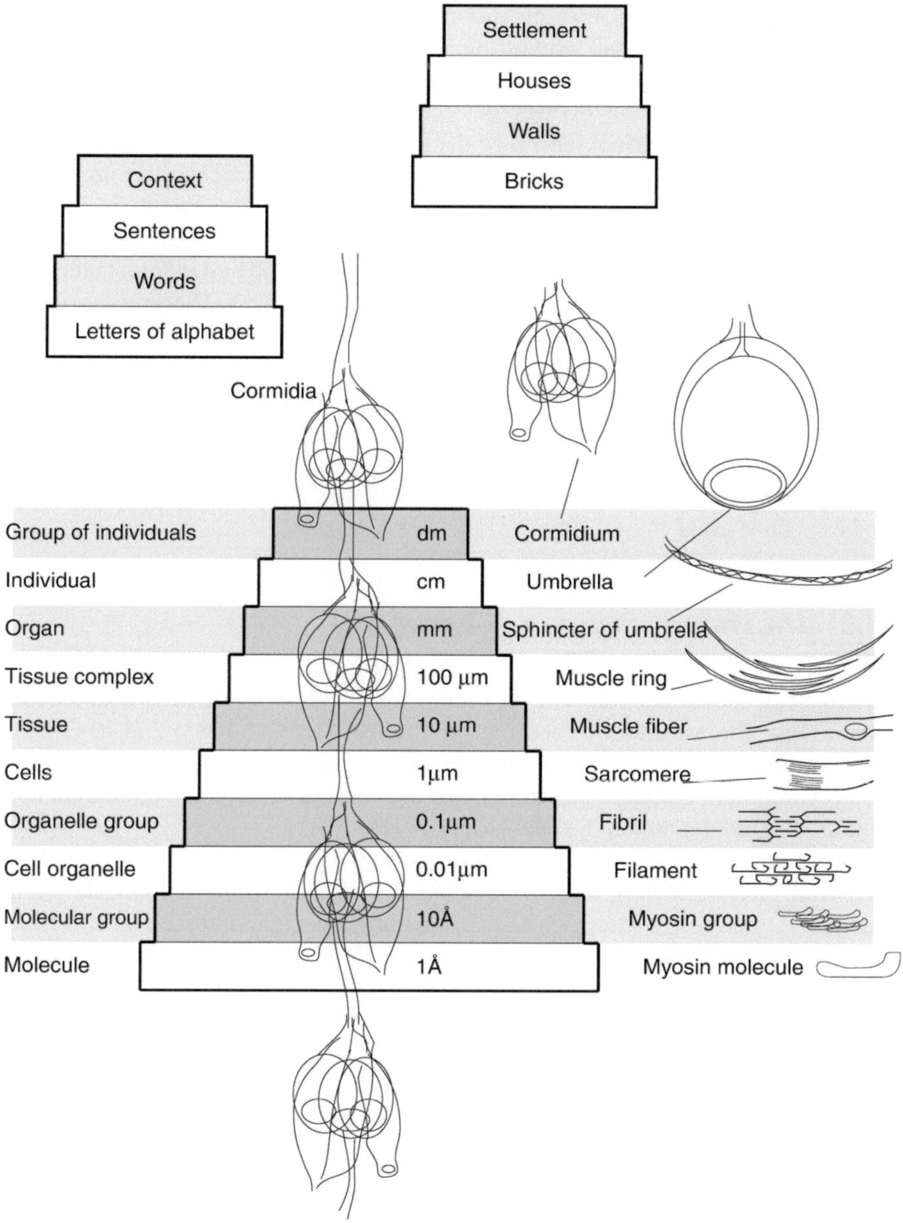

Fig. 4.1 The hierarchic structure of organisms, exemplified by two artefacts and the parts of a siphonophore (jellyfish). On left, tier designations, on right a constituent subsystem, in the center the approximate orders of magnitude

require an average 500 attempts if done by trial and error. If the amount is successively divided in half, then a mere ten attempts would be sufficient. The winning panelist of ‘guess my occupation’ TV game shows is the person who consistently and cleanly divides the occupations in half.

Life cannot be perfect. This is evident in examining how it arose (Sect. 4.1.3) and distanced itself from conditions of stability and physical equilibrium. Life’s balancing act will always be disturbed by misunderstandings, accidents, illness and death. Accordingly, the most obvious expectation is that, under competitive pressure, hierarchic structures in both the physical world and in human thought are more economical and can better withstand disturbances. The structure of the world not only programs our hierarchic thought processes but actually promotes them based on identical underlying principles.

Such interrelationships are introduced as ‘homoiologs’ in Sect. 4.3.1 (e). This refers to analogies (adhering to environmental demands) based on homologs (having the same history/origins). Note, however, that this discussion is anchored in systems of assumptions, even if they are the most plausible ones based on state-of-the-art information.

4.1.2 On Transformation and Emergence

A second consideration underlying any reflection about the world involves transformation. Take the assumption ‘nothing pre-existed’ (Lorenz 1983) and reverse it: accordingly, the cosmos evolved based on internal conditions, successively triggering an evolution of chemical structures, life, societies, communication, thought and cultures, one building upon the next (Riedl 1976).

For ‘creationists’ this assumption reeks of blasphemy, for ‘evolutionists’ of platitude. Nonetheless, it triggers sub-assumptions, none of which are trivial. Three structural conditions behind such transformations must be addressed, even before the causes per se are sought.

Namely: (a) what is being transformed (b) under what conditions, and (c) what do such transformations involve.

(a) *What is being transformed*: First, not everything in the cosmos undergoes change. When matter first appeared, it was in the form of hydrogen, the simplest of elements. And, as cosmologists have taught us, the cosmos continues to remain a 90%-plus hydrogen world. Only on a few planets in its solar systems has evolution given rise to a wealth of elements, complex chemical compounds and, no doubt even more rarely, to life.

    At the same time, consider that life on our planet arose immediately after parts of its surface cooled down to below 100 °C and that, according to cosmologists, the cosmos probably contains as many planets as there are suns. Extrapolating from the untold number of suns, life must have evolved countless times.

Finally, not everything in the realm of living organisms has undergone further development. The earliest life-forms—the prokaryotes, bacteria and blue algae—not only continue to exist today, but their abundance far exceeds that of all the other more differentiated forms combined. And they continue to be vital by fueling material cycles in nature and by being key players in the metabolism of more highly developed species.

The genesis of consciousness and of advanced cultures follows a similar tack. Is there a principle underlying this relationship, which is dominated by conservatism and limited evolutionary processes? We do not know. In some cases the 'higher' did evolve at the expense of the 'lower'. Nonetheless, it is too early to refer to a general principle. Experience shows that only a small proportion undergoes further development, albeit in an astounding and unfettered manner (Riedl 1976).

(b) *The conditions of transformation*: A more fundamental issue here is the conditions under which something evolves. Anagenesis ('higher development') is a case in point. Phylogenetic trees, tiered structures, the levels of structural hierarchies—all tempt us to assume that, much like a tower construction, the higher levels build, story by story, on the lower ones.

This notion is misleading. It contains a kernel of truth, but not the full truth. More correctly, all new, differentiated tiers develop as 'inserts' that are inserted between their component parts and an environment that enables and 'approves' the emergence (but sets its own conditions). Once again, the terminology needs to be expanded.

The most useful approach is to describe (b1) the process itself and then (b2) the conditions that underlie it.

- (b1) *The process itself*: The facts about cosmological, biological and social processes are well known and treated here in that order:

  The current opinion is that the four physical interactions or forces first originated in the early cosmos and probably even developed from one another. The same apparently holds true for the weak and the electromagnetic interactions. The nuclear forces must have separated from the gravitational forces even earlier in time (Weinberg 1977).

  The four forces differ from one another in their reach and, in reverse order, in their strength. In the explosive expansion of the materializing cosmos, the strong and weak interactions defined the realm of microphysics (atomic structure), whereby the effects of gravity and electromagnetic interactions also determined the expanse of the cosmos. The material cloud expanded unevenly: disparate gravitational fields developed and amassed matter locally. Accordingly, all cosmic structure, i.e. galaxies, solar systems and planets, developed in an interplay between two conditions: the constituents ('amount of hydrogen material') and the milieu or environment ('strength and movement of the gravitational fields'). At a smaller scale, galaxies serve as the environment for their solar systems, and the sun as that of its planets (Fig. 4.2).

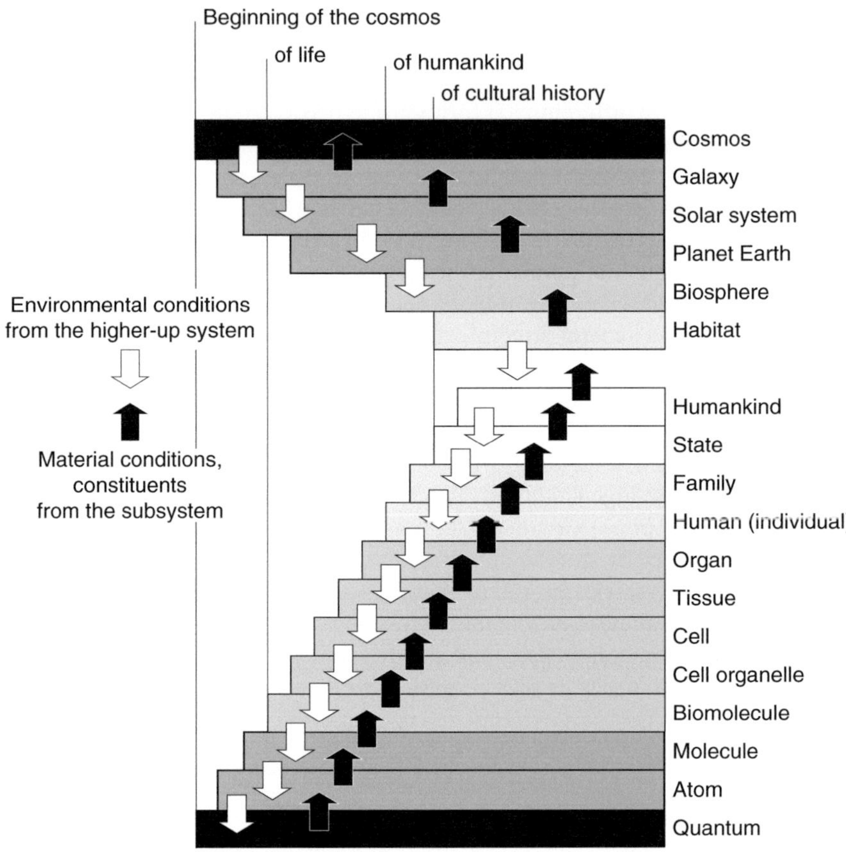

Fig. 4.2 Differentiating the world through insertions between the disposition of materials (material causes) and selection through the respective, next-higher environment (formal causes), highly simplified. Top: key time periods (details in Chap. 6)

The constituents for evolving life were nucleic and amino acids, and the environment was represented by energy-rich compounds, high temperatures and a quantum-flooded atmosphere above the 'primordial soup' of the early water bodies (Urey 1952). The concept of environment is anchored in biology and almost trivial in that discipline.

This all changes when considering that, at higher levels of development, the building blocks play a two-fold role: they represent both each other's constituents and their environment. The potential structural types of unicellular organisms depend on the available biomolecules and their functions (i.e. whether the organism forms a silicate or calcareous skeleton or can also produce chlorophyll). Equally, the types of organelles responsible for their fitness and survival are determined by the overall functional ensemble.

This becomes even clearer at higher levels of differentiation. Whether bone substance, calcareous shells, keratin or tunicin is available for skeletal support depends on the capabilities of the cellular components. The bodyplan, however, determines whether and especially where such support is positioned. Thus, tissues arise in an interplay between structure-determining cells and form-determining organs; organs, in turn, insert between tissue and bodyplan conditions (Riedl 1975).

Constituents and environment—in the form of individuals and milieu—are a more common concept in sociology, ethnology and cultural sciences. Note, however, that milieu here goes beyond the rainforest, a desert or a city to justifiably include the formative ambiance of a family, a clan, a state and its political structure.

As noted earlier, we need to expand concepts when pursuing a uniform principle in the context of a longitudinal theory. The concept of environment or milieu is valid up to the cosmos scale and proves to be applicable at the levels of histology, organology and cultural institutions as well: the same fundamental evolutionary conditions are demonstrable at all these levels.

- (b2) *Conditions of transformation*: The structural interrelationship outlined above becomes even more convincing when examining the conditions and effects of the bidirectional determination behind any new system. The two prove to be quite different and to retain this difference throughout the evolutionary process.

  The difference between these two conditions lies in the type of selective effect they exert. The constituents decide 'pre-selectively', the milieu 'post-selectively'. The selection concept becomes too narrow here because, traditionally, selection has been attributed solely to the exterior milieu. One thing is certain: the constituents exert a decisive (perhaps the more decisive?) selective effect. They do this before any new system can come into being, namely by determining whether, when and what materials are available for the new entity.

  This is true both quantitatively and qualitatively. The origin of cosmic structures provides a quantitative example. If too little matter is available, no sun can form. If too much is present, the sun will collapse. A qualitative example is building a bridge. If enough cement is available but no material is present for its mould or formwork, then the bridge cannot be built regardless how strong the demand. If only ropes or bricks are available as construction material, then the bridge—although the external, environmental constraints are the same—will take on a festoon or arcade form.

  The post-selectivity of the milieu can make its decision only after construction of the attempted novel entity has begun. Although that entity harbors a drive, many drives fail to reach fruition. The milieu ultimately decides about the system's survival—biologically via fitness or culturally via some added value provided. And it always has the final word! If, in the example above, the milieu requires the bridge to support a train, then the rope bridge would be eliminated.

  The most modest and reliable consequence of this insight? Considering only one side of the two conditions fails to shed adequate light on overall evolutive processes! Moreover, different types of causality are at work here (see Chap. 5).

(c) *The mechanics of transformations*: Transformation itself refers to a phase transition and involves the issue of emergence (for both phenomena see Chap. 3, Sect. 3.3.3). In short, cognitive problems are also involved. These are difficult to overcome due to the constraints our powers of imagination face.

The discussion here is restricted to the conditions behind structure formation and, to avoid jumping ahead, it suffices to simply recognize phase transitions as a phenomenon. This means accepting that emergences, even if the changes are solely quantitative, do not enable reliably foreseeing the novel qualities. Moreover, these qualities are not present in the constituents, even in traces. Finally, historical, non-repeatable events are involved: the crossroads negotiated by transitions cannot be reconstructed (see Chap. 5, Sect. 5.2 for details).

4.1.3 The Broadest Parameters

The parameters addressed here were introduced in Chap. 2, Sect. 2.3.2, and the task here is to determine which of them should be incorporated into our paradigm and how this can be done. Again, this effort is rooted a theory of the world. Moreover, the paradigm itself also has an empirically unresolvable background: no paradigm can validate itself.

The conditions behind every potential transformation in this world pertain equally to the theoretical frameworks of physics, of life and of human cognition.

This calls for addressing (a) the entropy problem, (b) levels of stability and (c) the energy-information relationship.

(a) *The entropy problem*: The tenet of entropy—the Second Law of Thermodynamics—is one of the main laws of physics. Like all physical laws it is universally applicable and states that, in closed systems, all temperature gradients will ultimately be eliminated and matter will become completely uniformly distributed everywhere. This is termed physical equilibrium. A steam-powered engine in an enclosed hall can run only until the temperature surrounding it has reached the same value as in the boiler. If we leave a flask of perfume open, then the effort required to return the odor molecules back into the bottle soon exceeds any conceivable effort.

What, however, marks a closed system? One might think of it as a container whose walls cannot be penetrated by either temperature or matter. No such container exists. Whether the cosmos itself can be thought of as a closed system also continues to elude us. This, however, is irrelevant for the paradigm discussed here, as is whether the universe will continue to expand or is doomed to future collapse.

The decisive fact is that evolutive processes produce differentiations where the law of entropy would lead us to expect a de-differentiation. This means that living systems circumvent rather than break that law. They adhere to it by delivering more de-differentiation into their environment than they need to build up and maintain their own differentiation. Any complex system that creates ordered internal structure where previously there was none must export disorder, at least in the form of heat. This can even be a steaming pile of manure due to the wealth of new life arising inside (Fig. 4.3).

As Schrödinger (1957) so aptly stated: "Life feeds on order". This truth is a major pillar in the theory of the world espoused in this book. Plants process photons into heat, animals convert both plants and other animals into dung. And it is self-evident that only lower organisms can profit from the remains of our digestive processes. We should never forget that our lives are based on degrading order, and that sustaining human society will ultimately require producing more differentiation than we destroy through consumption.

(b) *Levels of stability*: Life exists far from any physical equilibrium, making it a labile 'dynamic equilibrium' traversed by energy and matter, as so beautifully illustrated by Bertalanffy (1968). Ultimately, life proves to be very life-threatening.

From a physical, energetic perspective, this can be described as a stable steady state (Fig. 4.4). Of course, this stability is only temporary, and even that is merely relative. Compromising a single stabilizing barrier can cause the entire system to collapse.

A more interesting question is how such higher energy levels are created. This harks back to the issues of emergence and anagenesis, discussed in detail above. The logical assumption is that emergences give rise to new energy levels, although we know little about the underlying processes.

The question becomes even trickier when applied to anagenesis. The special conditions under which anagenesis prevails remain unknown.

Phase transitions yielding evolutionary advantages for organisms abound. A prime example is using photons to gain energy. This was a more effective

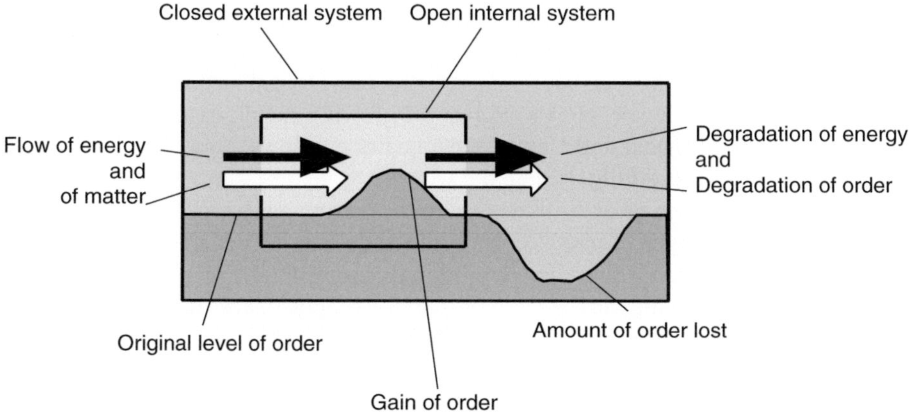

Fig. 4.3 Principle of 'open systems'. They can build up order if traversed by energy and matter—under the condition that they export more disorder into the environment

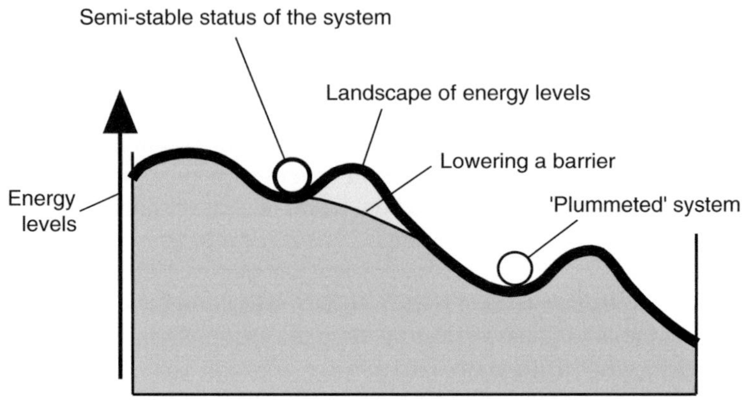

Fig. 4.4 Scheme of stable levels of energy and order, symbolized as a cross-section through a landscape containing stable positions. Lowering a barrier-forming elevation suffices to allow the entire system to slip

strategy and defined the step from the earliest chemoautotrophic organisms such as sulfur bacteria to plants. Equally, feeding on pre-formed energy resources marks the transition from plants to animals, and the transition from cold- to warm-blooded animals enabled mobility independent of outside temperatures.

Interestingly, the advantage never lies in a more economic use of resources. On the contrary, energy is used ever more wastefully over the course of higher development (Wieser 1989). The advantage is always only relative, i.e. within the framework of the inhabited environment. The development resembles a seesaw process. Once organisms in a habitat have reached a certain level of organization, some individuals will benefit by transcending that level through mutation and successfully standing the environmental test.

In principle, this is annoyingly simple. To paraphrase an old witticism: these driving forces alone have enabled a pond of amebas to ultimately develop the Paris Academy. This very same seesaw process—transposed from the organismic to the social sector—helps explain the growing standards and aspirations of human culture (Riedl and Delpos 1996a). Energy flows alone, however, prove to be insufficient to fully explain the world.

(c) *Energy and information*: With respect to the entropy problem, one aspect of the scientific paradigm remains to be addressed: how to generally describe the product of reducing entropy. Schrödinger himself (1957) already spoke of negentropy, meaning the opposite of thermodynamic chaos. In the present context this can be equated with 'order' (Riedl 1975, 1991). Unfortunately, physicists have failed to embrace the concept of negentropy. Our paradigm calls for addressing three sub-problems in this context.

These are (c1) the tiered structure of information, (c2) the information–energy equivalence and (c3) the redundancy content in the concept of order.

- (c1) *The tiered structure* of orderly systems is apparently the easiest notion to convey. A simple example suffices:

  Telling your children "your room is one big mess" refers to level one, i.e. the toys have taken on a random distribution. The toy clock, however, is still working. Adding "now you've broken the clock as well" (level 2) means that its components have lost their functional relationship. Nonetheless, all the wheels of the clock remain intact. The final stage would be "and now one of the wheels is broken too" (level 3—rarely induced by children's play). But even then, the metal alloy of the wheel (level 4) would remain intact.

  The ten or more hierarchic levels of living organisms (Fig. 4.1), along with all their interlinkages and interdependencies, are all the more daunting. This clearly underlines how difficult it is to arrive at a metric description in biology or even in the organization of human artefacts.

  And how do languages fit into this scheme? Can't their content be definitively quantified by the information unit 'bit'? Yes and no. The number of times a printer's letter case with 26 letters of the alphabet has to be halved in order to arrive at a particular letter (Shannon and Weaver 1949) can be readily specified, as is the degree of surprise when a rare letter appears. Nonetheless, even semantic contents are not necessarily comprehensible ('the volume at that traffic volume'). Not to mention syntax, where setting a comma can reverse the entire information content (e.g. 'A panda bear eats, shoots and leaves').

  Nonetheless, our paradigm leads us to expect comprehensibility in the sense of predictability (Riedl 1986), despite our poor success in dealing with complexity.
- (c2) *The equivalence problem*: The notion of an equivalence between energy and information has repeatedly been raised in the efforts to arrive at a solution. The brain teaser involving Maxwell's demon—inspired by Boltzmann (edition 1979)—is instructive. The demon is seated at a tiny door between two gas-filled chambers in a physical micro-world. He allows all the molecules crowding toward the left chamber through, but not those crowding toward the right. This enables the demon to create a pressure gradient. The energy that he gains must be equivalent to the information surplus that he has over us. This is directly related to our macro-world, especially in how the biosphere and its organisms function (Fig. 4.5). As determined earlier, building up differentiation must be 'bought' by exporting de-differentiation. Both processes can be described in units of energy (Odum 1971; Wieser 1989). They are also quantifiable in the sense of available information or predictability (Riedl 1975). This prompted a search for an information–energy equivalent (Riedl 1973, 1976).

  I remain convinced that an undeniable relationship exists between these two key parameters because information can never be transferred without expending energy, and an energy gradient always conveys something. At the same time, my conclusion is that such a relationship can only be determined on a case-by-case basis. The question "How many dollars does a bit cost?" can be answered. Nonetheless, that answer can also depend on the distance over which that bit

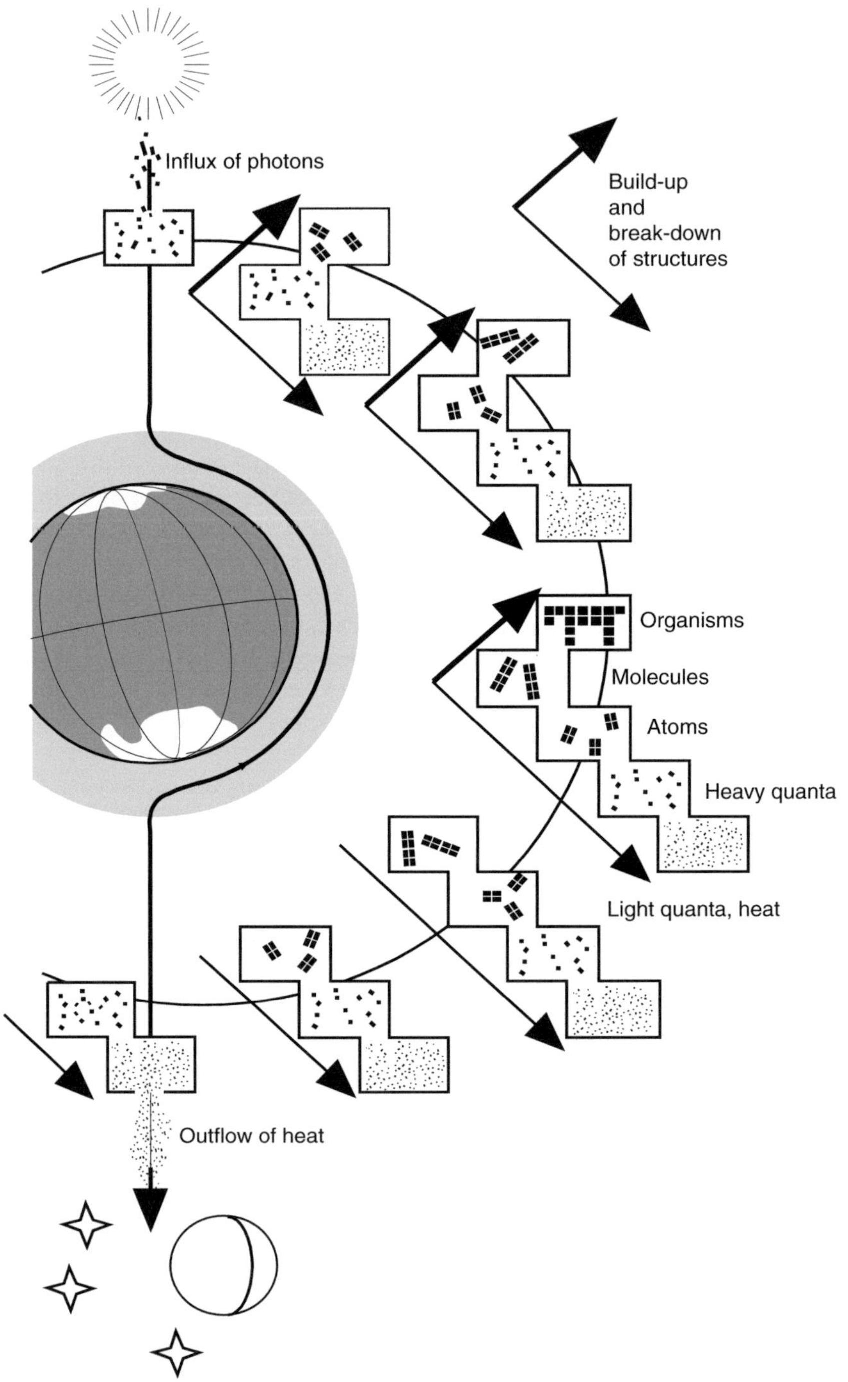

Fig. 4.5 Conditions for building up order in the biosphere (by exporting heat). Note the transformation of heavy quanta into atoms, molecules and biostructures; and the same compensationally back to light quanta and heat (details in Riedl 1976)

must be sent. Values also clearly play a role here. No generally valid equivalence can be expected.

This brings us to the far end of the paradigm—some might also say to the weak link in this 'theory of the world'. Here, cognitive dualisms hinder our perceptive abilities. The classical example of such a dualism is the wave–particle duality in microphysics. All experience points to it being our approach to the world (rather than the world itself) that is thusly split. This condition extends through every tier in the form of a function–structure dualism. It is even reflected in the characteristic separation of verbs and nouns in every human language.

The situation may be related to the space–time duality. This exposes our handicap in imagining a 'continuity' in studying megacosmic dimensions. A clear relationship between space and time is evident here as well, albeit without us being able to specify how many cubic meters of space can be given or obtained for a second.

- (c3) *The redundancy phenomenon*: A similar cognitive handicap, yet of a different sort, is evident in the phenomenon of redundancy. This insight again underscores the limits of human thought.

The redundancy in orderly systems is a clear prerequisite for any empirical gain in knowledge: the degree of certainty in gainable insight decreases with decreasing redundancy. At the same time, the construction effort or degree of differentiation (the perceived 'value' of a system) continues to increase until we can no longer differentiate a redundancy-free order from an absolute chaos of all possible states. A case in point: One hundred million bricks are to be arranged in perfect order in a warehouse with only minimal instruction. They can, for example, be stacked in groups of 20 high and 20 wide, 50 such groups in a row directed north, with a 10-brick distance between groups. For each brick, the position of 3-6 neighboring bricks can be determined. The redundancy of this order is high, but its 'value' is rather low.

Arranging the same bricks to form a cathedral requires enough instructions to fill an architect's office. Redundancy is greatly reduced but still present. Knowledge about the left side of the cathedral enables predicting the configuration of the right side, information on one window can help predict the structure of maybe nine others, and the left window halves provide insights into what the right halves look like, etc. As redundancy decreases, we perceive the value of the order to increase, typically in the same measure as we can estimate the effort required to produce that order, its degree of differentiation and its level of uniqueness.

If this value continuously increases as redundancy drops, does redundancy-free order necessarily have the greatest value? This level of order would be achieved if every brick were positioned such that it provides no information on the position of any other. One hundred million bricks would individually have to be given unique space coordinates. The instruction volume would increase enormously, the data potentially filling an entire library. For the uninitiated, this type of order would be indistinguishable from random chaos.

The data on all gainable knowledge are derived by comparing. Anything that is incomparable remains incomprehensible. If God cannot be compared with anything tangible, is it any wonder that he/she cannot be empirically verified? The relationship between structures and comprehension is treated in detail below.

4.2 The Order of Things

This section builds on the two hypotheses of 'the apparently true' and 'the comparable', which control our innate 'world interpretation' via *simul hoc*. Their interplay is examined here by presenting and validating the hermeneutic process addressed several times above.

The cognitive process remains the central issue, although the boundary to the explanatory process seems to narrow here. This is because the differentiated human cognition of complex patterns is intimately associated with developing explanations. All the more reason to cleanly differentiate the two processes.

Similar to the simpler forms of gestalt perception dealt with in Chap. 3, Sect. 3.3.1, the hermeneutic process also involves an interaction between our innate programs and the structures of extra-subjective reality. Our cognitive stance reflects the hierarchic patterns of the world around us, an insight that plays a role here as well and therefore served as a starting point (Sect. 4.1.1). Conversely, this process helps draw conclusions about types of similarity as well as about the basic patterns of natural order.

Three entities require consideration: Sect. 4.2.1 reciprocal enlightenment, i.e. hermeneutics (as the method), and, for weighting purposes, Sect. 4.2.2 the 'three fundamental forms of complex similarities' as well as Sect. 4.2.3 the 'four basic patterns of natural order' we perceive.

4.2.1 The Process of Reciprocal Enlightenment

Many colleagues have with good reason urged me to avoid the term hermeneutics. It has (through Gadamer 1960; Habermas 1970, among others) become the domain of philosophizing cultural sciences, lost some of its methodological edge and, in that form, admittedly gravitates towards the dilemma of the 'hermeneutic circle'. It is telling that, at the time of this writing, the entry on hermeneutics in the multi-volume 'Encyclopedia Britannica' is restricted to its use in the framework of biblical exegesis. Only more recently have the social sciences once again more systematically embraced hermeneutics (overview in Lamnek 1993), yet without addressing the alleged logical circularity in its strict, epistemological formulation.

Principally, the term describes a knowledge-gaining process: its precision was recognized early on and merits being re-formulated in the present context. I retain this historically matured term out of respect for those who coined the term and to help meet the challenges of the longitudinal theory espoused here.

Importantly, the relationship between hermeneutics and subsumption (addressed under the subsumption scheme, Chap. 2, Sect. 2.3.3 (b)) has gone unrecognized. This scheme of causal explanation, dating back to Hempel and Oppenheim (1948), focuses primarily on its 'deductive-nomological' character (see also Chap. 5, Sect. 5.3.1 (b, c)). In the present context, however, it also encompasses the interrelationships of theories. Both are anchored in the hierarchic structure of the world.

The two methods demonstrate that theories (based on cases from a particular tier) themselves become the cases of a next-higher theory. This yields the hierarchic arrangement of theories already illustrated in Figs. 2.11 and 2.12. Moreover, the 'three pathways' (Fig. 3.12) distinguished here show strong links between the processes of cognition and explanation. This all contributes to making the above-mentioned gap between the two appear so narrow.

This situation calls for (a) providing a synopsis of the history of the term hermeneutics, (b) demonstrating the underlying principle with examples, (c) validating the process based on the current paradigm's structure and, finally (d) rebutting the allegation of circularity.

(a) *A short history*: The term hermeneutics is derived from Hermes, the benefactor god, guide and patron of interpretation and oratory. Methodologically, the first distinction is the speculative *hermeneutica sacra*, as the interpretational art of religious texts. Later, beginning in the Renaissance, the *hermeneutica prophana* was introduced, dealing with the interpretation (still intuitionally) of profane, usually testamentary texts.

Methodological developments are documented at the turn of the eighteenth to the nineteenth century and shortly thereafter. Three positions characterized the differentiation at that time: those of Goethe, August Boeckh and Schleiermacher, although each was unaware of the others' contributions.

Goethe focused on the problem of the 'morphological type', i.e. on clarifying the cognitive process potentially yielding a canon on comparative approaches. Inspired by the debate between Cuvier and Geoffrey Saint-Hilaire (on whether the direction taken in comparative anatomy should proceed from the particular to the whole or in the opposite direction), he developed a recursive principle involving synthetic-analytic cycles.

In 1795, he called for extracting the typical from the similar and dissimilar features of the entities being compared. The resulting expectations or theory behind the type concept (Goethe referred to 'idea') could then be tested on case examples. This approach, after several rounds—referred to above as iterations—would yield the general character. Applied to systematics, he wrote (p. 235): "The classes, genera, species and individuals behave like the cases in a law: they are contained therein but do not themselves contain or yield that law." This was very farsighted and anticipated iteration and subsumption.

August Boeckh focused—even in the late Goethe era—on a formal theory of philology. "Where the grammatical understanding is insufficient to determine the objective word meaning," he opined, "the historical interpretation must be consulted. Whether the grammatical understanding is insufficient can only be determined when one knows the individuality of the author and the genre of the linguistic work." Importantly, evaluating an author and genre of literature calls for an additional theory, namely that of "the historical framework of the linguistic work" (p. 14 of the 1966 edition: the original edition was first published posthumously in 1877 by his students).

This even presages the interlinkage of levels that Goethe already recognized but did not elaborate on. The next section presents the linkage of the tiers based on a simple example.

Schleiermacher, in turn, forwarded a hermeneutic theory that was differentiated methodologically into several fields of application: theology, jurisprudence, philology, archeology, literature, art history and musicology. Goethe's comparative anatomy finds no mention even though Alexander von Humboldt was closely befriended with Goethe, and his brother Wilhelm von Humboldt was versed in the topic of hermeneutics. The relationship to the natural sciences was not recognized and remains so to this day.

(b) *Examples for the principle*: The underlying principle is basically identical in both the natural sciences and humanities.

The first step is to (b1) identify the commonality and then provide one example each from the (b2) cultural and from the (b3) natural sciences.

- (b1) *The commonality* of the methods has already been broached more generally (Chap. 2, Sect. 2.3.3 (b)). In both cases a knowledge-gaining process is involved: perceptions, tempered by reinforcement or disappointment, lead to generalizations (Fig. 2.10). The increase in achievable certainty is based on iterative, reciprocal controls in a hierarchic network of expectational stances. This hierarchy itself develops from such an iteration. The expectations adjust themselves to the structural hierarchies of the respective complex entities and develop hierarchies of classes. Equally, the enhanced prognoses improve the arrangement of the items. This turns the theories of one level into cases of a theory at a next-higher level. The result is a contradiction-free system of prognoses that continues to be confirmed in every new case examined.
- (b2) *Language and writing*: When in unfamiliar terrain, the best strategy is to initially proceed based on plausibilities. This calls for an example that everyone will have experienced: decoding a complex artefact, namely deciphering a letter in very unusual handwriting.

  This example is chosen for its didactic insight in helping trace the process. Note that the heading 'structuring the perceived' in Chap. 5 selects a process (treated in more detail under Chap. 6, Sect. 6.4.2 (b)) conventionally subsumed under understanding. In fact, understanding and cognition are sufficiently related to justify this didactic approach.

  The first step in tackling a poorly legible letter is based on expectations. A salutation and signature will point to a complete missive. We expect that the type of language used and the overall sense can be determined, and that the body of the text will be based on a structural hierarchy involving letters of the alphabet, words and sentences, all arranged into a context. Although readers typically assess all these levels simultaneously, the individual letters are a good starting point.

A commonly occurring symbol appears to be a 'v', although its might equally be a 'u' or an 'n'. The next step is to form words based on individual cases (Fig. 4.6; tier 5) and analytically develop a theory of symbol meaning (tier 6). In words containing only three letters, that symbol is often the first. Additional symbol meanings in such short words are recognized. The expectation: the first letter must be an 'n'; the symbol groupings support this—they yield 'not' and 'nor'. The reader composes them based on other cases of deciphered symbols. The language turns out to be English.

One deciphered word is apparently 'swallow'. Whether it refers to a bird or to a bodily function cannot be inferred from the isolated word. The next step involves using cases based on sentences to, again, analytically develop a theory of word meaning. Conversely, we—synthetically—assemble the sentence meaning from the interpreted words. Whether a sentence is meant ironically or not needs to be determined analytically from the context (tier 3), just as, in a reverse process, the context is assembled synthetically based on the deciphered sentences. The solution is attained when all consistently predicted symbols, words and sentences yield coherent sense. Importantly, the process can be initiated from any level in this structural hierarchy.

- (b3) *The natural sciences*: The complex field of biology provides good examples. The first example is readily understandable even for non-biologists, the second is designed for closer scrutiny by specialists.

One example stems from (i) familiar systematics and the other from (ii) the process of discovery in comparative anatomy.

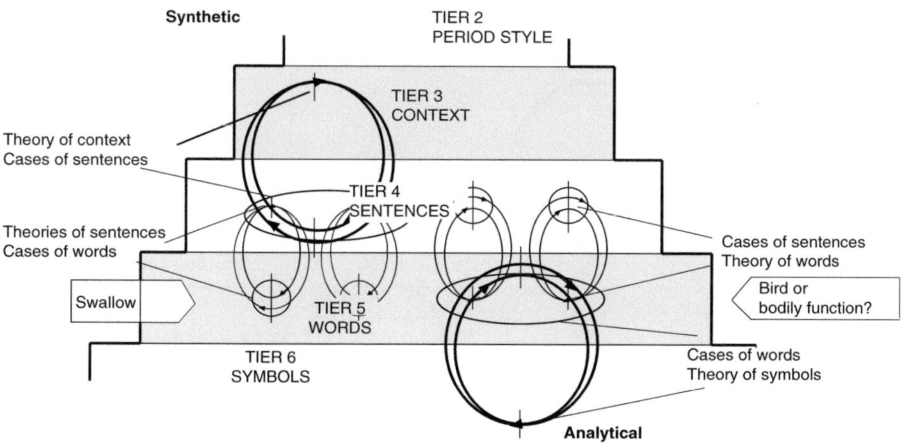

Fig. 4.6 The tiered and reciprocal interpretation of a text ('swallow'), exemplified by the relationship between symbols (letters of the alphabet) and context. The interplay between the applied theories (expectations) is outlined in the text. Scheme as in Figs. 2.10 and 2.11

(i) *In systematics*, most readers will be acquainted with the rough ranking of the mammals. The term 'mammal' (tier 4 in Fig. 4.7) clearly comprises a very specific content. Analytically it encompasses the orders falling under Mammalia, for example the hoofed animals (tier 5). Breaking this down further yields a series of families, each featuring a similar body plan. One example is the horse-like animals, which in turn contain species such as the domestic horse along with its many breeds and then the individual animals themselves. This justifies the content of the term analytically.

The same must also be valid in reverse, i.e. synthetically. The term mammal itself is based on the clustered similarities of the other orders (tier 5). At the same time, it also gains its 'sense' from its membership within the vertebrates (tier 3). That term, furthermore, attains its meaning within the framework of multicellular organisms (2), of animals and ultimately of organisms in general. This founds the term mammal from a synthetic perspective as well.

(ii) The example from comparative anatomy involves deciphering a complex natural entity, namely resolving an as yet unknown bodyplan.

Biologists are often faced with the task of determining a new species or the representative of a new genus. Experience shows that such relatively simple cases fail to optimally illuminate the underlying principle. The example selected here therefore focuses on the process behind the discovery of a new animal phylum, one that later became known as the Gnathostomulida.

Based on didactic considerations, the selected starting point lies at the upper end of the structural hierarchy. The description is more detailed here to help acquaint readers with the methodology and necessary expertise.

The researcher is confronted with an about 2-mm-long, wormlike 'something' pinned down in a drop of salt water between a slide and cover glass under

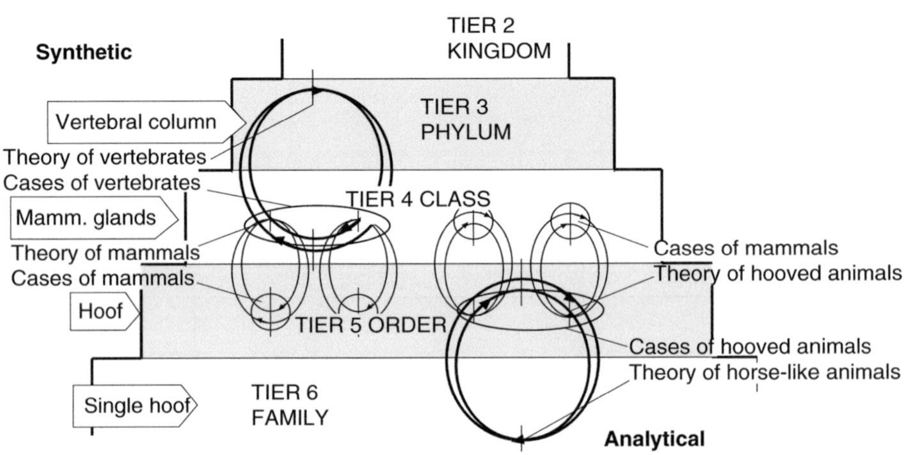

Fig. 4.7 The tiered and reciprocal interpretation of kinship, illustrated based on the relationship of systematic categories (between family and phylum). This interplay is outlined in the text; scheme as in Fig. 2.11

the microscope. The first question is: Is it a complete organism or, as is often the case, a fragment of a tentacle, for example from a bristle worm or from a cnidarian polyp (which can, once detached, continue to crawl about for days)?

The inspection fails to show any ruptured or torn ends. And, much like in deciphering the above-mentioned letter, the researcher now expects a coherent context. All the available background knowledge is mustered to start the process of developing hypotheses that incorporate all potentially comparable classes and all levels of their structural hierarchies. For the sake of clarity, Figures 4.8 and 4.9 (tier 2) present this in a linear manner. The second question:

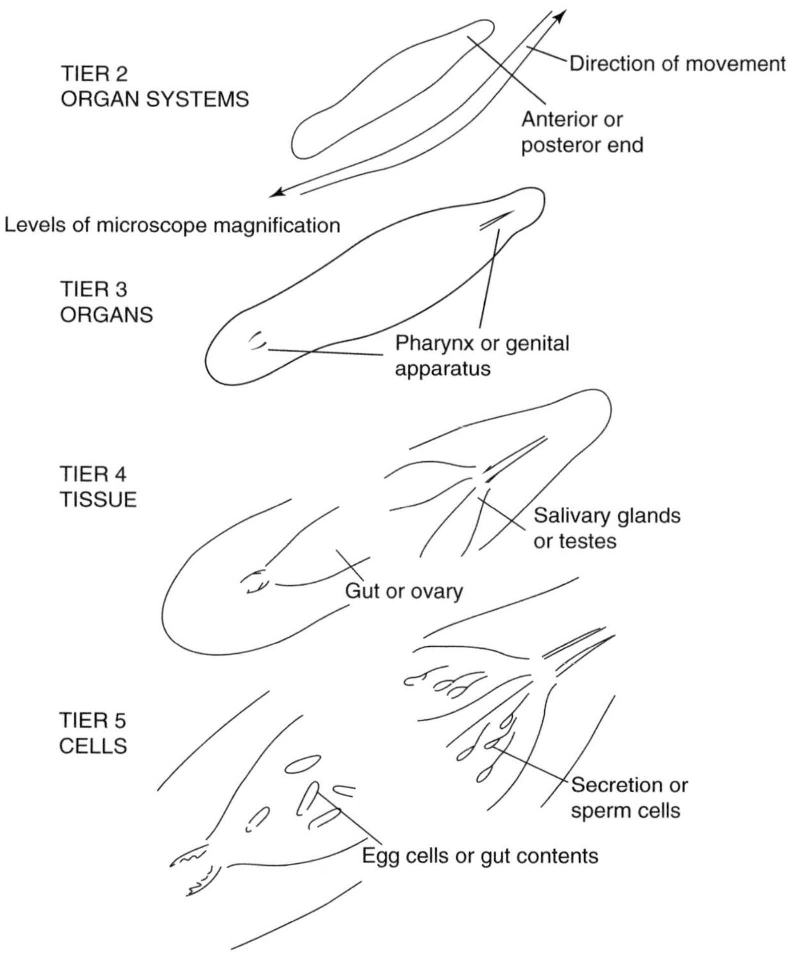

Fig. 4.8 The stepwise interpretation of an organization, exemplified by the discovery of a new organism group (the structural hierarchy of the millimeter-sized Gnathostomulida). The successive questions reflect the respective microscope magnification; see text for interpretations based on allocation to class hierarchies (after Riedl 1998; see scheme of theory formation in Fig. 4.9)

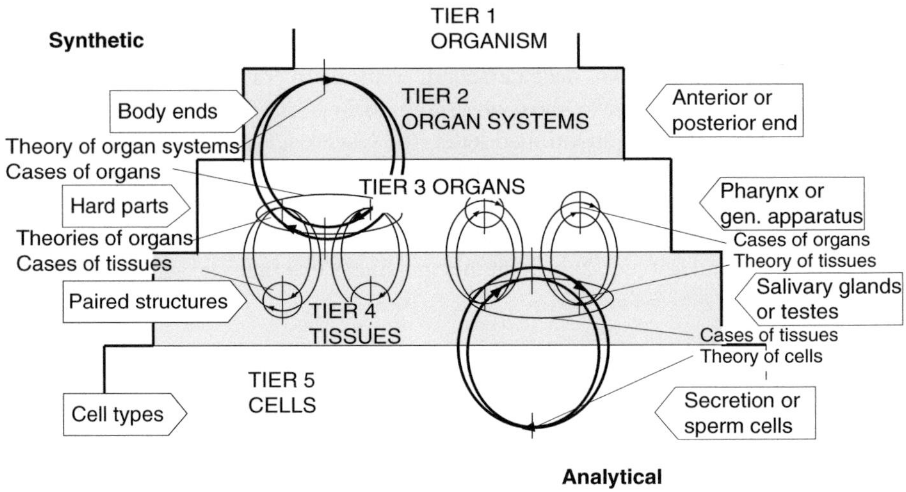

Fig. 4.9 The interplay between tiered theory formation, based on the example and the alternative interpretations of the case study in Fig. 4.8 (of a gnathostomulid). Left: the objects awaiting interpretation; right: the alternatives. See text for discussion and solution

What is the front end of the enigmatic organism? (compare Figs. 4.8 and 4.9, tier 1).

The answer is typically revealed by the direction of movement. Let's say, however, the animal is no longer moving (or, as in subsequently discovered, related species, it can move in both directions; Figs. 4.8 and 4.9, tier 2). This calls for falling back on anatomical criteria.

The next higher magnification (Figs. 4.8 and 4.9, tier 3) reveals presumed organ systems, hard parts, a probably 'cuticularized', rod-shaped element at one end of the body, paired elements at the other end. Rod-shaped parts are known in the pharynx of parasitic rotifers (Seisonidae) but also in the penis of various 'lower worms'. Forceps-like cuticular elements, in turn, are known from the jaws of certain rotifers, from the anterior end of certain flatworms (Kalyptophnchia), and less frequently from scattered other groups in connection with the genital apparatus.

Where does this leave the effort? The researcher must put his or her trust in the next higher level of magnification. Based on knowledge about the organs of other worm-like animals, this can provide clues enabling a theory to be formulated about the animal's orientation. Conversely, case examples of anterior and posterior ends of potentially related organizational types lead to the expectation of a pharynx at the front end, a genital apparatus at the hind end. Such types can be viewed as paralleling the potential languages of the above-mentioned letter. Trying to determine the animal's orientation parallels seeking the sense of the letter's overall context.

The remaining uncertainty is comparable to the sense of the sentences.

The next higher magnification reveals the tissue level (Figs. 4.8 and 4.9, tier 4) and sets the framework for theories between the tiers describing organs and tissue structures. We trust that the cases involving tissues will help formulate a theory about the organs they make up.

The observation process directs the researcher to the tissues, revealing that the paired hard structure is associated with an unpaired tissue sac, the single rod-shaped structure with a paired sac. Paired tissue sacs could be either salivary glands or testes, the unpaired one an ovary or a gut. The meanings of the words in the letter example offer similar alternatives.

The final step is the highest magnification. This leads to theory development between the tiers describing tissues and cell structures. We trust that case examples of cells will help formulate a theory about tissue function, just as case examples of tissues provide us with prognoses about the participating cell types.

The microscope (Figs. 4.8 and 4.9, tier 5) reveals that the unpaired sac contains material resembling food particles, whereas the paired sacs contain sperm. Even this, however, does not mean the organs are definitely testes: they could equally well be foreign sperm, and the organ could therefore be a receptaculum or a bursa seminalis, i.e. part of the female genital apparatus.

Despite the remaining uncertainty, however, the detective work begins to yield a coherent picture. And, much like in the letter example (where the degree of certainty of interpreted symbols cannot be derived from the symbols themselves), it is the overall context that helps to identify the roles the cells play. The food particles confirm the gut, this in turn the jaws and anterior end. Similarly, the sperm confirm the 'testes' theory, this in turn the penis stylet and the posterior end, which ultimately confirms the anterior end.

This biological example omits many additional characters in order to avoid overburdening the story with complexity and to highlight the basic principle. The principle of reciprocal enlightenment or insight was followed in resolving both the letter and the worm example, regardless of whether we term it the subsumption of explanations or the hermeneutics of cognition.

(c) *Validation based on structure*. The principles behind the epistemological paradigm of this book help understand how unique the above achievements are in four respects.

These are (c1) their recursive structure and (c2) their two-sidedness along with the fact that (c3) all of this is pre-consciously controlled and (c4) takes considerable effort to reconstruct consciously.

- (c1) *This recursivity* reflects our innate, knowledge-gaining mechanism. That mechanism must be able to start off in complete ignorance and gradually arrive at the attainable degree of certainty based on arbitrary expectations and their correction through experience.
- (c2) *The two-sidedness* of operating reciprocally between levels in the structural hierarchy—and transcending the phase transitions separating them—is related to

how things unfold. Recall the expectation that differentiation involves insertions: insertions between the constituents of a new system and an environment in which they must assert themselves. The logical conclusion: each newly inserted system can only be understood based on the interplay between two conditions.

The sounds of spoken language and the intentions of communication no doubt existed before the particularities of semantics were developed and before any syntax. Sounds and intentions are their prerequisite. Equally, the cells and the overall individuality of every multicellular organism existed prior to its tissues and organs, providing the framework for their development.

Chap. 6, which deals with the systems of explanation, underlines that this reciprocality helps reconstruct the developmental conditions behind complex systems.

The phase transitions linking or bridging the levels are necessary because, among other things, the details behind the emergence of new systems typically cannot be reconstructed.

- (c3) *The control exerted by our faculties*: The most astounding fact of all is that such complex processes are controlled entirely subconsciously. Impossible, one might say. The undeniable proof is provided by comparative anatomists and systematists, who have used this approach to arrange millions of bits of data into an entirely correct 'natural system'—without ever having been aware of (i.e. expressly defining) the underlying process. Moreover, the newest applications of the method in the social sciences have also failed to provide true validation—despite the indisputable successes.
- (c4) *Establishing the insight*: The awkward effort required to convey the process, to render it intelligible, reflects the nature of human language. Chapter 2, Sect 2.3.2 (c) already underlined language's definitional character. Clearly, a transitory manner of speech—to better convey the transitions—would help to order things, especiall because terms can only gradually attain their full contours.

  Moreover, the linear, sequential form forced upon spoken (and therefore written) language is clumsy in rendering recursive relationships and processes (see d2 and d3 for condensed versions).

(d) *Untangling the circularity*: Finally, the allegation of circular argumentation needs to be countered. Ever since the hermeneutic method has been applied in practice, it has been clouded with the suspicion of circularity, of harboring 'circular logic'. Even, Galileo in his 'Dialogus' (1632) made fun of the process by having the literary figure 'Simplicio' assert the impossible, namely that in order to understand Aristotle one would "have to have at one's fingertips every word he ever wrote."

This calls for (d1) a short overview of the problem coupled with (d2) the conditions for solving it.

- (d1) *The nature of the problem*: The problem itself was already recognized by Francis Bacon (1620): "From all words we need to extract the sense, in whose light every word is to be interpreted." That approach, as summarized by Popper (1973a), "is the famous problem that Dilthey and others termed the 'hermeneutic circle': The whole (a text, a book, the work of a philosopher, an era) can only be understood if one understands its components, but these are only comprehensible if one comprehends the whole." The recent efforts made by sociologists have made little headway in resolving this dilemma, although their special forms of subjective and structural hermeneutics have clearly opened new and important avenues in social research (see overview in Lamnek 1993).

Nonetheless, how can—according to the harshest criticism—A be understood based on B, when B needs to be understood based on A. Stegmüller (1979) stated: "If the hermeneutists, in the thesis they forward about the indissolubility of the hermeneutic circle, are referring to the insurmountability of this dilemma, then this cannot lead to a statement about the workings of humanistic cognition. Rather, it can solely conclude with the demand that all disciplines affected by this dilemma close their gates because their activities represent a hopeless undertaking."

- (d2) *The solution to the dilemma* lies in demonstrating that a validated and reliable cognitive process—rather than circular logic—is involved. Recalling the parallels between hermeneutics and subsumption, the solution requires the following conditions.

  First: Cognitive processes form a double-pyramid of case examples and gainable perceptions about structures and classes. These are ordered based on the hierarchic, tiered complexity of the examined systems, and all operate across the phase transitions between the individual tiers.

  Second: Iterative spiral processes are at work. The spiral represents the knowledge gain based on comparable case examples, an operation involving observation disciplined by reinforcement or disappointment. The resulting theories never originate in isolation. Rather, they themselves always constitute case examples of next higher-order theories. They form their content and are themselves then contzrolled by that content.

  Third: The examination can be initiated at any level. Setting the sight on the overall system makes the examination 'synthetic', setting it on the constituents makes it 'analytical'. The structural and class hierarchies therefore intersect at all levels, enabling reciprocal testing for a fully coherent insight.

  The solution is illustrated in Fig. 4.10. The supposed 'circularity' (A), when viewed from 'perspective' (B), proves to be composed of a synthetic and an analytical component that, upon their 'separation' (C), can be examined individually. The decisive step for the 'validation' is that the analytical and the synthetic examination, rather than standing in isolation, are embedded in a hierarchic framework of theories (D). The essential point is that both of these theory-systems can stand on their own and legitimize themselves. The sole requirement is that they mutually confirm and complement rather than contradict each other.

  This raises the demand that case examples of tissues yield our concept of an organ, just as case examples of organs yield our concept of a tissue (Fig. 4.10d).

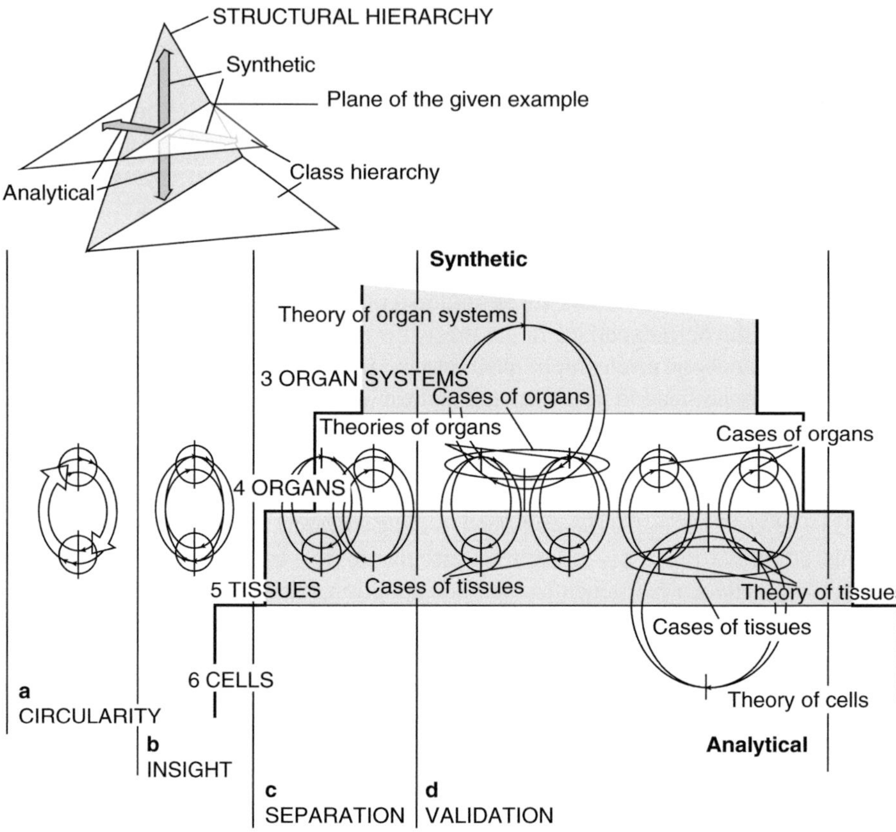

Fig. 4.10 Disentangling the hermeneutic circle, exemplified by a tiered constellation of anatomy (structural hierarchy refers to the preceding case in Figs. 4.8 and 4.9). How should (**a**) organs be interpreted based on tissues when tissues are interpreted based on organs? Because (**d**), both synthetically and analytically, hierarchies of theoretical frameworks initially develop interdependently of one another. The expectation is that, when combined, they confirm one another. The above sketch outlines the positional relationships of the considered class hierarchies (details in Figs. 3.11 and 3.14)

The parallel situation: examples of words promote the interpretation of a sentence, and sentences help interpret a word (Fig. 4.6). Equally, taxonomic groups such as hoofed animals (ungulates) contribute to the definition of mammals, while case examples of mammals help define hoofed animals.

Circular reasoning in the sense of Stegmüller—fishing in murky waters—can crop up with insufficient insight into and information about the interrelationships. In reality, circular logic is not involved. Quite the opposite: the 'structure' of the development of complex systems is concurrently being depicted.

4.2.2 The Three Fundamental Types of Complex Similarity

There is no such thing as false similarities. This unusual conclusion is based on two considerations. Recognizing similarities, i.e. comparable entities, is initially a purely ratiomorphic achievement. A root can remind us of a witch, a stalagmite of a pieta. Human perception, which Popper (1973b) so aptly compared with a spotlight that continuously scans our reality, searches for recognizable and classifiable things. This is a prerequisite for successfully orienting ourselves in our surroundings.

What can be—and often is—incorrect in a perceived similarity is the attached explanation. The *propter hoc* and *simul hoc* are simply two different things (see Chap. 2, Sect. 2.3.3). This necessitates more strictly differentiating the explanations of similarities (Chaps. 5 and 6). The focus remains on the cognitive process.

Supposed similarities can well turn out to be entirely different things: a person on the horizon might actually be a stone manikin, or a fried egg on a sandwich a plastic imitation. Such party gags, along with run-of-the-mill magicians, entertain us with such veiled realities.

Three types of similarity already crop up at the subconscious level. It pays to examine them separately: (a) supposed identities, (b) analogies and (c) metaphors.

(a) *Supposed identity*: The apparently identical things in this world deserve special attention for two reasons. Firstly, it is astounding enough how unsurprised we are at clearly identical objects in nature and, secondly, how challenged we are in confronting supposed identity when that identity has been modified.

The first step is to examine (a1) our lack of astonishment and, conversely, (a2) what triggers astonishment and research.

- (a1) *On astonishment*: We accept with great equanimity that the chick hatching from a chicken egg is almost always perfect—as if it were simply serially rubber-stamped on paper. This despite the fact that the program behind the embryogenesis of every higher organism is perhaps the most astounding and yet-to-be deciphered process that the cosmos has ever brought forth. Perhaps our trust in the precision of such highly regulatory processes is rooted in the lawful order in nature, namely the very order we rely on to survive: the cave bear, human and bee will remain a cave bear, a human and a bee.

  Identical replications also abound, in various patterns, within complex systems. They are often taken for granted. Some are well-defined, others not.

  A well-defined case is the phenomenon of symmetry in its various forms. Bilateral symmetry characterizes most organisms (i.e. designating the Bilateria as a group), our vehicles and many architectural styles. The underlying principle is an identical doubling of a structural principle in a mirror-image configuration. Some of it can be explained functionally, some also based on esthetic criteria. 'Cheap order' also plays a role.

  The planes of symmetry in organisms have decreased during the course of evolution. Six, 12, 24 and 48 planes of symmetry can be laid through a sea anemone. Here, minimal instruction yields a great amount of order. In a spherical planktonic organism, the number of possible symmetry planes is actually unlimited. The term 'differentiation' is used to describe the evolutionary reduction of such symmetries.

In contrast, we lack precise terms to describe the serial arrangement of identical structural elements such as the legs of a millipede, the vertebrae of a snake or the curbstones along a sidewalk. The same holds true for spatial arrangements such as the cells of an epithelium, a street paving, the hairs on a furry animal or the repetitions of a wallpaper design. And we clearly lack a terminology for the spatial packaging of identical elements, whether they be the grains of a sandy beach, the cells of a liver, or the bricks stacked in a construction warehouse. Nonetheless, in all these cases the term differentiation can be applied to describe their evolutive modifications.

- (a2) *On the origin of scientific study*: Interestingly, our attention tends to be drawn to cases in which identity is suspected rather than clear-cut—the more difficult the confirmation, the more meticulous the effort. Borderline dissimilarity helps trigger all automatic reflection and, subsequently, thought. This holds true both for magical thinking (Lévi-Strauss 1968) and scientific thinking, with their full ranges of background knowledge, underlying assumptions and personal doses of fantasy.

  The reason behind this faculty? It is vital for proper orientation, and making correct prognoses improves our chances of survival. Or, as Wilhelm Busch (1982) ironically phrased the question, "what kind of bird a maybug was."

  Our sciences swing into action as soon as a method is agreed upon or suggested. This holds true across all fields of inquiry, be it asking whether the 'humps' that orbit around Jupiter might be moons after all (Galileo), whether dolphins and cows, apes and humans, lido and spit of land with lagoon, or 'father' and 'père' are not principally 'one and the same'.

  This method is always a morphological one along with its subsumptive and hermeneutic aspects. Chap. 4, Sect. 4.3.1 discusses the specifics of its application in the framework of the 'homology theory'. That includes other pre-programed cognitive faculties involving gestalt and fields of similarity, along with the concepts of type and metamorphosis.

(b) In contrast, *analogy*, in today's use of the word, does't lead us to expect identity. It represents a different type of comparison, yet remains a decisive source of knowledge. This insight was important enough to have been the focus of Lorenz' (1974a) Nobel Prize speech. When some kind of similarity is surmised in a dissimilarity, then analogy drives the comparison and basically involves the same automatic correlational process discussed earlier.

Initially, analogy was also used in the sense of homology (e.g. by Goethe 1795; 1824). And for good reason. After all, a differentiation between analogy and homology first becomes necessary when seeking the cause behind a similarity, and this was not Goethe's focus. For a detailed treatment of the cause behind similarity, (see Chaps. 5 and 6).

Evidently, sorting supposed analogies into (b1) chance analogies and (b2) functional analogies already takes place at the pre-conscious level. This raises the issue of causality.

- (b1) *Chance analogy* represents the first and most immediate category. A dumbbell shape, for example, can describe the outline of a molecule, a plant spore, exercise gear, an island or even a galactic nebula. It is interesting how quickly we lose interest in such perceptions—probably because we expect no common cause behind these similarities in form.

  Bell-shaped forms elicit more lively interest (Riedl 1976). A diving bell and a glass dome used to cover cheese may well be somehow related to a bell jar, for example. The relationship becomes murkier when comparing a church bell, a flared skirt or a bellflower, quickly causing our interest to wane.
- (b2) *Functional analogy*: The category comprising true analogies includes those similarities whose function is recognizably the same. The one decisive caveat: a high probability that these similarities arose independently of one another. Examples include the streamlined shape of fishes, aquatic dinosaurs, aquatic mammals (Fig. 4.11) and even of submarines. These represent 'functional analogies'.

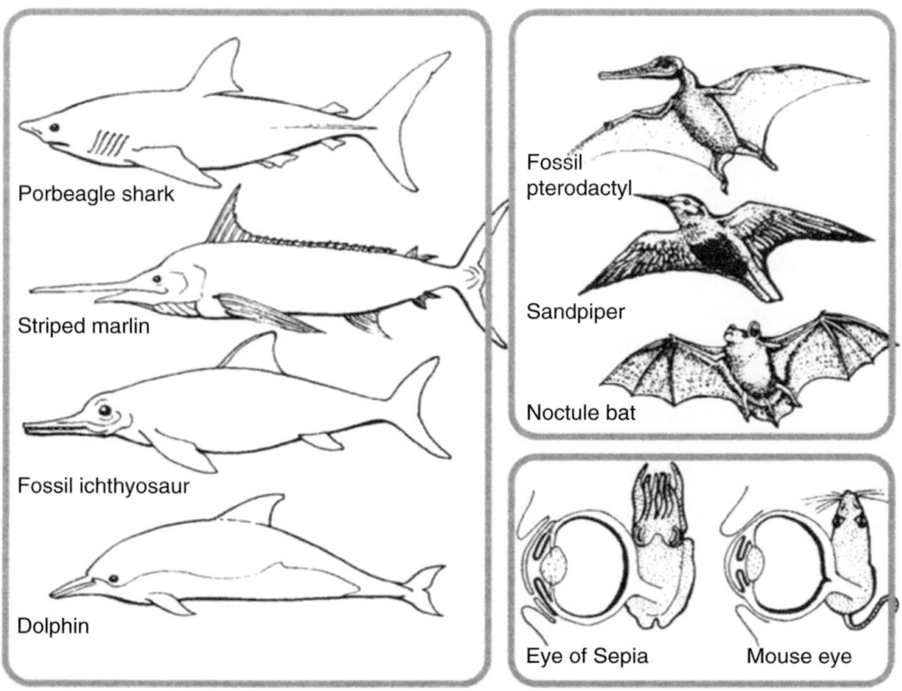

Fig. 4.11 Forms of functional analogies, illustrated based on streamlining, the flight apparatus in tetrapods, and on the vertebrate and squid eye. The analogies are expected based on the type of similarity, confirmed by the dispersed occurrence of such convergences in an otherwise harmoniously divergent field of similarity (see also Fig. 4.12; after Riedl 1980, 1987a)

The interesting aspect of this perception is that it invites explanation, and that acting on this invitation calls for recognizing the underlying patterns of similarity. Importantly, we already order these patterns pre-consciously, and no special concept of causality is required to establish a high probability of independent origin. This is because (see Chap. 3, Sect. 3.4.2; Figs. 3.8 and 3.9) 'harmoniously divergent fields of similarity' are a pre-conscious phenomenon. They minimize contradictions and disharmonies by re-arranging the respective items, satisfying esthetic considerations more than any need for explanation. At best we attribute this harmony to nature following its own internal principles. In today's terminology, such principles are 'system imminent'.

The features pointing to analogy are those that are dispersed or randomly distributed in such a harmoniously divergent field. They are also important clues in helping to recognize convergences (Fig. 4.12). Such constellations support the interpretation that the principles driving these similarities lie outside (environment) rather than within the system (organism).

Let's simplify these abstract considerations by using an explanatory approach. Accordingly, harmonious divergences are attributed to modifications of a trait, convergences to adaptation to the same environmental conditions. This involves the process of explanation, not of perception: similarities must be perceived before any explanation can be forwarded.

We recognize analogies in the cultural sciences in much the same way. This is valid for every facet—languages, instruments, art—whose harmonious divergence is attributable to genealogies, much like in the organismic realm. Such geneologies may be difficult to determine because information and knowledge are transmitted much more freely in human cultures. For example, the similar streamlining of a submarine and an airplane is based on the widely known principles of flow dynamics, even though the first flying machines were far from being aerodynamic.

(c) *Metaphors* are the third type of expression for perceived similarity. On the literary level, they probably established themselves as extensions of analogy. In such cases a third element, the *tertium comperationis*, is inserted between the two items being compared: 'golden' instead of 'admirable', moving 'gazelle-like' instead of 'gracefully'. Metaphors are occasionally used in science to illustrate a point, but are never confused with analogy as such.

4.2.3 The Four Fundamental Forms of Complex Order

This discussion also serves as a retrospective on 'the order of things'. Firstly because two of these four basic forms have already been treated from another perspective; secondly because a generalization is still lacking; and thirdly because it echoes the approach of the chapter '*Denkordnung als Folge der Naturordnung*' ('Ordered thought as a reflection of ordered nature') in Riedl (1987b).

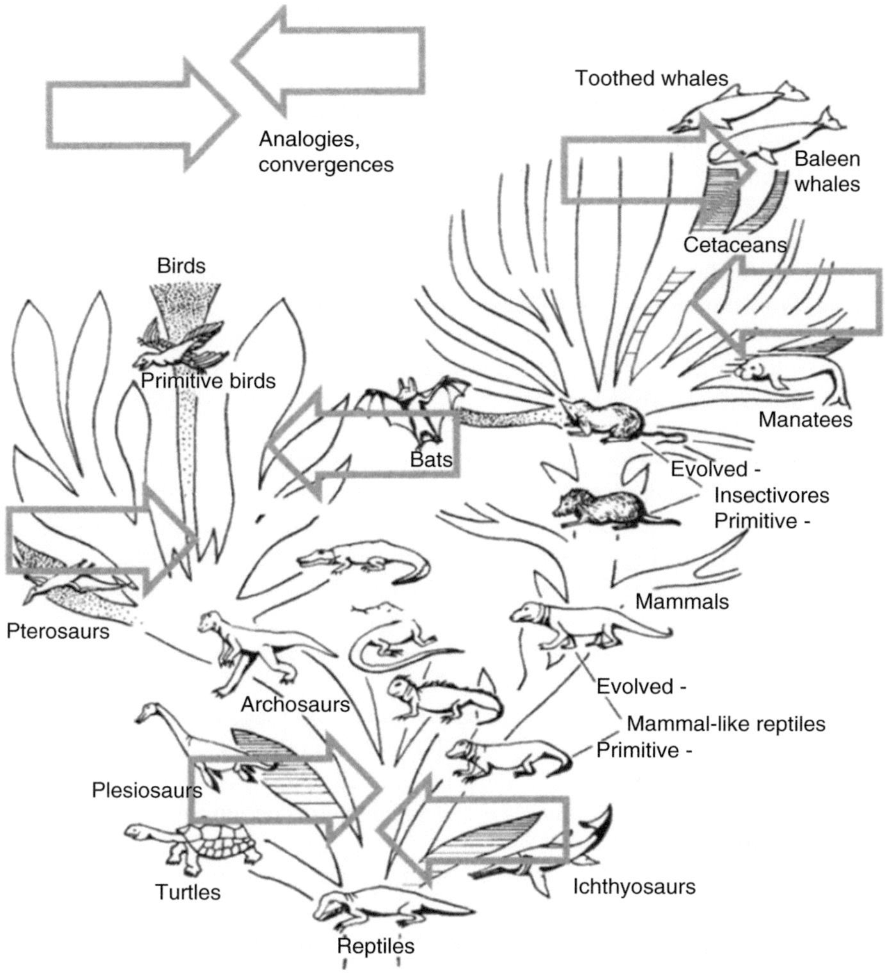

Fig. 4.12 Dispersed distribution of functional analogies, depicted as convergences in a harmoniously divergent field of similarities that is later explained as a phylogenetic tree. Proving analogy requires demonstrating that the common stem form lacks the trait (here the streamlined form or the wings) (after Riedl 1987a)

This condensed presentation is warranted for several reasons. A monographic treatment is already available in Riedl (1975). Moreover, I myself was often lucky enough to recognize the principle behind a phenomenon before that phenomenon began to ramify. Finally, this approach takes on the challenge to causally resolve—in the framework of 'explanation' (see Chap. 6)—key issues that require revisiting these four fundamental types. The focus here remains on the cognitive processes.

Readers with a bibliographic bent may note that recognizing the four patterns of order (in my book 'Systems theory of evolution') prompted the notion that structured thought must be an adaptation to structured nature. This insight was subsequently brilliantly confirmed by Lorenz' ethological research (1973a) and spawned the concept of 'evolutionary epistemology' (Campbell 1974; Vollmer 1979).

Importantly, the innate hypotheses were introduced and treated relatively late (Riedl 1980) (see also Chap. 3, Sects. 3.2 and 3.3). Accordingly, perceiving orderly patterns behind the 'norm' is governed by the 'hypothesis of the apparently true', whereas perceiving the remaining three patterns of order is governed by the 'hypothesis of the comparable'.

The four fundamental patterns are (a) norm, (b) interdependence, (c) hierarchy and (d) transmission of historically gained features or tradition (German: *Tradierung*). The discussion of each basic form leans on organismic examples, then draws comparisons with artefacts and finally presents the cognitive tools that help us recognize the underlying patterns.

(a) *The norm* as a phenomenon has already been raised in connection with the catchwords 'standard building blocks', 'cheap order', 'identity' and 'redundancy'. Adding the term 'iteration'—used earlier here in connection with the basic conditions for human cognition—underscores the connection.

Standard building blocks provide the basic structure of the entire cosmos. They extend into the living world as well, from biomolecules to colonies and populations. And such building blocks themselves serve as the substrate for all individualization and differentiation.

Standardized elements also underlie spoken language and every manner of written record, from letters of the alphabet to entire dictionaries. The same holds true for other artefacts: most have, beginning with the bricks that built Babylon, undergone continuous development and experienced a renewed surge in modern industrial society.

This redundancy, this economy of fit and of instruction at the lowermost rung of order, consists of 'law times (thousand-fold) application'. Such redundancy is also the driving force behind all associative learning. This is because 'cognition' equals 're-cognition', which is also rooted in conditioned reflexes (Fig. 2.6). A conditioning of identities is involved. Unsurprisingly, we not only perceive, build and think in norms, we also perpetuate this principle in our artefacts (supported by its economic advantages).

(b) The term *interdependence* is newly introduced here, but the phenomenon itself is omnipresent in the form of mutual or reciprocal dependence. Ever since the paradigm of the 'solitarily traveling particle' of early physics has had to be abandoned, interdependence has proved to be a governing principle in the cosmos. This ranges from the micro-world, in which matter originates, to the meso-, macro- and mega-worlds. The catchword is the so-called 'butterfly effect', the metaphor in which the mere wingbeat of a butterfly can turn out to be the ultimate cause behind the collapse of a solar system.

Living organisms are composed entirely of interdependent structural elements. In fact, all complexity is characterized by a non-random combination of parts and therefore by mutual dependencies. While this statement may appear trivial, the decisive consequences of such dependencies are less trivial. In developing explanations for complexity (Chap. 6, Sect. 6.3.2) these consequences will be introduced as 'genetic' and 'functional burdens or constraints' . They prove to be the factor channeling all developmental processes—and ultimately serve as the foundation of 'natural order' itself.

Interdependencies also turn out to be the precondition for all forms of communication. They are among the oldest programs governing social systems, are active in every facet of human activity, and exhibit considerable differentiation in all our artefacts. The more complex the social system—along with its languages, arts and technical products—the more complex the interrelationships.

This prompts the conclusion that our pre-rational interpretation of the world, and ultimately our thinking, must have developed under the expectation of interdependence.

(c) *Hierarchy* is the cornerstone of every issue raised in this book, determining how we perceive and deal with structures and classes. All differentiated cognitive gain presupposes hierarchic structures, regardless of whether this process be understood as subsumption or hermeneutics.

The previous chapters have underlined the hierarchic order of the world with all its levels and phase transitions, each with its unique set of qualities. This explains why physics, chemistry and biology all require a tailored terminology. Human languages also proved to be hierarchically organized: comprehension involves storing the phoneme in order to interpret the syllable based on the word, then the word based on the sentence, and the sentence on the overall context. Finally, all human artefacts are also hierarchically structured.

This calls for further examining our cognitive predisposition for perceiving hierarchies. A justified question is whether the very fact that we think hierarchically might not enable us to see the world as anything but hierarchically arranged. That would make the structure we perceive behind extra-subjective reality a mere artefact, a projection of our own thought patterns. My friend Bernhard Hassenstein drew my attention to this fatal possibility. The crux of the matter was the reality content of the class hierarchies in the 'natural system' of organisms (Hassenstein 1951, 1958). This actually prompted my interest in epistemology. It turns out, however, that our thought patterns are so complex, and that the patterns in nature mirror this so perfectly, that chance can be ruled out as an explanation. This means two things: first, one must be the cause of the other and, second, the older the cause of the younger. The next step—arriving at a potential explanation—was a small one for a biologist. The conclusion can only be that this correspondence is a product of adaptation. Of all the conceivable operational modes our brain structure would be capable of, those that correspond best to the underlying structure of the world must be those that ultimately prevail (Riedl 1975, 1980, 1987b).

(d) *Transmission of tradition* (German: *Tradierung*) is a phenomenon that has received little discussion. It refers to the passing on of instructions, structures and functions across generations. In principle this represents a special case of maintaining lawful order. A more interesting issue is the tenacity of that preservation and transmission—even when a structure has become superseded or functionless or, as is often the case, when such traits have retained their function despite major transformations.

In the organismic kingdom, every degree of order behind structures and classes is based on transmission. This is so self-evident—anchored in genetics—that we take the spectacular process for granted. This also explains why cases of atavism or rudiments are all the more conspicuous and surprising. A case in point: the auricular or Darwinian tubercle on the upper edge of our ears, a relic of more pointed ears. 'Spontaneous atavisms' are an even more impressive example of traits that suddenly emerge from deep in the history of our genetic make-up. In some persons this is manifested as a tiny tail, an entirely furry face or superfluous nipples. Even 'neck fistules', representing remnants of long-gone gill slits, have been documented (Fig. 4.13). In fact, ontogeny is replete with such transmitted 'palingenetic' traits. The arteries of the gill slits, which are still present in the embryos of all mammals, are a classical example (compare Fig. 4.43).

Human culture, much like organisms, is also anchored in the transmission of established forms of order (helping explain the derivation of the German expression from "tradition"). Interestingly, cultural atavisms that persevere despite altered function are more common than thought. Otto Koenig (1970) used this phenomenon to develop the discipline 'cultural ethology', which he

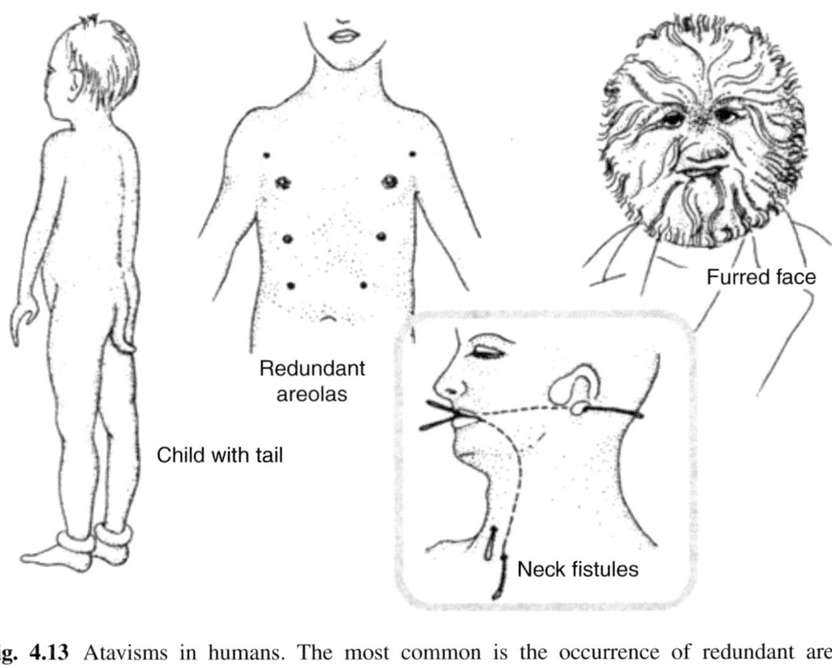

Fig. 4.13 Atavisms in humans. The most common is the occurrence of redundant areolas. Physicians and anthropologists have even reported neck fistules, interpreted as remnants of gill slits (from Riedl 1975)

substantiated with numerous examples (some of which are reproduced here in Fig. 4.14).

The question about our cognitive predisposition for this order-producing pattern (and its cause) remains. Importantly, the differentiation achieved by even a ‘primitive culture’, along with all its social rules, can be destroyed in a single blow. Equally clearly, creating that differentiation from point zero requires more than the wave of a magic wand. This highlights that, from the onset, we must have been drilled to accept the things we encounter and to measure ourselves against them. In many publications, Eibl-Eibesfeld (e.g. 1978, 1984) demonstrated that social forms of behavior are innate human traits. Species survival itself therefore dictates being geared to perceive transmitted, historically rooted conditions.

This chapter has presented various forms of experience demonstrating the high degree to which our ratiomorphic faculties have prepared us for the order behind things, for perceiving their similarities and for deciphering complex systems. Chomsky (1970) and Lenneberg (1972) concluded that children mostly need to learn a vocabulary, not a language as such. This supports the hypothesis that we are innately prepared for the basic structures of extra-subjective reality: we merely need to insert the case examples.

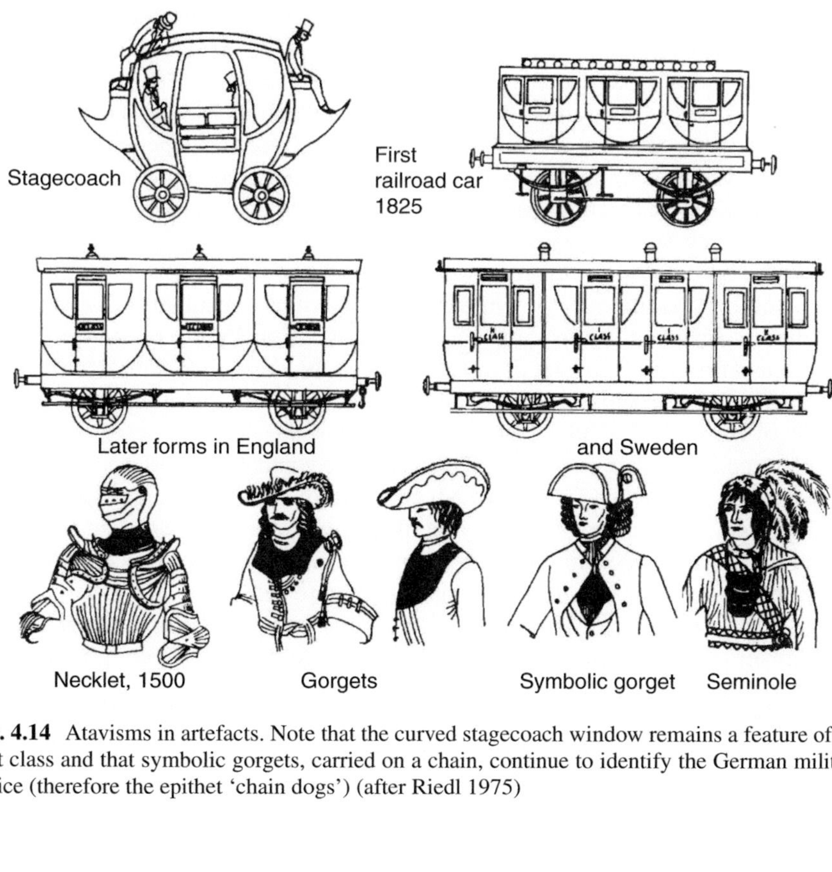

Fig. 4.14 Atavisms in artefacts. Note that the curved stagecoach window remains a feature of the first class and that symbolic gorgets, carried on a chain, continue to identify the German military police (therefore the epithet ‘chain dogs’) (after Riedl 1975)

This is the point to introduce the principles behind research in this field. ‘Morphology’ represents the traditional framework. It is the foundation for comparative anatomy, systematics and therefore for our interpretations of evolution and phylogenetics. The distinction between morphology and systematics is merely a concession to today's parlance.

4.3 The Principles of Morphology

In general, the same principles underlie both morphology and systematics, although the modern take is to treat comparative anatomy and systematics as separate disciplines—as if any clear conclusions could be drawn by treating the hierarchies of structures and classes as separate disciplines. In fact, as outlined above, the two are interrelated, with the difference lying solely in the viewing direction (i.e. in the primary interest).

Doing justice to morphology requires (i) examining the term itself, (ii) discussing ideas and experience as concepts and (iii) presenting the content of the discipline.

(i) *‘Morphology’* combines *morpho...* (‘pertaining to gestalt’) and *logos*, a term that already featured prominently in the cultural history of the Greeks. It can be best understood as 'the teaching of...’, namely in the sense it was coined by the comparative anatomist Karl Friedrich Burdach in 1800 and subsequently embraced by Goethe, who also analyzed its application. The resonating idea was to validate the methodology behind a comparative approach that yielded ‘the typical’ (in Goethe’s sense the ‘type’) or even the ‘archetype’ of all organismic structure.

The earlier users of the term were well aware of the empirical nature of their method. Nonetheless, the influence of ‘German (metaphysical) Idealism’ already spawned the first misunderstandings. A lucky circumstance enables citing a first-hand account of this divergence. In a first encounter with Schiller, Goethe (1817) reminisces: “I in an animated manner related the metamorphosis of plants and, with a few broad strokes of the pen, generated a symbolic plant before his very eyes. He listened intently and examined it all with great interest...; yet, when I had finished, he shook his head and said: ‘that does not represent experience, but merely an idea’. I was dumbfounded and somewhat vexed, because the standpoint that separated us had been so succinctly delineated.” This calls for going into somewhat more detail.

(ii) *Idea or experience?* That was the question from the very start. And many thinkers consider it to be unresolved or believe that morphology remains an idealistic philosophy. For many natural scientists it represents an anachronism.

Goethe considered Schiller to be a sophisticated Kantian thinker who had actually critically examined Kant’s critical writings, but who had—in the slipstream of Plato's concept of ideas—become an ethical-educational moral philosopher. Namely of German Idealism. Schiller set out to illuminate harsh nature with intellect, whereas Goethe set out to illuminate the searching soul with nature. For Schiller, nature was made for the soul, for Goethe the soul for nature.

Goethe did, in fact, use the word 'idea' in developing his morphology. Nonetheless, as becomes clear in the outlined method, in the sense of an expectation or theory based on experience with case examples—mammal skulls. The distinction between 'idea' as understood by Plato and a product formed by experience is difficult to establish, especially at this crossroads. For Plato, ideas of the world are a given. Things, along with our souls, all participate in them, much like in a memory. While Goethe's morphology also involves the expectation of discovering principles behind things, the process is primarily inductive, requiring substantiation by experience. In Plato, the process is primarily deductive, preceding experience and not validated by it.

Sloppy differentiation does in fact create confusion here. And this confusion only increases without adequately appreciating the hermeneutic or subsumption method, or by simply casting the suspicion of circularity.

(iii) Morphology is *the scientific, experience-based discipline of comparing*. It must be grasped and internalized as the foundation for both comparative anatomy and systematics. In enabling us to recognize harmonious systems of similarity, it also becomes the prerequisite for their explanation, namely for evolutionary theory and phylogenetics.

This has largely been brushed under the rug today, with 'morphology' being taken as synonymous with 'anatomy'. This interpretation has met little resistance because, as outlined earlier, *simul hoc* processing already occurs subconsciously. This also explains the establishment of university departments titled 'Comparative Morphology' or 'Functional Morphology', clearly ignoring that the former is a pleonasm, the latter a contradictio in adjecto. Namely, morphology already incorporates the comparative aspect, and the approach itself excludes functional explanations.

If the differentiated comparative processes are already intuitively anchored in our make-up, why, then, make the effort to illuminate or render intelligible a discipline involving comparisons? After all, we were able to construct the phylogenetic tree of organisms without such background knowledge. This fundamental question requires attention, and delving into these differentiated processes helps make progress.

Three factors merit mention. The first is the apparent difference in the hemispherical preferences, in the talents, of our brains. This directly influences our interest in comparing gestalt and determines the significance we attach to such comparisons. Second, morphology takes practice. The 'innate master teacher', as Lorenz named it, must be prompted to teach us. This explains Ernst Mayr's view that the subtle intuition of an experienced systematist deserves primacy over 'numerical taxonomy'. Third, this discipline is necessary because we would otherwise fail to transcend intuitionistic discussion and attain the status of scientific argumentation.

The comparative approach is essential for all polymorphic systems, particularly when these exhibit a genealogy. This holds equally true for geomorphology as it does for artefacts, for linguistics and literature studies, ethnology, art history and cultural history. In biology, this approach is by no means restricted to anatomical characters. Ever since Lorenz (most recently 1978), it has also been applied to the temporal facets of ethology, helping establish 'comparative ethology'. It is also valid for genealogical phenomena in genetics and ecology.

The reason why the comparative approach is elaborated here based on anatomy reflects the unparalleled wealth of accumulated experience in this field. Two million species plus a half million system groups, multiplied times up to twenty specific 'homologies' in each, yields fifty million individual terms—ten times the number of words in any major language. The basic framework, however, is and remains the general theorem of comparison.

The first step is to (Sect. 4.3.1) introduce the homology theorem and then use that to derive (Sect. 4.3.2) the terms 'type' and 'bodyplan' and, more holistically, (Sect. 4.3.3) a theory of the character or trait.

4.3.1 The Theorem of Homology

The word homology was coined by Richard Owen (1848) and juxtaposed to the term analogy (which it had earlier been equated with). At that time it was already recognized—based on the continuity of similarities—that a particular type of identity was involved in the respective structural elements. This was later referred to as 'similarities of character'. In Chap. 3, Sects. 3.4.2 and 3.4.3 this was derived more precisely based on the form of harmoniously diverging fields of similarity. This insight needs to be applied here.

The first step is to (a) separate homologies from analogies. Only then can (b) the criteria of homology be determined, (c) synthesized, (d) attributed to a theorem of probability and (e) validated for all forms of homology. A discussion (f) of the various standpoints on the homology concept is briefly appended.

(a) *Differentiating and separating analogies from homologies*: Sect. 4.2.2 (b) outlines that chance analogies and functional analogies (Fig. 4.11) can be recognized as the dispersed distribution of convergences in otherwise harmoniously divergent fields of similarity (Fig. 4.12). Specifically, this means that the common ancestors did not possess the same specific character or trait. Such cases call for explanation, bearing in mind that cognition precedes explanation, is its precondition and remains independent of it.

In contrast, those traits whose modifications recognizably define the harmonious and divergent aspect in a field of similarities are termed homologies. We again seek an explanation. Superficially, such similarities are explained by kinship. This, however, is again no substitute for the cognitive process and, as will become evident, is also insufficient as an explanation.

The first step in examining homologies is to focus on individualizable (namable) homologous body structures that can easily be counted (!) as units. Additional forms of homology and the linkages with analogies are outlined further below.

(b) *The homology criteria*: We owe the first correct presentation of the homology criteria to Adolf Remane (1951, reprint 1971). Remane, well versed in comparative anatomy and systematics routines—from the lower Bilateria to the primates—resolved (once again, intuitively) two key conditions for the process of recognizing homologies. Unfortunately, his book was never translated into English, and leading English-speaking researchers are therefore unaware of its content.

Remane's terminology is used here, differentiating between (b1) principle criteria and the rather unfortunately designated (b2) auxiliary criteria. Paragraphs c and d then provide the epistemological validation of this categorization.

- (b1) Remane distinguished three principle criteria: (i) positional, (ii) structural and (iii) transitional.
  (i) *The positional criterion* is based on the fact that no homolog stands in isolation. Each is embedded in a matrix of structural elements that can be homologized. Experience shows a high consistency regarding position. If the correspondence of an element in two different species is not apparent based on its structure, then its position can providc the decisive clue. Importantly, this operational step clearly already requires (Fig. 4.15) recognizing the correspondence of the respective adjoining elements. The link to the underlying cognitive process is evident.

  The starting point, as in the example given here (Fig. 4.15), is at 'focal positions', e.g. the incisors of the upper jaw and at the occipital foramen. Then, a process of stepwise reciprocal confirmation coupled with the theory of positional constancy helps to gradually resolve the correspondence. (If you do not know what the carburetor in the new motor looks like, but you do recognize the air filter, the fuel pump and the cylinder head, then the carburetor must be the device at the junction of the three.) This illustration of the substantiation process is good preparation for the upcoming treatment of probability.

  The positional criterion provides an even more universal access, namely based on the hierarchic arrangement of the homologies. This is because the vertical or nested component of the hierarchy also incorporates the horizontally adjoining elements. The term 'framework homologies' is introduced here: it proves useful in defining the relationship to the subsequentg structural criterion and (in paragraph c) to additional forms of homology. Importantly, homologies not only adjoin one another, but are also positioned in and atop one another, frame within frame (Fig. 4.16).

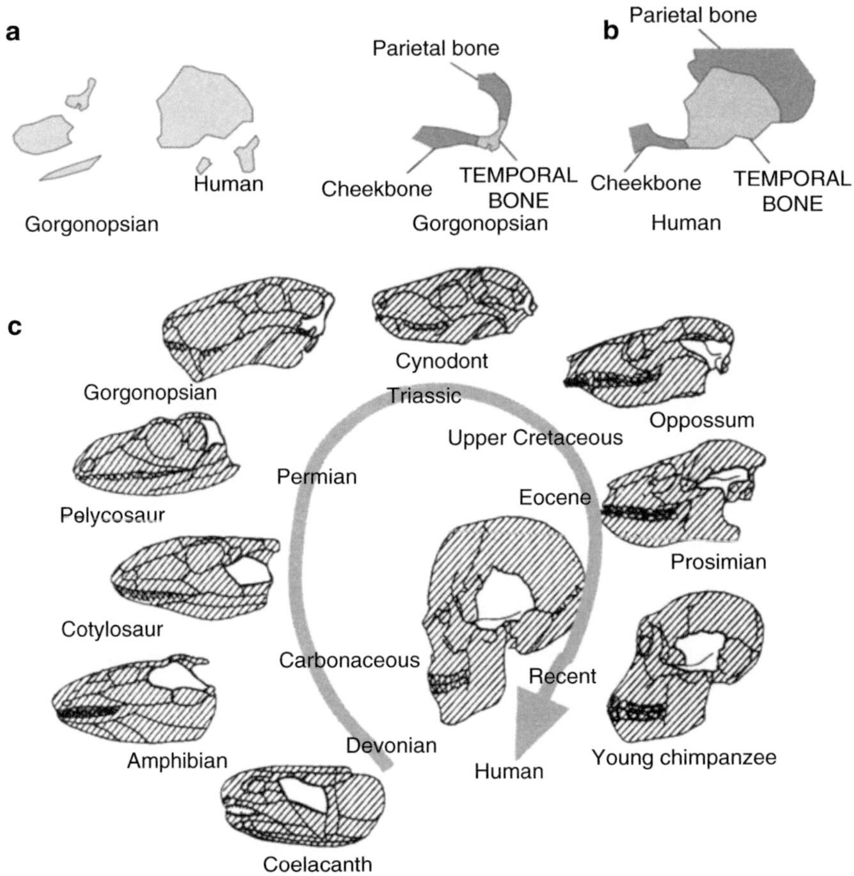

Fig. 4.15 The positional criterion of homologization, illustrated using the temporal bone. Based on shape alone (**a**) no corrspondence of three bones (of a representative from the Permian and a recent human) would be recognized; the consisent positional relationship (**b**) makes this correspondence likely and ultimately (**c**) reveals the continuity of the relationship (sketch of underlying principle in Fig. 4.26; after Gregory 1951 and Riedl 1975, modified)

Figure 4.16 illustrates such a hierarchy involving the human backbone. The homolog 'backbone' or 'spinal column' is the framework for five subframes. The first is the cervical spine, which forms the frame for the seven homologizable cervical vertebrae. The second is the epistropheus, forming the frame for the vertebral body, vertebral arch and 'tooth' (the dens epistropheus). This, in turn, encompasses three individualizable homologs, including its ventral articular surface, the facies articularis ventralis dentis epistrophei. The position of the latter homologous structure is termed 'minimum homolog' here because any further breakdown yields a new set of perspectives. Clearly, however, every superordinate framework homolog defines the structural features of the homologs in its subframe.

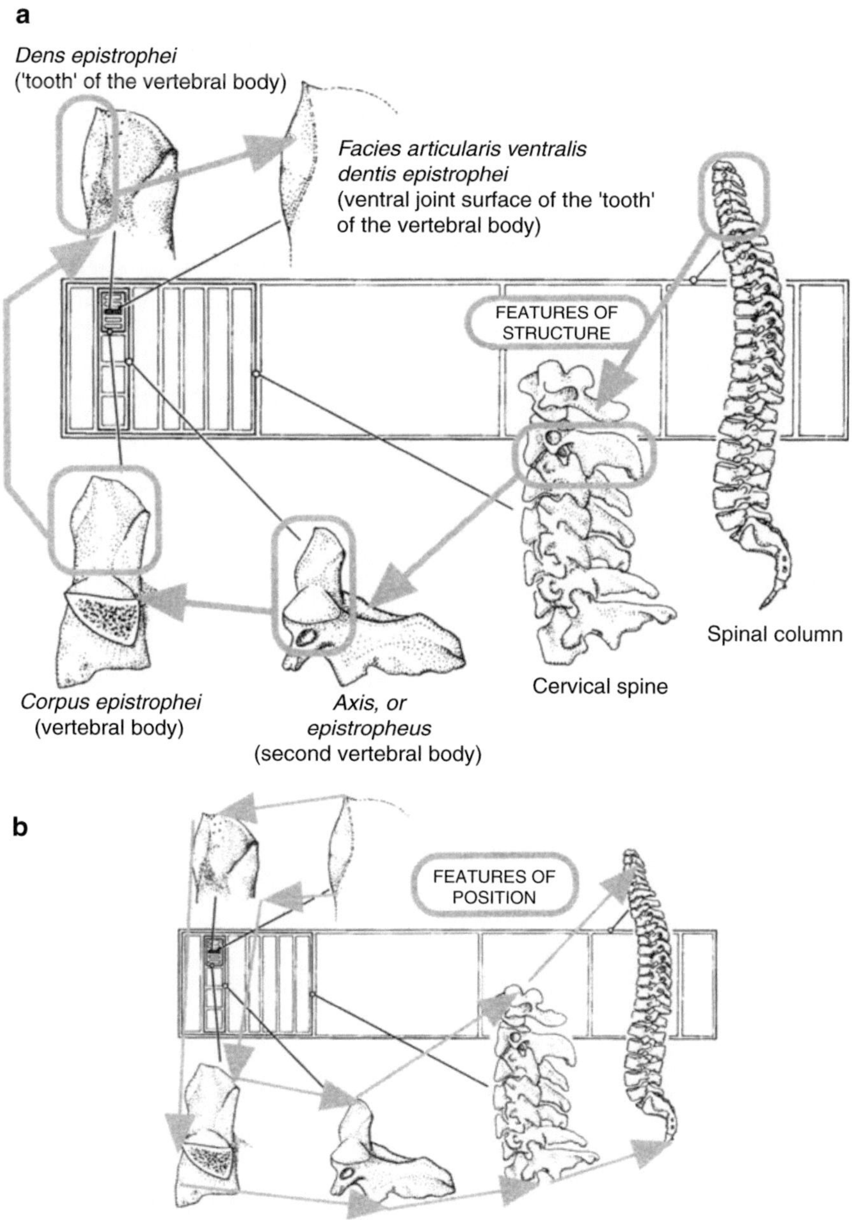

Fig. 4.16 The hierarchy of homologs and the interplay of structural and positional features, illustrated based on six hierarchic levels in the human spinal column. Note that (**a**) the structural features are identifiable from the upper to the subsystems, (**b**) the positional features in the opposite direction

In ethology, the positional criterion is equated with the 'serial position' of a behavior, in ecology with the 'spatial position' of a community, in molecular biology with the position on the 'gene map'. In the cultural sciences, this criterion plays the same role as in biology, ranging from the study of syntax to the history of architecture.

(ii) *The structural criterion* is based on the conservatism of the individual building blocks and their substructures. If the correspondence of a structure in two different organisms cannot be determined based on its position in the respective matrix, then its internal structure can provide the decisive clue. Again, this presupposes knowledge about that structure. (Knowing what a spare tire looks like helps you find it under the floor of the car trunk.) The biological example chosen here (Fig. 4.17) is the testes of two chordates, the lancelet and a deer. Even if the transitions in their position remain unknown, their internal structure, above all the stages of sperm development, cements the correspondence.

Structural features also do not stand in isolation. Structures are always themselves structures of structures. Simply reversing the perspective on the

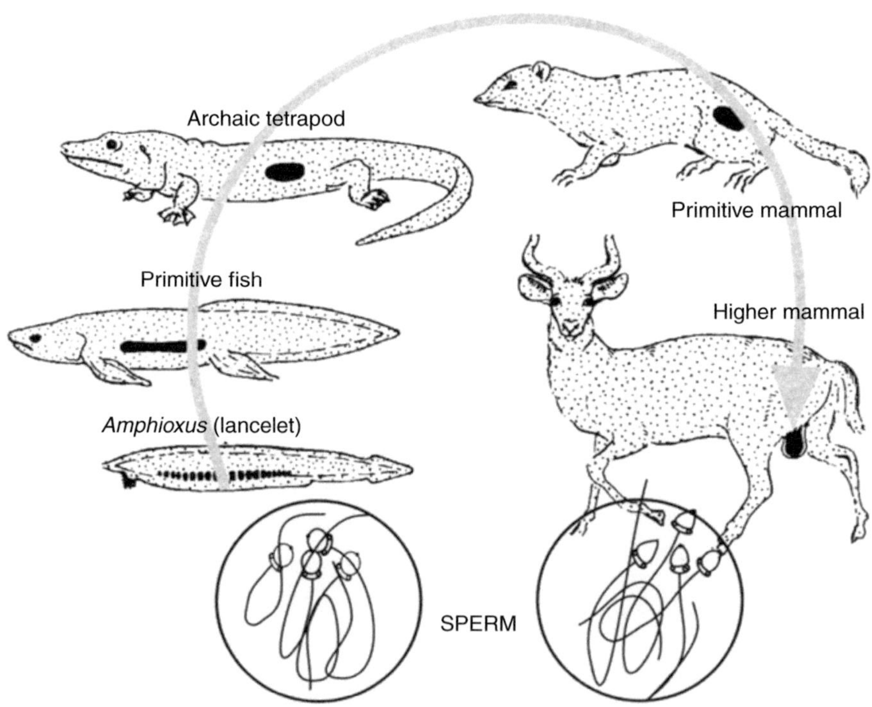

Fig. 4.17 The structural criterium of homologization, exemplified by the different positions of the testes in the lancelet and elk. If the internal structure of the organs is highly similar, as the sperm in the lower inserts shows, then they correspond regardless of their different positions (after Riedl 1975)

hierarchy of homologs (based on the example in Fig. 4.16) reveals that a cervical spine—if it were unearthed separately in a dig—would be identifiable based on the structures of its vertebrae. Equally, the characteristic 'tooth' (dens epistropheus) would suffice to identify an isolated epistropheus.

In ethology, the structural criterion is equated with the makeup of an activity, in ecology with the structure of an 'association', and in molecular biology with the composition of a 'peptide chain'. In the cultural sciences the structural criterion is important in prehistory and throughout the history of literature and art. For example, structural principles alone already clearly define what linear pottery is or, in sacral architecture of the Gothic period, a 'window tracery' (Fig. 4.24).

(iii) Remane's *transitional criterion* begins to stray from the idea behind his principal criteria. This is because the positional and structural criteria, in their pure form, should already substantiate identity when comparing two structures. Nonetheless, any comparison of two structures automatically invokes the overall field of similarity.

This criterion is more of a didactic tool: the transitions between structural forms can provide important, sometimes even the crucial, clues. This can be illustrated based on examples of form and position modifications that would leave researchers clueless were only the two comparable forms available. The first example (Fig. 4.18) is that of the three auditory ossicles in our middle ear. Studying the transitions reveals that they are in fact traceable back to, i.e. homologizable with, three massive cartilage elements in the jaws of sharks. Moreover, no one would ever, in a comparison restricted to a sea squirt and a hummingbird, come to the conclusion that both organisms belong to the chordates (both have a dorsal support, the chorda dorsalis—the former in the larval tail, the latter in the embryo).

This is a step beyond binary comparisons, opening up an entire chain or field of similar forms in envisioned space. Transitions are broken down into series of positional structures. This marks the shift to Remane's auxiliary criteria.

In ethology the transitional criterion is crucial for our understanding of changes in behavioral components. In ecology it helps understand 'successions'. In molecular biology it is important for the 'consonant shifts' that govern changes in the coding of large molecules. This criterion is also essential in the cultural sciences: consider the diversity of 'idioms' behind every style, or the fact that words such as 'père' and 'father' are harmoniously derived from 'pater', or that Daimler's first automobile (Fig. 4.18) is so far removed from a Formula 1 racing car that the connection between the two can only be determined (albeit quite simply) by tracing back the history of the automobile.

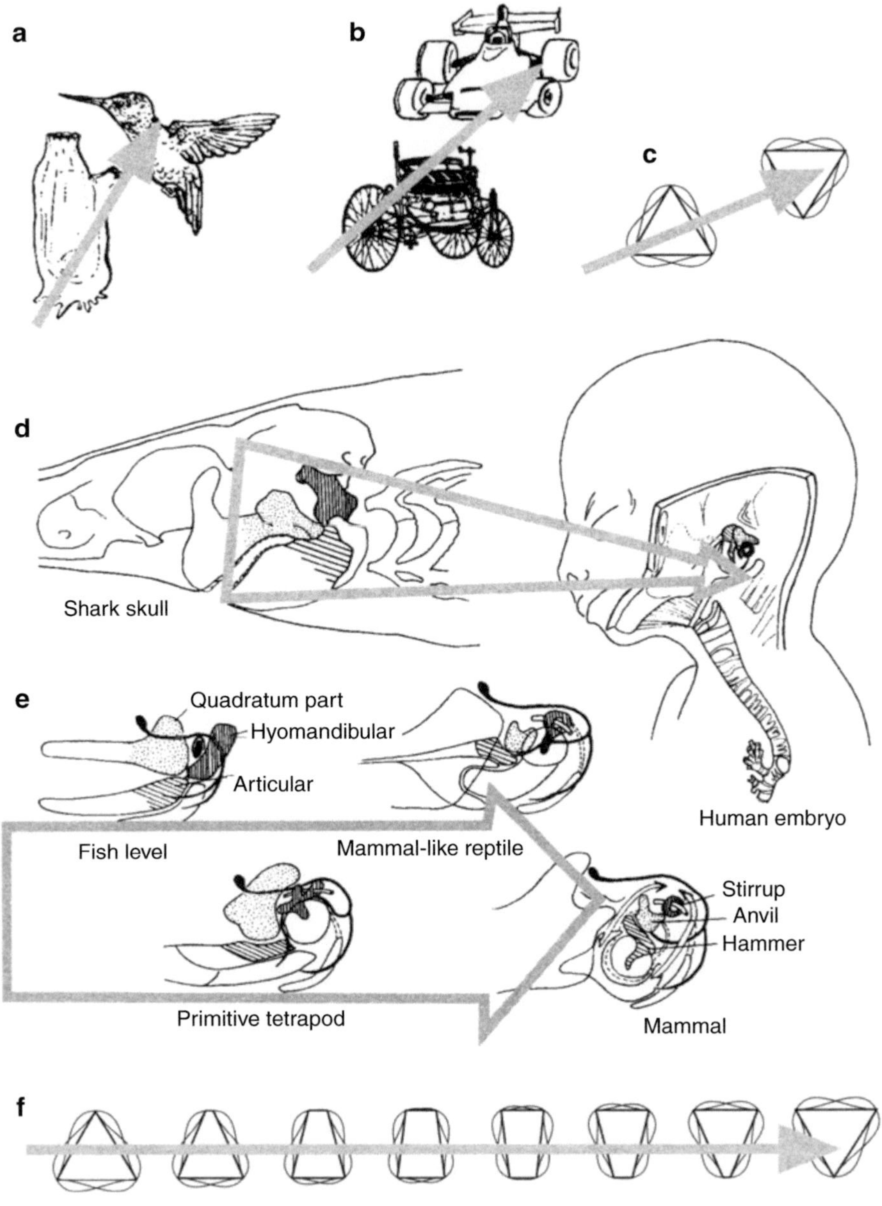

Fig. 4.18 The transitional criterium of homologization as exemplified by the evolution of the shark jaw into the auditory ossicles of humans (**d**) and their embryonic development (**e**). Note that without knowledge of the transitions, the relationships would be unrecognizable (for example between a sea squirt and a hummingbird) (**a**, chordates), between the first 'Benz-Mobil' and a Formula-1 race car (**b**, automobiles), or between two figures (**c**). The latter first becomes evident based on the series of transitions (**f**) (after Riedl 1975, 1976, amended)

- (b2) Remane distinguishes three *auxiliary criteria*. The discussion here is restricted to a 'coincidence' and 'anti-coincidence criterion' based on their information content. In principle, the two represent a reversal of one other. Their message, however, is so polarized that they are treated separately. The idea is to adjust or rectify Remane's description as 'auxiliary' and emphasize their fundamental importance.
  - (i) *The coincidence criterion* also enables homologizing, in Remane's sense, very simple structures when these occur regularly in a kinship circle or, in the present context, in a field of similarity. This is valid only within that particular field. A case in point is the chorda, a trait after which the Chordata were named. Its remains valid regardless of whether that feature is present only in the larval tail of sea squirts (and then disappears when that tail is lost) or whether, in vertebrates, it is reduced to remnants in the spinal discs, the so-called nuclei pulposi.

    This criterion clearly also helps homologize inconspicuous traits when the positional and structural criteria are insufficient. Nonetheless, it also plays a role for all traits, including the most prominent ones (see below).
  - (ii) *The anti-coincidence criterion*, in contrast, raises a red flag against homologizing any trait showing an entirely scattered distribution in a field of similarity. The classical example is a 'red throat spot', a feature distributed among widely separated vertebrates including coral reef fishes, lizards and birds. The same holds true for the crista sagittalis, a bony ridge on the skull of hyenas and gorillas.

    This criterion plays an equivalent role in all the natural and cultural sciences. No isolated object can give rise to a concept in ethology, ecology or molecular biology. Coining a term requires many confirmations of that concept in a field of similarity. The same holds true in establishing a tangible cultural concept. Isolated finds remain puzzles in cultural history, for example the 'man(?) from Manopello' (southern Italy) or the stelas on the Easter Islands. Although homology research is epitomized in biology, it is only one facet in the 'general theory of comparability' (German: '*Allgemeine Vergleichslehre*') that Goethe (1817) already strove for. Goethe meant this in the full empirical sense, even if he presented it only as a sketch (Riedl 1995b, 1998).

    The argument here is that an intuitionistic solution is available, and a complete one at that (see below). Cognitive dualisms are at fault for causing us to perceive position and structure as being different things, and to juxtapose them with transitions and coincidences. Comparisons involve nothing else.

(c) *A synthesis of the criteria* is appended here. It is justified to ask why in particular there are five of them. Remane's intuitional and apparently largely correct categorization provides the answer. In fact, the literature on this issue forwards up to twenty-one criteria of homology. All of them fail the test of rigorous inquiry.

This topic—if the reception by colleagues and feedback from classroom teaching is a barometer—involves more effort than merely grasping the criteria themselves. A two-step approach is best:

The first is to (c1) synthesize the principal criteria and then to (c2) juxtapose the principal and auxiliary criteria.

- (c1) *Synthesizing the principal criteria*: The framework homologs are hierarchically nested. Accordingly, the difference between the positional and structural approaches is merely one of perspective. Looking in the direction of the upper systems reveals the linkages to be a positional matter, looking toward the subsystems reveals the structural composition. Importantly, this flip in perspective can be initiated from any hierarchic level.

  Some of the confusion may reflect this system-inherent duality. To stick with the spinal column example in Fig. 4.16: the epistropheus is equally a structural trait of the cervical spine and a positional trait for the dens. Conversely, the dens is the structural trait of the epistropheus and the cervical spine concurrently its positional trait. This may seem trivial but needs to be absorbed in order to proceed with confidence.

  The result is a matrix of reciprocal control and insight, much like discussed in connection with subsumption and hermeneutics. This is the second hurdle in understanding the underlying process.

  Remaining within this framework of positional and structural comparisons, consider only two forms (for didactic purposes two related forms): the spinal column of a human and of a gorilla.

  Now closely observe what automatically transpires when comparing the two: prognoses are made from element to element and at each tier in the framework. The context can be paraphrased as: "I wonder whether that element will be present there as well?!" In the case of related species, there is confirmation aplenty. Namely equally as many as the number of cases in which individual details are found to correspond in position and structure (i.e. homologies are determined). Thirty-four comparable vertebrae multiplied times at least 15 homologous structures, i.e. situations that are individually recognizable and individually named by anatomists, yields over 500 homologs.

  The process then involves studying whether (and presupposing that) these separately designated sites are predictably present in all yet unexamined adult representatives of a species. Those features that are characteristic but not necessarily consistently present are subsumed under so-called 'epigenetic' traits, which are understood to reflect 'channeled degrees of freedom' in ontogeny (compare Fig. 6.21).

  Note that, as in the letter-deciphering example (Sect. 4.2.1 (b2)), we subconsciously begin at all levels of comparison and nearly simultaneously cross-process the whole object and all its many details. Conveying this process requires systematically pursuing it level by level, in the present case based on the sections of the vertebral column, individual vertebrae, along with their main and sub-parts.
- (c2) *Synthesizing the auxiliary criteria*: The task here is to synthesize the principal and auxiliary criteria. Unraveling this process of cognitive gain calls for introducing 'simultaneous' and 'consecutive' coincidences. The first aspect is

'simultaneous cognitive gain', i.e. the experience gained in comparing only a single pair of items.

(i) *A simultaneous cognitive gain* corresponds to the example just given above. Clearly, the didactic advantage of having selected two very similar forms already requires knowledge about the respective field of similarities. At this level of investigation such knowledge can be presupposed, but ultimately needs to be examined. Recall that we think in fields of similarity and that the net needs to be recast more widely only when an entirely new structural form is encountered. The fact that we recognize a spinal column can be taken for granted here.

(ii) *A consecutive cognitive gain*: Science has tested the term 'spinal column' on a wide range of case examples. This is by no means trivial because the spine of a shark and that of a frog show surprisingly little similarity (Fig. 4.19). Determining the correspondence requires combining the tools of transitional criteria and coincidence criteria and by excluding anti-coincidences.

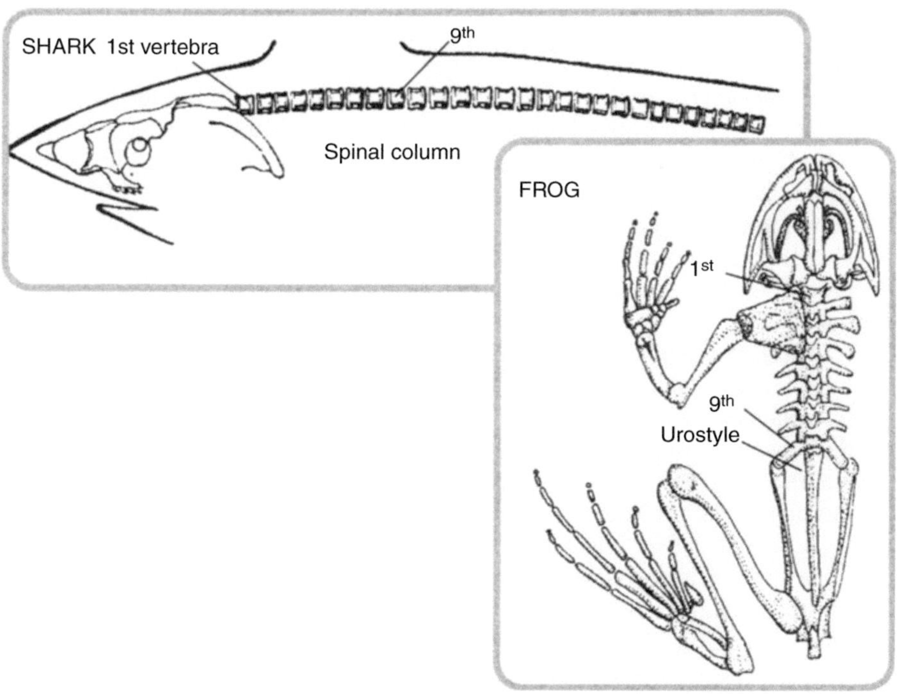

Fig. 4.19 Differentiation of initially equivalent structural elements, exemplified by the transition of the spinal column from the shark to the frog. In sharks the vertebrae are still uniform ('interchangeable'), in frogs differentiated into nine 'individualities' and a rod, the urostyle (from Riedl 1975, 1976)

The study of homologies is therefore based on a 'one-after-the-other' principle, i.e. a 'consecutive cognitive gain' in which species after species is compared. In mammals alone, in which the above-mentioned homologies re-occur in the 22 recognized orders, this amounts to 5000 case studies (recent species only). Conservatively assuming that only half of these homologies are detectable in the tedious comparison from species to species, from the platypus to humans, this still yields one hundred million confirmed prognoses.

Although the spinal column—our example—is complex, it nonetheless represents only part of the skeletal system. That system in turn, along with tendons, muscles, blood vessels and nerves, is part of the musculoskeletal system, and this is ultimately only one part of the organism as a whole. In mammals alone, the species-by-species approach yields an estimated one hundred million confirmable prognoses.

Positional and structural experience is related to experience in transitional and coincidence matters in the same manner as the perception of simultaneous and consecutive coincidences: both combined yield full experience.

Importantly, homologies are quantifiable. Every structure that can consistently be found and identified (and thus earns a separate designation) counts. The scientific discussions about the remaining borderline cases play only a negligible role (see also paragraph C3).

(d) *A theorem of probability*: Tracing the homology theorem back to the parameters of probability and explaining it based on those parameters requires five insights, namely into: (1) the comprehensiveness of the homology criteria (as demonstrated above); (2) in the quantifiability of confirmed prognoses (predictable homologies; see back-of-the-envelope numbers above); (3) the probability-based character of all natural laws; and (4) the relationship between confirmable prognoses and theorem reliability. In the following, the four will be interlinked.

The fifth insight is somewhat different: it pertains to the delimitability of traits. The first and second insights presuppose this, but the insight relies on the interplay between optimizing the concept of trait and that of field. Accordingly, this 'theory of the trait' bridges the gap between the principles of morphology and the principles of systematics (therefore discussed last in dealing with the morphological problem; paragraph C3).

As noted above, understanding the synthesis of the homology criteria is more difficult than merely determining the criteria themselves. Their subsequent validation proves to be even more difficult. This is rooted in three considerations, namely in accepting the insights 3, 4 and 5 below.

It is best to start with (d1) the issue of probability and (d2) the weighting of prognoses and then to show (d3) what this means.

- (d1) *Probability of theories*: It is common knowledge that natural laws are laws of probability. The respect afforded to the formalizable laws of physics is an obstacle here. Many bio- and cultural scientists tend to think that—contrary to their own field of expertise—the supposed 'eternal' validity of physical laws elevates those laws above the restrictions of probability. This is clearly not the case.

Conversely, many "inorganic" scientists believe that the relative recency and descriptive character of organismic laws inevitably reduce them to indeterminate rules—which is also not the case. The reliability of a discovered correlation—incorporating the hermeneutic and subsumption schemes—can be measured entirely based on the degree to which predicted cases are confirmed.

- (d2) *Weighting of prognoses*: Accepting this, then the main problem lies in how to assess such a degree of probability. A simple brainteaser: Assume an equal probability of a prognosis reflecting a lawful, rule-based event or a random occurrence. Since the exact proportionality of the two outcomes is not the issue here, tossing a coin can serve as a model. How often can my opponent win the toss in succession before I become convinced (and to what degree) that intention (swindle) is involved?

The process is recognized to involve the base 2 logarithm. Of course, it is possible that the 'heads' betted on lands ten times in a row. The probability of this, however, is a mere 2^{-10}, i.e. on average once per 1024 cases of 10 tosses each. If 'heads' lands 100 times in a row, then the probability of chance is reduced to virtual impossibility on any earthly scale ($2^{-100} = 1/1.3 \times 10^{-30}$). Even giving the possibility of randomness more leeway, let's say of being one hundred times more probable than any common-sense insight, then the explanation based on randomness is still an astonishingly low 10^{-28} (explaining why exact proportionalities are largely irrelevant here).

This justifies raising the question of how often a prognosis about a surmised homolog needs to be confirmed until the probability of having detected an organismic law is high enough. The prognosis that every healthy-born vertebrate exhibits a chorda dorsalis in its embryonic development can serve as an example. How many individual cases can we rely on to date? That would be about 50,000 species multiplied times more than 100 million generations times an average of 20 million individuals: 5×10^4 times 10^8 times 2×10^6 yields 10^{19} realized cases for which the presence of a chorda dorsalis can be assumed.

The probability that our prognosis is incorrect has therefore been reduced to 6.4×10^{-58}. This is equivalent to the probability that a grain of material on our hand would suddenly cool down to near absolute zero and, defying the laws of gravity, shoot up to the ceiling at relativistic speed—namely if all its molecules would by chance simultaneously move in that direction.

The above discussion has focused on isolated homologs. Returning to the spinal column example, specifically to the narrower circle of modern mammals, yields a similar picture. Comparing the recent species yields 500 *simultaneously* confirmable multiplied times at least 4000 *consecutively* confirmable prognoses. The probability (2×10^6 cases here) that the same lawfulness is not involved is a mere 7.5×10^{-37}. Viewed in the context of the 100-million-year-long mammal era, with on average 2000 species multiplied times about 50 million generations times 10 million individuals, this yields 2×10^3 multiplied times 5×10^6 times 10^7. Accordingly, based on 10^{17} realized cases in unbroken series, the probability that a mammal would appear without a spinal column is a mere 6.7×10^{-52}.

- (d3) *Probabilities of homologies*: The take-home message from the above is that biologists deal with true laws derived from nature. The idle talk about a 'merely descriptive natural history' has no epistemological bearing whatsoever. Laws are either involved or not. Haeckel's Law—stating the ontogeny represents a simplified repetition of phylogeny—has been downgraded to a mere rule. Yet it reliably fulfils justified expectations anchored in historically developed, lawful order. Returning to the chorda example: ontogeny encompasses 'palingenetic' and 'cenogenetic' laws. Palingenetic refers to the repetition of stages once experienced, for example that every mammal exhibits not only the chorda in its embryonic development but repeats—and then dismantles—many other traits such as the system of gill arteries that characterize fish (Fig. 4.43). Cenogenetic, in turn, involve adaptations to larval or embryonic life, just like the embryos of all higher mammals are nourished via an umbilical cord.

  An important step forward for morphology is to transcend its intuitionistic stage by weighting homology expectations. This demonstrates the degree of certainty behind statements but also reveals when such expectations may be less reliable or are still at the stage of purely mental constructs.

(e) Combined, *the forms of homology* introduced below show that there is no third category beyond analogies and homologies. The forms themselves largely reflect our cognitive approach, the degree to which analogous structures are based on homologies, and the step from perceiving a similarity to expecting a homology. The above probability-based justification remains valid for all forms.

Four forms or types of homology are presented here: (e1) homonomy and symmetry, (e2) homodynamy, (e3) homoiology and (e4) the transition of isology to homology.

- (e1) Homonomy refers to those homologous elements of an organism that, in contrast to the singly individualizable homologs, repeat themselves in the same organism (often as no longer individualizable standard building blocks). Symmetries represent a special case in this category.

  Each framework homolog usually bears its own specific name. Nonetheless, pursuing this in descending order ultimately hits a barrier. In the spinal column example above, it was a joint surface on the second cervical vertebra, the *facies articularis ventralis dentis epistrophei* (Fig. 4.16). Breaking this articular surface down a step further reveals the bone trabeculae, which actually do still have a name. Extracted from their matrix, however, it becomes impossible to determine from which part of what bone they come from. They are simply standardized, indistinguishable building blocks.

  Such 'final' homologs in the descending hierarchy of framework homologs, for example the above joint surface, can be termed a 'minimum homolog' to indicate how far homologs can be recognized. This represents a 'homonomy boundary': beyond that limit, homologies must be dealt with in a different manner (Riedl 1975).

Equally pertinent is the fact that these homonomous elements are themselves again hierarchically structured. The bone trabeculae, for example, are composed on bone cells, these in turn of cell organelles and so on, down to the last still homologizable biomolecules (compare Fig. 4.1). Interestingly, this homonomy boundary is quite high in plants: in a spruce tree, for example, only the trunk still represents a homolog—the branch whorls are already exchangeable homonoms, all the way down through the branches, twigs, and needles, further to their stomata, their cells, etc. Early German systematists tellingly referred to sea anemones as '*Blumentiere*' (= 'flower' animals) because the situation is similar in corals, sea anemones and their kin.

This takes us back to *symmetries*. These also involve homonomy, i.e. identical repetitions of the same building instructions. As a rule, such homonomous building blocks become more abundant, and more conservative, the deeper they are embedded in the hierarchy. This ranges from 10^{12} equally formed "tiny gray cells" in our brains, to the thousands and millions of homonomous organelles and biomolecules they contain. Again, the organization of organisms exhibits high redundancy, i.e. 'cheap order'.

At higher levels of homonomy, this is represented by the uniform extremities of ancient arthropods or the teeth of ancient reptiles. Our fingers and even the principally uniform design of our arms and legs are additional examples. Reducing such similarity, i.e. differentiation, represents the transition from the homonomous to homologous structural elements.

In ethology, homonomous building blocks play an important role as repetitive, homologizable movement sequences and repeated verses of songs. The same holds true for isomorphic molecules (to the extent that these prove to be homologous; see also Sect. 4.3.1 (e4)). In artefacts, assemblages of identical structural elements reflect the redundancy phenomenon. The term homonomous is particularly applicable to those elements that exhibit a genealogy. This ranges from the repetition of terms in different languages to the form of capitals atop pillars or of the ribs supporting vaults in architecture.

- (e2) *Homodynamy* refers to the transfer of homologous instructions within the framework of the epigenetic system, i.e. ultimately gene interactions. So-called induction processes immediately come to mind: a blasteme—an embryonic cell complex—transfers special tasks of differentiation to another complex (compare Fig. 6.20).

  The chorda of vertebrates once again serves as an example. It is the source of an instruction to the dorsal mesoderm, the future primordium of the musculature, to arrange itself segmentally. This, in turn leads to induction of the primordia of the vertebrae and subsequently to that of the spinal ganglia, etc. Removing the chorda from even the most primitive vertebrate, for example from a lamprey embryo, and implanting it under the abdominal skin of a chicken embryo triggers segmentation of the ventral mesoderm even there.

The message is therefore 'universally' understood in all vertebrates and initiates the development of homologous structural elements. This justifies terming the phenomenon a homology.

Similarly, the term homodynamy is applicable to functions, specifically those that extend down into the molecular realm via metabolic processes and energy transfers. Among human artefacts, this term applies to heritable 'codes of conduct' and even the functions of semantics and syntax.

- (e3) Homoiologies, in contrast, refer to analogies that build upon homologous foundations. The classic example is the crista sagittalis, the bony skull ridge (introduced in Sect. 4.3.1 (b2) above) that grows over the suture of the parietal bones. It develops, independently, in those animals whose temporal musculature is strongly developed (and needs an expanded area of attachment) for a particularly powerful bite. Hyenas and gorillas are a case in point. Similarly, the powerful pectoral musculature of birds and of moles give rise to a bony ridge along the sternum.

  The categorization here is correct because the positions at which the analogous structures develop are clearly homologous. Homoiologies are much more common than generally assumed. This is because they can arise whenever homologous structures, independently from one another, are subject to the same demands. The classical example is the fins of ichthyosaurs and dolphins, but the principle is also highlighted in marsupials—they have developed nearly every conceivable ecological type known in higher mammals.

  Finally, difficult-to-recognize homoiologies are detectable behind many behaviors and human artefacts.

  (e4) *Isology*, in contrast, is a term borrowed from chemistry to describe structurally identical molecules. These have nothing to do with homology because they can have different origins. In some cases, however, one and the same large molecule can be determined as having had the same origin and undergone the same transformational steps. Here, homology becomes highly probable. The reliability of this interpretation is further cemented if such molecular transformations clearly correspond to the overall trends the organisms show in their field of similarity. An early and convincing example (Fig. 4.20) is the cytochrome-c molecule.

  The transition from isology to homology is an issue going beyond large molecules, becoming particularly pertinent in the context of genetic coding. This is because homologizing (with its rigid criteria) becomes increasingly difficult as the number of traits decreases.

  More need not be said here about cognition. As far as explanation is concerned, however, the above boundary actually marks the beginning of the problem (see Chap. 6).

(f) *Interpretations of homology*: The discussion of the homology concept is marked by two features. First, it has failed to yield a uniform interpretation. Second, and related to this, it has sought to support or replace the pure cognitive process—which is mistrusted—with explanatory processes. Importantly, this lack of trust

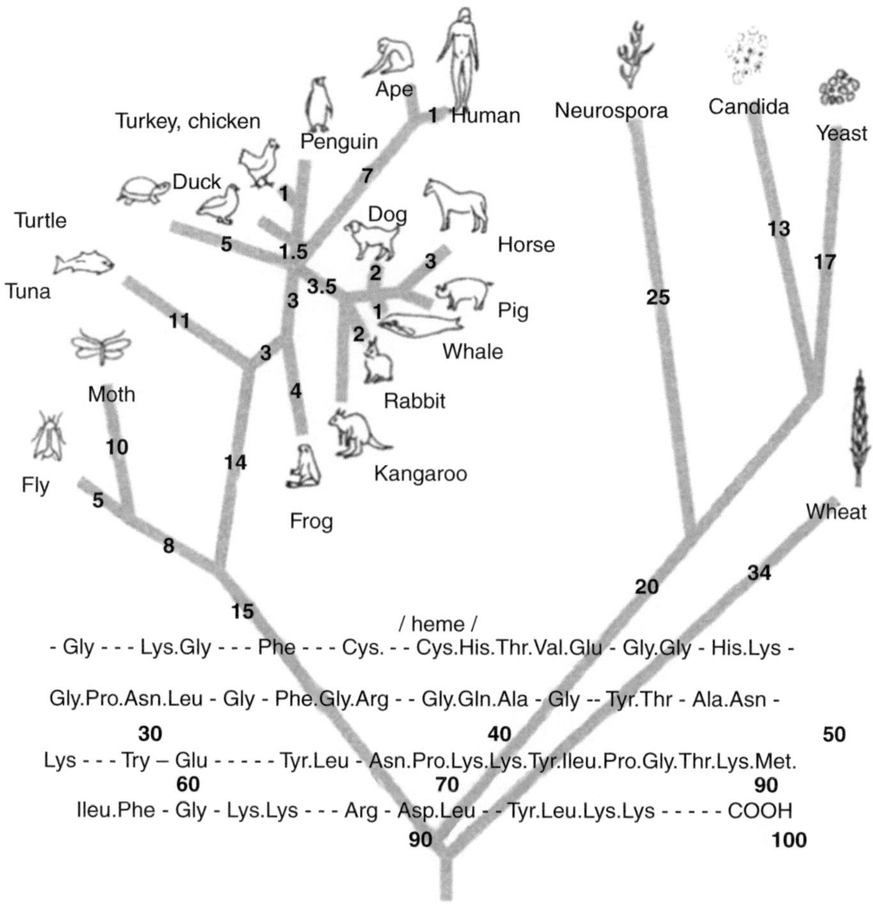

Fig. 4.20 The homology of isologies, exemplified by the cytochrome-c molecule. Applied to the phylogenetic tree of organisms, this enables determining how many (mutative) changes can be postulated in the long chain of amino acids from representative to representative. This reveals the chemical similarity to be a homology. Bottom: the amino acid sequence that has remained the same from yeast up to humans (compiled in Riedl 1975)

is based on an insufficient understanding of cognition, as outlined in this chapter. Attempting to substitute cognition with explanation, however, inevitably runs the risk of becoming mired in the circularity mentioned earlier: no explanation can be better than the preceding description of the entity being explained. These considerations were termed 'idealistic' or 'classic' in the 1980s, other monikers being 'historical', 'evolutionary' or 'biological', 'phenetic' or 'cladistic' as well as 'utilitarian' (Rieger and Tyler 1979; Patterson 1982; Wagner 1989). 'Idealistic' or 'classic' are positions restricted to the cognitive process. 'Historical', 'evolutionary' and 'biological' refer to different explanatory models (treated in Chap. 6 and less pertinent here because explanation has little influence on perception and cognition).

'Phenetic' and 'cladistic' are standpoints that seek to circumvent homology as an issue. They are more informative when applied to systematic and are therefore treated—along with the term 'operational homology'—in Sect. 4.4.1 (a).

Finally, 'utilitarian' was coined to encompass positions seeking to avoid homology as a term, regardless whether it involve cognition or explanation. This approach replaces homology with the term 'character' or 'trait'—as if the kinship used to explain similarities can itself be attributed to these very similarities. The 'fields of similarity' presented earlier lie at the heart of the matter. They presuppose an unreflected, intuitional solution, which again obscures the necessity of fully understanding the process. They fail to move us forward on this issue.

4.3.2 Type and Bodyplan

The type concept has been addressed in relation to its development (Sect. 4.2.1 (a)) and empirical determination. Bodyplan (German: *'Bauplan'*), in contrast, remained a colloquial metaphor in earlier chapters. In the broader sense, both can be used synonymously, but it is useful to distinguish the two. Contextually, and reflecting the way we use language, 'typical' is better when referring to the adjectival sense. In hierarchic contexts the plural 'bodyplans' is preferable to denote the interlinkages of levels.

The best way forward is to address (a) the issue of type followed by (b) the phenomenon of bodyplans.

(a) *On the type concept*: The dictionary definition of the modern use of the word 'type' refers to "the ability of humans to extract, from a series of similar circumstances or entities, that which is 'typical', a process that from a perception psychological point of view is already pre-programed by gestalt perception" (Brockhaus). Progressing from the discussion in Chap. 3, we need to determine what is typical in the type and what expectations lie behind determining that which is typical.

    The concept of the 'taxonomic type' is a given. It refers, for example in the designation *Genus typicus*, to the first discovered genus of a family. This alone demonstrates how removed researchers already were from an understanding of type.

In principle there can be only one type, namely (a1) the 'morphological type' along with modifications based on (a2) the aim of the investigation or on simplifying (a3) the method or (a4) the manner of depiction.

- (a1) *The morphological type* can be defined very simply based on today's state of knowledge as 'the degrees of freedom and of fixation demanded of and enabled by the structural components of a system'. Degrees of freedom and fixation are systemic conditions. This malleability and rigidity are recognizable in harmoniously divergent patterns, i.e. in the respective field of similarity. The hand skeleton of mammals can again serve as an example (Fig. 3.8). The conservatism of the central axis is immediately apparent, as are the reduction patterns of the lateral toes, along with the freedoms in both the absolute lengths and relative proportions of the metacarpus and fingers, etc.

  What possibilities does a system harbor? What constraints are unsurmountable? From a comparative anatomy perspective, this can be determined only by examining the demands that were placed on that system. We would be unaware that finger bones can elongate in a spider-like manner or, in contrast, further subdivide were in not for bats and dolphins.

  The morphological type yields key clues about the system properties inherent to a particular structural forms. This goes beyond any individual hierarchic level. To stick with the above example: the hand skeleton type of can be variously represented tofocus only the even-toed animals (a narrower depiction) or encompass all tetrapods (a broader depiction).

  The shortcoming of this type concept is that it defies a single, cohesive depiction. This is because the metamorphoses of several qualities are superimposed on each other. While these can be quantified individually, they modify themselves in different directions. The result is structural transformation in three dimensions, a meshwork of positional shifts involving changes in form and even novel structures.
- (a2) *The systematic type* as a concept is prompted by the desire to derive the original form based on representatives of a kinship group. Tracking the peripheral representatives in a field of similarity, i.e. the end-points of opposing trends, back to a common hub can actually help approximate the unspecialized, original structural form (see Fig. 3.8 in support of this approach).

  This strategy is by no means foolproof. The problem lies in the direction in which series of similarities are 'read' (see Sect. 4.4.3 (a) for more details). This is an issue in systematics and is bolstered by fossil evidence. The ancient bird Archaeopteryx, for example, would have been difficult to reconstruct based solely on recent representatives. We humans, as another example, are highly unspecialized mammals—without, however, resembling the earliest mammal representatives.

  Returning briefly to Goethe's '*Urform*' (archetype) and his '*Urpflanze*' (archetypical plant): despite sounding similar to the above concepts, these terms focused on the prerequisites behind a structural form rather than on phylogenetic aspects.
- (a3) *Simplifications of the method*, in contrast, are designed to arrive at a 'generalizing type' and a 'central type'. In the former, backtracking to 'the general'

requires omitting all special modifications. In the latter, something akin to the geometric center at the hub of a particular field of similarity is sought by retracing the series of metamorphoses. The early discussions already lamented that this approach did not yield fully functional organisms. While that is correct, it again highlights that the critics must have expected the result to be something like an original form or an archetype.

In fact, a morphological type need not represent a viable creature. Rather, it embodies the 'possibilities of a particular structural form'. It corresponds to a system's functional and epigenetic degrees of freedom, delimited by the opportunities afforded to or demands placed on that system.

All simplifying methodologies designed to establish a type cannot yield anything beyond what that simplification allows. Whether they operate by rounding off or by averaging, they can never capture the full potential of a structural form.

- (a4) *Simplified depictions* lead to the 'diagrammatic type'. Depicting an insight or experience in a simplified manner is entirely legitimate and sometimes a technical necessity (e.g. graphic representations). Recall, however, the obstacles facing graphic renditions of morphological types. Accordingly, diagrammatic representations tend to restrict themselves to a few selected characteristics of a type. Generally, two figures are generated, one depicting those structural elements that are consistently present, the other the most common positional relationships (see Fig. 4.21 for an example). Such illustrations are misleading because they tend to suggest proportions that were not necessarily meant to be depicted in the first place. Considering the range of mammal skull forms, for example, there is no reason to specifically represent the 'type' based on a dog's head (Fig. 4.21). Nonetheless, some form must be chosen before any positional relationships can be depicted.

  Everyone is familiar with what each discipline dealing with complex systems considers 'typical', whether it be geomorphology, ethology, ecology or the full range of cultural sciences. Nonetheless, the term tends to used colloquially even though it would be more instructive to consider the various type concepts.

  Importantly, gestalt perception and gestalt comparison already steer us very adeptly to the morphological type presented here. The underlying principles, however, have rarely been elaborated. The process remained intuitional, condemning the various proponents and schools to argue intuitionally. In fact, the essence behind the type is the key to clarifying the actual metamorphoses in this world.

  The type is and remains a multidimensional affair. And by no means in the abstract sense. Quite the opposite, it represents truth anchored in reality: everything it encompasses defines real functional limits and has given rise to real, living representatives.

(b) *The term bodyplan* or bauplan is typically taken to be a metaphor, but in a certain sense the analogy is fitting. This is rooted in Aristotle's four *causae* and extends into the interdependencies defining all hierarchic structure.

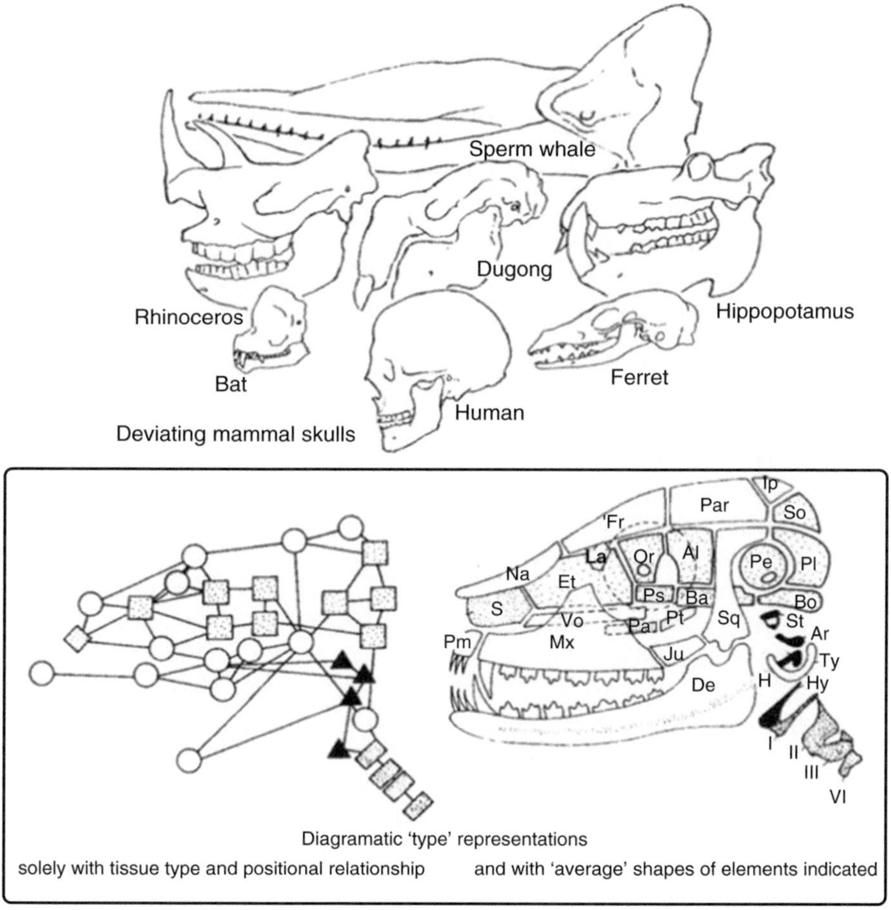

Fig. 4.21 The diagramatic 'type', exemplified by the mammal skull. Left: only the types of structural elements and their positional relationships (after Riedl 1975); right (nach Kühn 1955): including their 'average' shape, which already goes somewhat beyond the type concept. Seven deviating mammal skulls are juxtaposed for comparative purposes (after Gregory 1951)

The discussion has again arrived at the boundary between cognition and explanation, hovering between the transition from *post hoc* to *propter hoc* as characterized by David Hume. Practically put, no theoretical concept is necessary to recognize the effectiveness of a stone axe or fur clothing.

'Good old common sense'—the ratiomorphic processing of the *simul hoc*—continues to help us (b1) grasp the links bridging the tiered structural hierarchies and (b2) define the framework of the class hierarchies behind these structural hierarchies.

- (b1) *Structures of bodyplans*: Examining the structural hierarchy of bodyplans is the stepping stone to cognitively 'differentiating effects'. This refers to the four forms of discernible causes. Our understanding of this has been modified in the history of science (see Chaps. 5 and 6). The focus here is on the double cognitive dualism structuring our perception of causes.

Building a house, to pick an example championed by Artistotle, first requires forces, sweat, money or power—the *causa efficiens*. The second requirement is suitable material—*causa materialis*. The third is a plan that specifies which materials are to be positioned where, i.e. a selective, form-giving or 'formal' principle, the *causa formalis*. The fourth and final requirement is a certain intention, goal or program—the *causa finalis*. All four are indispensable (see Chap. 5, Sect. 5.2.2 (a) for a more detailed account of the relevant sections in Aristotle).

This can be directly transposed to the structure of organisms. This is trivial in the case of the first two, driving forces and material causes. The latter two, the *causa formalis* and *finalis*, are recognized selection principles in biology, designed to ensure species survival. The four causes do not quadrisect the world. Rather, our cognitive approach appears to recognize two alternatives.

(i) The first type of symmetry in these alternatives was already recognized by Aristotle. In simplified form, he considered the *causa efficiens* and *finalis* to act from the outside, *materialis* and *formalis* from within. Today's understanding of things confirms this. The driving forces behind the structure of every organism are derived from the sun, and the conditions for species survival are dictated by the respective environment. Material transformations and form-determining selection operate from within.

This cognitive symmetry reflects human experience: we experience the result of the *causa materialis* and *formalis*—the material and form-giving aspects within the systems—as structures through gestalt perception, i.e. as cells and bricks, as organisms or houses. In contrast, we experience the *causa efficiens* and *finalis*—the driving forces and 'purposes'—as being functions that are alike in their immateriality. This is reflected in the syntax—the separation of nouns and verbs—in all languages (Fig. 4.22).

(ii) A second symmetry crops up as soon as the hierarchic tiers of complex systems enter the equation. It becomes apparent that the *causa efficiens* and *materialis* operate through this hierarchy from below, the *causa formalis* and *finalis* from above. Again, this may seem trivial for the drivers (initially always quantum forces) and for the materials. The fact that selection and purpose always operate from the higher levels is evident: an organ determines the form and purpose of its tissues, just as a room helps define the optimal furniture.

The cognitive aspect of this symmetry is also experience-based and again reflected in syntax—as the passive or active voice. We generally feel we can actively control the material and formative process, but we perceive ourselves fatefully subjugated to the forces from above (Riedl 1998). In fact, our lives depend on knowing whether we are the movers or are being moved. This is anchored in our neural pathways ('reafference principle'; Lorenz

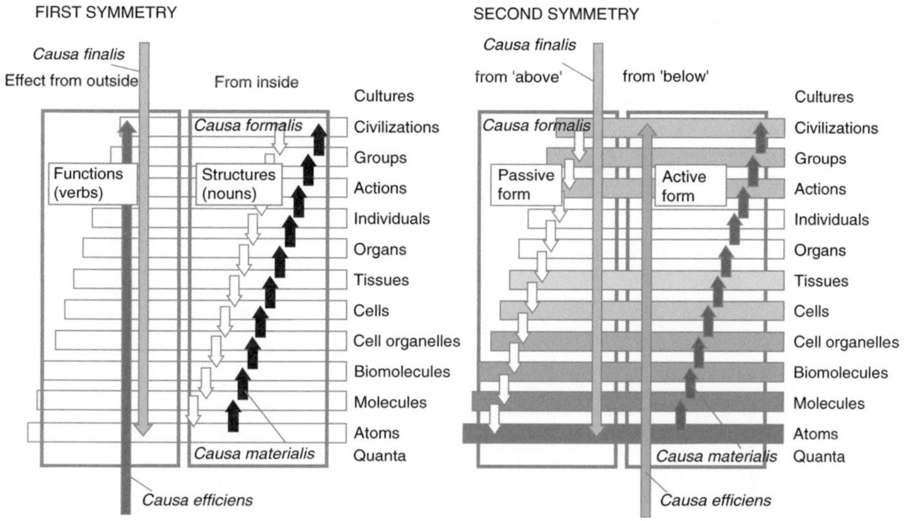

Fig. 4.22 The two cognitive symmetries of our perception (interpretation) of the causal patterns in complex systems. Exemplified by the hierarchic structure in the organismic realm. The four categories of causes and the symmetries were already recognized by Aristotle (but not yet the hierarchic structure). The second symmetry (Riedl 1978/79) is based on the studies presented here (see Chap. 6 for details; from Riedl 1998)

1978), i.e. whether we think we are shaking something or being shaken. Whether we are eliminating something or being eliminated ourselves is fundamental.

- (b2) *Classes of bodyplans*: We need to extend our view to the concept behind structural and class hierarchies. The analogy between the construction types of organisms and of artefacts raises the terms 'conditions and pre-conditions'. This highlights the hierarchic relationship even more strongly because it involves 'the bodyplan for bodyplans'.

  The systematics addressed in this book is based on the systems of prerequisites specified in that construction plan for construction plans. An illustrative example: The bodyplan of crustaceans (Fig. 4.23) is based on a rigid architecture involving the outer skin and turgor, was well as on metamerism—the serial subdivision in largely similarly built body segments. If early stages of true extremities develop, then these adhere to that serial construction.

  This stage is already represented by annelid (segmented) worms. The development of their extremities is accompanied by reinforced skin, which helps anchor these paddle-like parapods. In the arthropods these take on the form of true extremities, and the necessary enhanced exoskeleton rigidity ushered in a new stage: the formative step to an entirely new group was taken. Importantly, this bodyplan serves as the foundation for additional bodyplans: the original

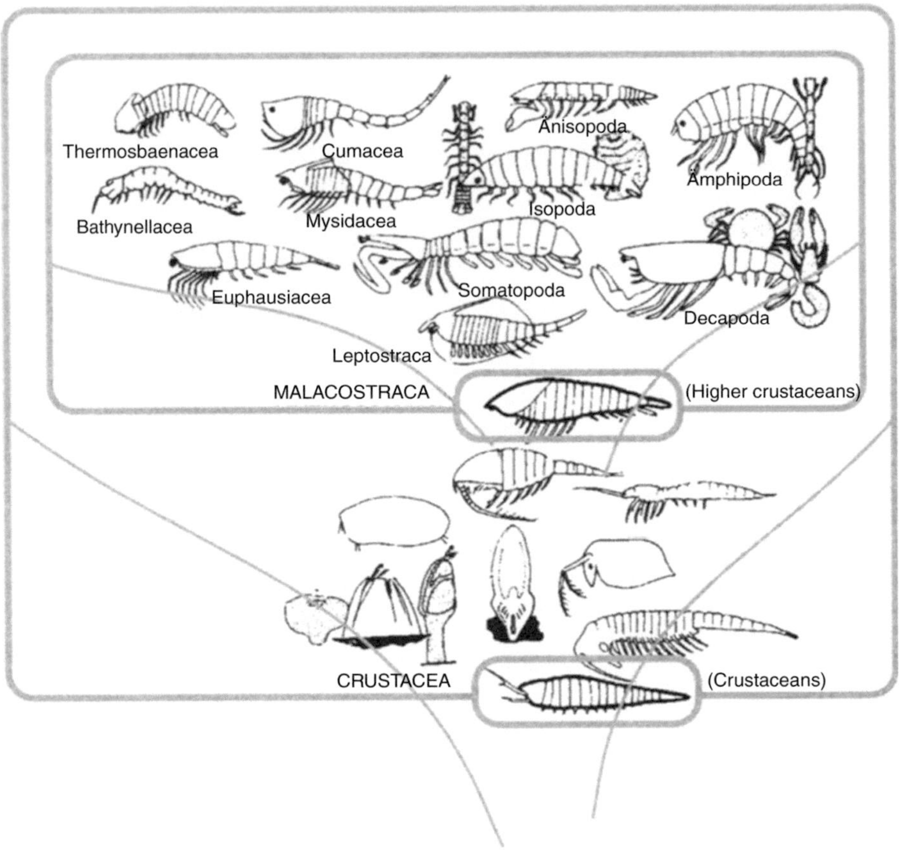

Fig. 4.23 The bauplan of the bodyplans, exemplified by the 'higher crustaceans' in the framework of the crustaceans. Note how the differentiation of the body shapes in the main groups of crustaceans is maintained in the representatives of the 'higher crustaceans'; subseqeuntly, only the trunk and leg shapes vary (modified after Riedl 1987a)

principle can either be further perfected such as in millipedes or, in the case of more highly differentiated forms, the metameres can become variously fused into regions. The latter strategy is manifested in the crustaceans, insects, spiders and their kin, and in a number of smaller groups.

The basic bodyplan of crustaceans exhibits eight variants, one of which is illustrated in Fig. 4.23 (i.e. the 'higher' crustaceans with eleven additional specified limb and carapace configurations). Importantly, there is no deviation from the principle involving an external skeleton, metameric regions and articulated extremities along with all the implications of that overall construct. This principle of pre-conditions of pre-conditions is a feature defining every hierarchy within such fields of similarity.

The same principle is evident in human artefacts, whether they involve a style such as Gothic (even in its reiterations such as Classicism or Neo-gothic) or, more broadly, the manner of thought and speech characterizing a particular epoch. All are governed by some set of pre-conditions. Figure 4.24 provides an example. Accordingly, the floor plan of cathedrals of a particular time period determines their vault construction, which in turn defines the overall ornamentation, which then extends to the tracery of the windows.

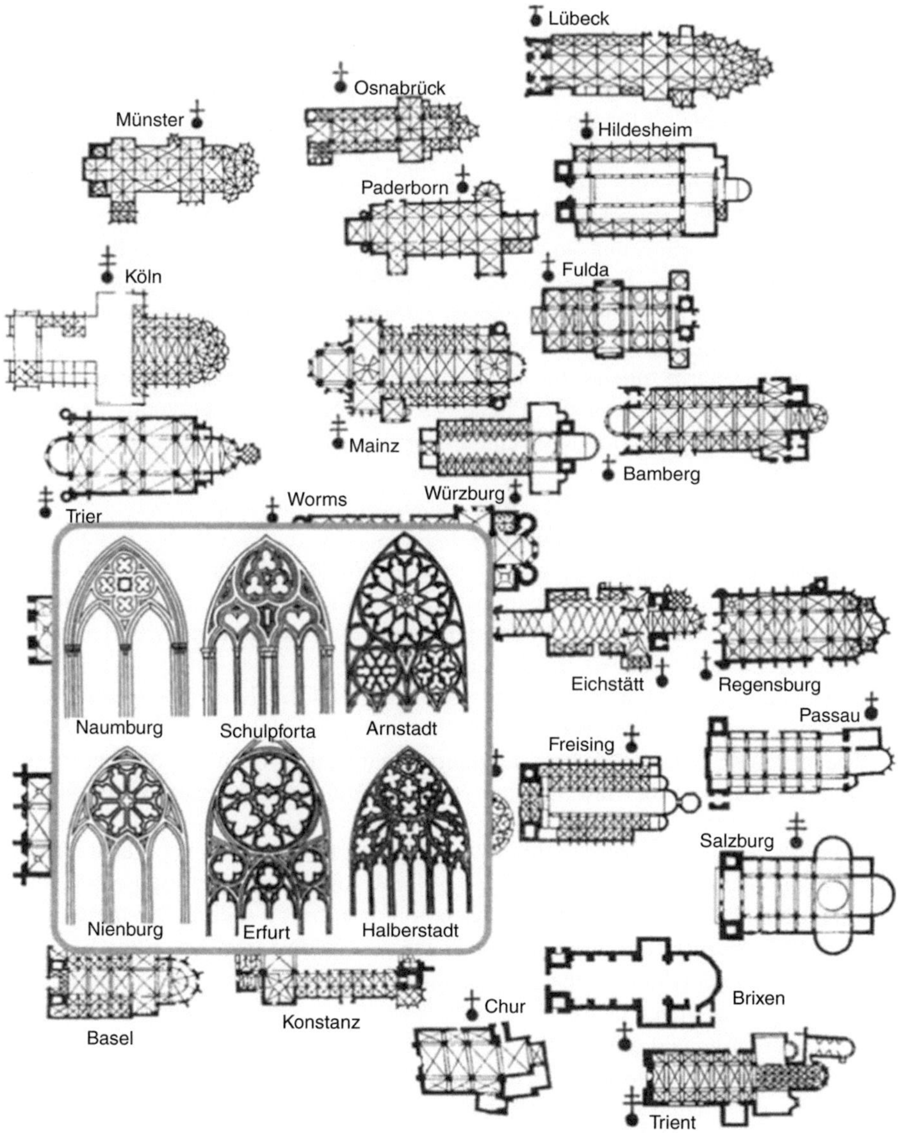

Fig. 4.24 Architectural styles as the establishment of principles, exemplified by church floor plans from the Gothic period. Insert: window tracery of German cathedrals from the North Sea to South Tyrol (after Braunfels 1980; Möbius and Möbius 1978; details in Riedl 1987a)

4.3.3 A Theory of the Phene and Character

The above discussion has repeatedly referred to characters or traits, more in a colloquial manner, without more precisely defining the term. As is often the case when grappling with systemic relationships, the definition was presupposed. This can no longer be deferred because it turns out that accurately defining 'character' or 'trait' requires practical experience. This involves a reciprocal optimization within the respective field of similarity: a trait must be viewed in its interplay with the higher-level structural and class terms. This addendum to the principles of morphology proves to be a foreword for the subsequent principles of systematics.

In the often very sensitive German etymology, the word for 'character' ('*Merkmal*') combines the roots of the words '*Merk*' ('realize') with that of '*-mal*' ('mark') to yield the notion of something 'marked' or 'remarkable'. Moreover, the English 'realize' implies that 'real'-ity is involved, i.e. that sensory data have something to do with reality. This semantic framework was also used by Uexküll (1937) to distinguish what an organism perceives ('*Merkwelt*') from the physical environment as such ('*Umwelt*').

Although the term 'phene' has not yet been introduced here, the related 'phenomenon' has come up repeatedly. Phene is a technical term conceptually similar to 'character/trait'. In preparing to discuss explanatory processes, they can initially be used synonymously.

Phene is an abbreviation of the Greek 'phainesthai', meaning 'becoming visible' or 'appearing'. It was introduced by geneticists, who felt that it was important to distinguish between gene and phene, between genetic make-up and its manifestation. At that time, this reflected the invisible versus the visible. In the current context the term presents little problem. In the subsequent explanatory framework, however, it gains gravitas and becomes theory-laden.

The systems context dictates initially behaving as if the fields of similarity framing the characters are already optimized (this optimization is treated in Sect. 4.4.2).

The first step is (a) to determine what a character is from the perception and epistemological point of view and then (b) to append the optimization process.

(a) *Perception of characters*: Asking what a perception (but not yet a character) is helps delineate the problem. A flash of light, for example, is not a character—except in the framework of an expectation or a theory such as "Is the airplane actually approaching through the night-time cloud cover?" Equally, if all observed objects are elongated, we colloquially understand that as being their quality rather than a character or trait as such—unless we expect the existence of a non-elongated object.

    A character can initially be understood as being part of a differentiating perception or as a notion of coincidences, one that harbors the expectation of being able to gain foresight.

The way forward is to examine (a1) expectational contents followed by (a2) qualities, polymorphisms and metamorphoses, then to (a3) discuss the relationship between similarity and probability and, finally, to (a4) summarize the practical aspect.

- (a1) *Expectations and their contents* are both the focus of and delimit what we experience as a character. Characters are per definition restricted by our sensory and mental faculties. Less trivial is the fact that alertness, interest and knowledge minimally influence the process.
  - (i) Every branch that our eyes glance over during a walk through the woods leaves some trace in our brains. Nonetheless, this information is impossible to retrieve willfully. However, if some recurring feature crops up that even only marginally attracts our interest, it gels as a 'peculiarity'. Conversely, the notion of 'being on to something' often leads us fully astray. The subconscious once again plays an important role here.
  - (ii) The structure of an expectation harbors a two-fold reciprocity (both already introduced above). The first involves the 'categorizability' of a character within a structural as well as a class hierarchy. The second in both cases is the predictability of the categorization based on character content and character affiliation. This bidirectional perspective progresses toward the inside and the outside. In structural hierarchies the context is structure and position, in class hierarchies lower- and higher-level classes.

    An unusual feature guiding our expectations is the 'infinitesimal' aspect: we expect the tiniest object to exhibit further, ever smaller components, and the most expansive object to again have some kind of even larger framework. This compels us to search further at both ends. And within this spectrum we expect to find characters of all sizes. In biology the range extends down to the atoms and their kinetics (as characters of a molecule) on one end, and up to life as a phenomenon and the biosphere as a whole (as characters of our planet).

    Finally, the difference between individuality (uniqueness) and exchangeability plays a role in structuring our expectation of characters. In structural hierarchies, the term 'homonomy boundary' (paragraph C1e1) is applied to the actual divide between the two. That boundary also changes our expectation of being able to determine position. The homolog has only one position, the homonom many (albeit, in the bony trabeculae example, limited to bones). Differentiation can make homologs out of homonoms. Equally, differentiated perception and interest can turn an exchangeable person into a unique individual. Small children do not yet distinguish between the concept of class and of individuality (Piaget 1978). In adults, that boundary shifts depending on interest, inclination or opportunity and can range from humans to domestic animals and other creatures, down to the grasses making up a meadow.
  - (iii) What an expectation promises, consciously or not, can differ considerably. The consensus is that it must be something useful. Accordingly, the effort put into turning a perception into a character meets our need for orientation—

even if only to satisfy our curiosity or to achieve inner peace (regardless how zany the underlying interpretation or notion).

- (a2) *Dealing with qualities, polymorphisms and metamorphoses* leaves the level of pre-conditions to address the question of how we conceptually handle perception.
  (i) Our penchant for counting and measuring has enabled great progress in a few select sciences, above all physics. That approach, however, is based entirely on assuming the exchangeability of entities considered to be identical.

  Test persons tasked with setting boundaries in a cyclical continuum do so in a carefree and largely unreflected manner (Riedl 1987a). Moreover, the boundaries are cyclical and reflect randomly assigned coordinates (compare Fig. 3.5).

  In fact, polymorphism is unavoidable even in the simplest geometric variables. Halving a distance or an angle leads to qualitatively different results. In surface areas (Fig. 4.25), simply halving individual variables yields a wide range of highly dissimilar transformations.

  Stating that the world is replete with qualities may sound trivial. Nonetheless, reducing everything to quantities makes sense only under special conditions. This is reminiscent of the question how many tomatoes are in the grocery store's four-pack of tomatoes if one is half-rotten and the other a 'twin': for the housewife the answer is three, for the botanist five.

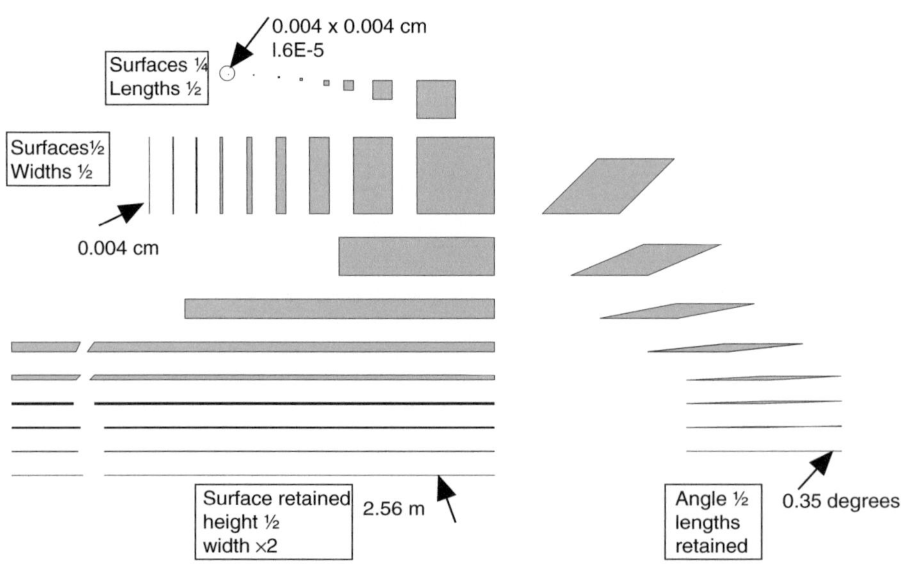

Fig. 4.25 Quantities and polymorphism, exemplified by the stepwise halving of one measure of a square. This yields entirely different sizes (after Riedl 1987a)

(ii) The issue boils down to the phenomenon of polymorphism. In fact, characters can always be assumed to be polymorphic. Even a simple color spot, which can be quantitatively described with a number representing its wavelength, has a certain size, is based on certain substances and is positioned somewhere, for example on the breast of a bird—making it useful in species recognition for an ornithologist.

This also holds true for exchangeable molecules: even a single guanine molecule can be understood as a character only in the context of its position in the chain and when embedded in a specific genetic material.

Polymorphism also underlies every conceptual framework: a clearly quantifiable spot of light on an oscillograph attains character status only in connection with the experimenter's expectation. The 'logical positivists' of the 'Vienna Circle' had hoped that the so-called 'protocol sentence'—for example that codifying the data pertaining to the above light spot—would already contain empirical certainty. It then became clear that this presupposes a theory about instruments along with a theory about the observer. Everything proves to be theory-laden.

(iii) If it is true that all characters are polymorphic at least conceptually, then the phenomenon of metamorphosis—the transformation of all complex systems—becomes important. After all, our current understanding is that objects change 'seamlessly'. The description 'continuously' is misleading because periods of constancy and comparatively rapid phase transitions clearly exist. In phylogenetic time spans, the latter can be aptly described by the metaphor 'fulguration' (Lorenz 1973a). Upon closer examination, however, they all prove to be smooth. This process goes beyond the transformations of individual parameters, which complicates the issue. In fact, change involves different parameters undergoing different transformations, which is coupled with the waning of old and waxing of new qualities. These issues lie at the heart of the problem behind optimizing the concept of a character or trait.

- (a3) *The relationship between similarity and probability* is a measure of the degree to which something is burdened with theory or expectation (Riedl 1987a). A simple example: What corners of a right triangle correspond with the corners of another (Fig. 4.26)? In open space, no decision as to which of the six potential solutions is correct can be reached. Knowing the position of both triangles in a system of characters, however, reveals the solution. This corresponds to the positional criterion of homology (compare Fig. 4.15).

The same holds true for the structural criterion of homology. If a positional relationship fails to provide a clue (Fig. 4.27), then six solutions are once again available. If one corner is determinable as a character, then only two possibilities remain; if two characters are fixed, the solution is unequivocal.

As simple as this consideration is, it yields three conclusions on dealing with supposed similarity. First, a theory on what a character is must be formulated, and this theory depends on knowledge about the character's surroundings.

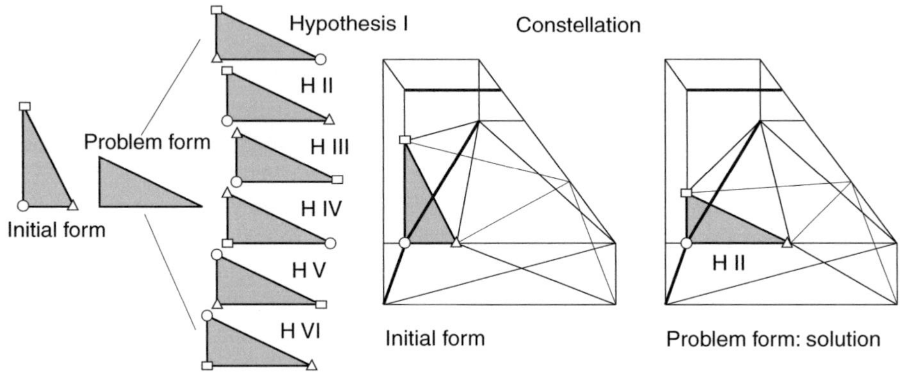

Fig. 4.26 Comparability based on position. An abstract example based on two triangles. Initially, no decision can be made between hypotheses H I to H VI. If the triangles are positioned in a constellation, however, then hypothesis II proves to be the solution. Compare the example 'positional criterion' in Fig. 4.15 (after Riedl 1987a)

Second, positional and structural criteria again play a role. Third, the probability of a homology being present can first be determined as of this point; it involves comparing the character with the full range of differentiations and other representatives (compare Fig. 4.17).

- (a4) As far as *practical aspects are concerned*, we note that characters are present at all levels of systems—from the homologs to the homodynamies to the homonoms. Moreover, these homonomous structural elements extend from the organ, tissue and cell level down into their organelles and large molecules. Both homologous and homonomous gene loci are definable. Finally, experience shows that the homonoms become increasingly conservative with their depth in a particular level. This makes them informative for ever wider fields of similarity.

The character richness of structural elements decreases along with their size. This, in turn, reduces the probability of determining individual homologs. Accordingly, the positional criteria gain importance through the entire hierarchy up to the organism's overall bodyplan. The reciprocity involved in critically evaluating homology becomes decisive here. The solutions lie in the overall constellation rather than in the depths of the system. Were we to discover a raven whose feathers were not composed of keratin (like in all other birds) but of tunicin (a material known only in sea squirts and other tunicates), we would still not order that bird among the sea squirts.

The phene concept has also been modified accordingly. We are increasingly able to fully describe individual genes, the process of transcription, and the translation into chains of amino acids. In many cases we know how the proteins are folded into large molecules with specific spatial structures. This has opened a further window of 'visibility'. Originally, the distinction between gene and phene corresponded to that between the invisible and the visible. Today, the invisible is being reduced or is shifting into the realm of the epigenetic system, the gene-gene

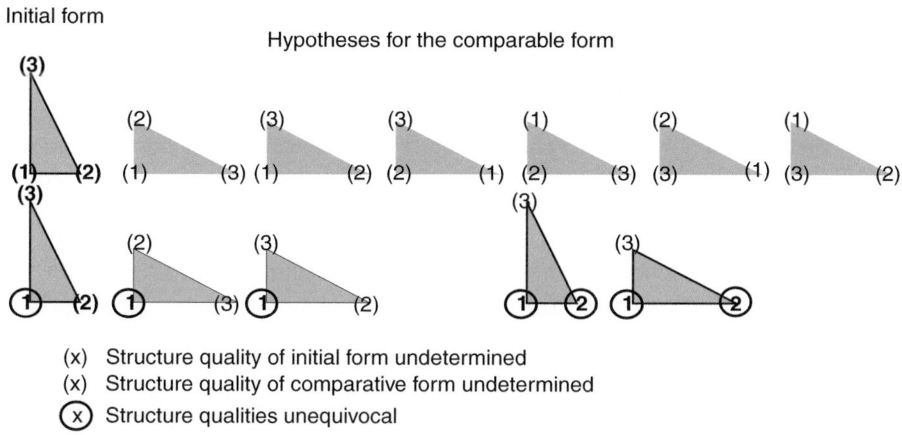

Fig. 4.27 Comparability based on structure, abstractly illustrated by two triangles. Their structure reveals either no definitive vertex, or only one or two. This yields six competing solutions, of which two or one is unequivocal. Compare the example 'structural criterion' in Fig. 4.17 (after Riedl 1987a)

interactions. Even there, inroads are being made, for example regarding the effect of 'homeobox genes' (Ruddle et al. 1994). The interrelated processes first manifest themselves (in the sense of the Greek 'phainesthai') when induction processes reveal the homodynamies.

Characters or traits play a role in all sciences. Accordingly, the gene–phene differentiation merely reflects a broadly understood cause-and-effect relationship. The trait concept has undergone further differentiation in traditional logic, in linguistics and in statistics. This is conceptually related to the phene issue. The commonality is that 'trait' refers to the determining feature or sub-feature of a term, of a symbol, sound or logical predicate.

(b) *Optimizing character concepts* is a somewhat different issue. This becomes evident when meanings shift. The discussion up to this point has focused on what makes up a character and how to grasp it. The question now is what predictive powers such characters offer.

In practice, a set of three complexes is involved. These encompass (b1) trends and discontinuities, (b2) optimizing a delimitation hypothesis and (b3) weighting a character correlation.

- (b1) *On trends and discontinuities*: Characters in complex systems are never exactly the same. A quick look at the veins on the backs of our left and right hands confirms this. The underlying principle, however, remains preserved and can be predicted based on a lengthy series of homologies. The principle behind even an isolated character needs to be determined. This is because even the simplest of characters are polymorphic in structure or in position.

The same holds true for delimiting a character, i.e. seeking a definition that encompasses its full range of variability. Figure 4.28 presents a simple diagrammatic scheme of such a potential range.

When arranged in this manner, the variations of the square and the different trends they take are clearly recognizable. Whatever names they might be given, the changes they undergo are conspicuous. And the experiments on partitioning continua (Fig. 3.5) reveal a strong conformity in how we subdivide such trends. This gives us great confidence in our ability to detect boundaries in a 'trend'. In fact, we are good at recognizing 'discontinuities' within phases of transition.

Importantly, character shifts never occur in isolation. Rather, characters harbor contents and/or are themselves the contents of additional characters, all displaying their own trends. Figure 4.29 illustrates such a framework in diagrammatic form: the trends within trends are evident, and it would present little difficulty to arrange the individual forms in the correct order were they jumbled. Figure 4.29 also underlines that the axes of trends can split, whereby some trends end and others begin. This also marks the beginning and end of characters and the terms applied to them.

These strategies of perception and arrangement define how we deal with all polymorphic systems. The results are perhaps most visible in biology, but are equally recognizable in human artefacts, especially those involving an interplay between transformation and transmission such as in the history of language, style or art.

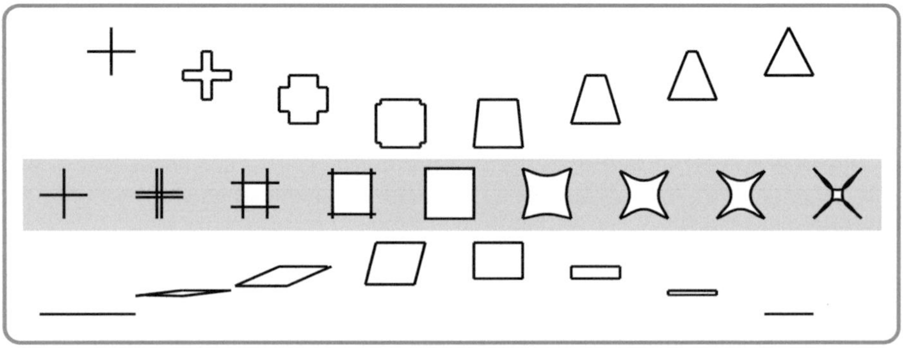

Fig. 4.28 Trends and discontinuities, exemplified by six different transformations of a square. Note that the mode of transformation of the figures in (**a**) cannot be determined, but that (**b**) reveals how, for example, an X can turn into a +

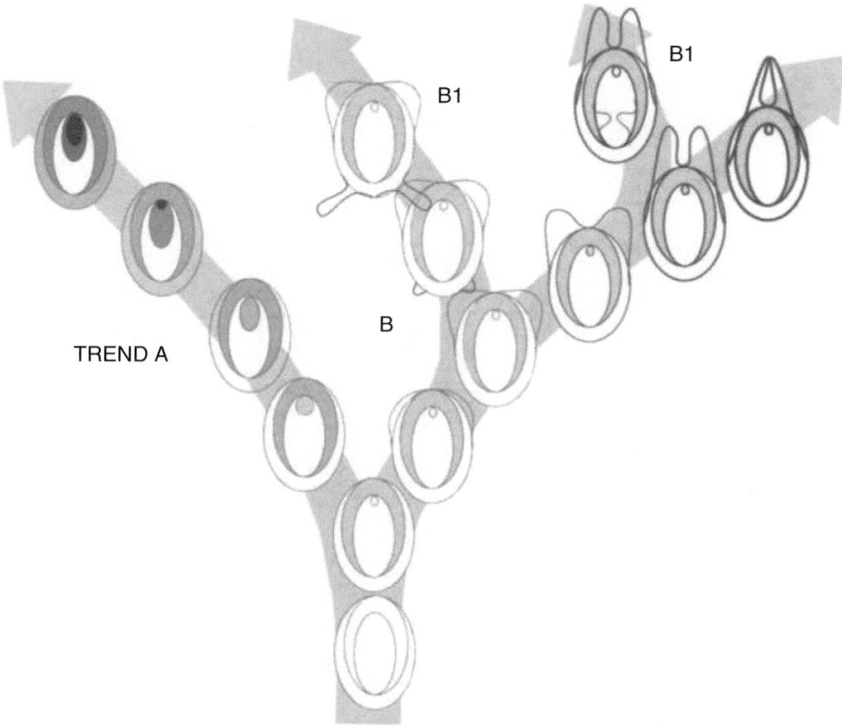

Fig. 4.29 On trends within a trend. The transformation process from figure to figure is clearly discernible in this graphic, including the sites at which the transformation processes branch (A versus B, as well as B versus B1 and B2). Test persons have little difficulty forming such series

- (b2) *Optimizing a delimitation process* must deal with character polymorphism. The issue revolves around where best to set a boundary within a series of forms. This can be difficult considering that a character is composed of other characters, all of which can shift differently and begin or end at different positions in such a series. Simply put: where does the most distinct discontinuity lie?

  The principles of probability dictate focusing on reciprocal confirmation—on a correspondence or co-occurrence of discontinuities—rather than on the most conspicuous sub-characters of a character. Again, gestalt perception drives the process, but can easily be led astray by hidden topological inconsistencies (Fig. 4.30) (see also Sect. 4.4 for more details). Like in the type concept, an isolated character cannot serve as a measure of the whole. Rather, the overall constellation is required to interpret individual characters.

  Two examples, one biological, one cultural, can serve as an illustration. What sub-characters of the leaf margin of oak leaves (Fig. 4.31) and what sub-characters of the protuberances on the skull of horned dinosaurs (Fig. 4.32, bottom)

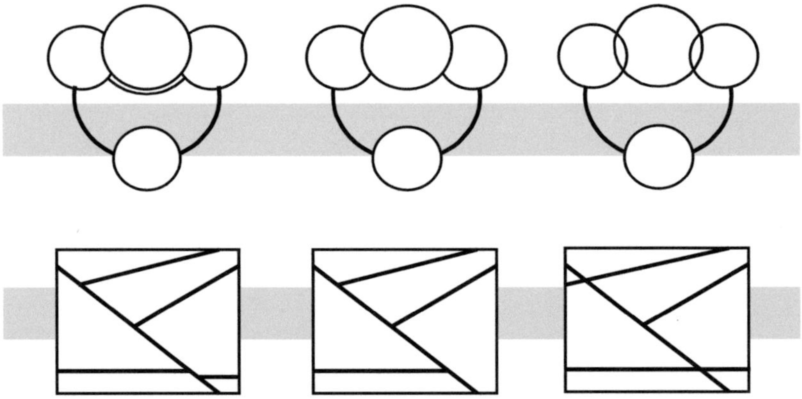

Fig. 4.30 Topology versus gestalt perception. The former involves the equivalency in the number of nodes and connections (the constructability of the transformation of figures), the latter involves the perception of forms. This also helps explain the emergence of two types of sciences. It is difficult to recognize that the three top figures are topologically equivalent to the three bottom ones, but the three clearly differ from one another

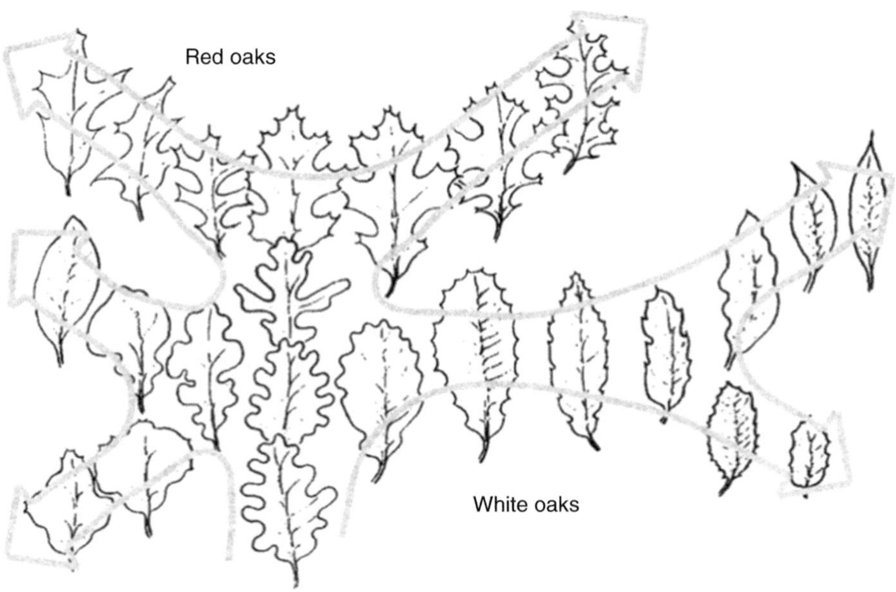

Fig. 4.31 Trends in the variation of similarities, exemplified by leaves of North American oak species (after Brockman 1968; Riedl 1987a)

best help to delineate trends? Shouldn't they all be compared to one another, i.e. in the leaf the overall tooth number, the teeth themselves, tooth clusters and lobes; in the dinosaur the number, width, length, curvature and location of the horns on the skull? Once again, structure and position must be addressed in a back-and-forth manner.

- Examining human artefacts calls for the same approach. This is equally valid for purely physical objects such as the development of cutting and thrust weaponry over the course of five centuries (Fig. 4.32) as it is for the development of Cartesian logic up to the present (Fig. 4.33). The positional criterion here lies in the framework of the respective time and culture.
- All four cases are already depicted in a manner intimating the interrelationships. Practical scientific investigations, however, involve a different strategy. Sixteen

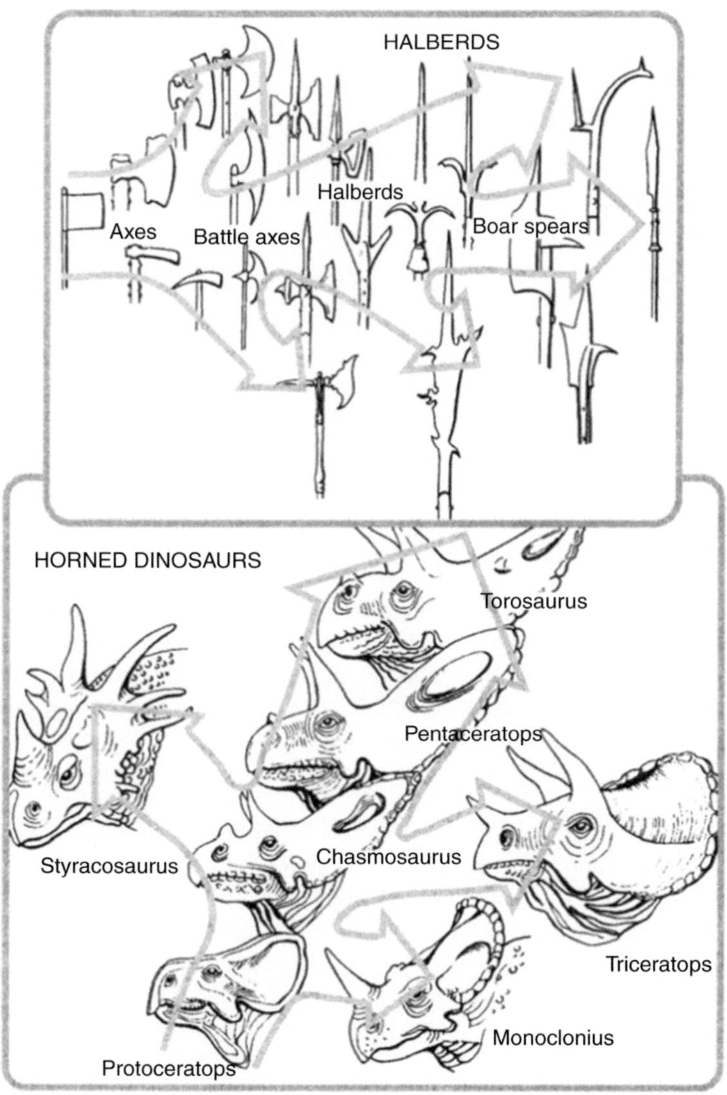

Fig. 4.32 Trends from the history of form transitions, exemplified by halberds (battle axes in the fifth to 'boar spears' in the twelfth century) and the skulls of horned dinosaurs (after Viollet-Le-Duc 1875; Thenius 1972; from Riedl 1987a)

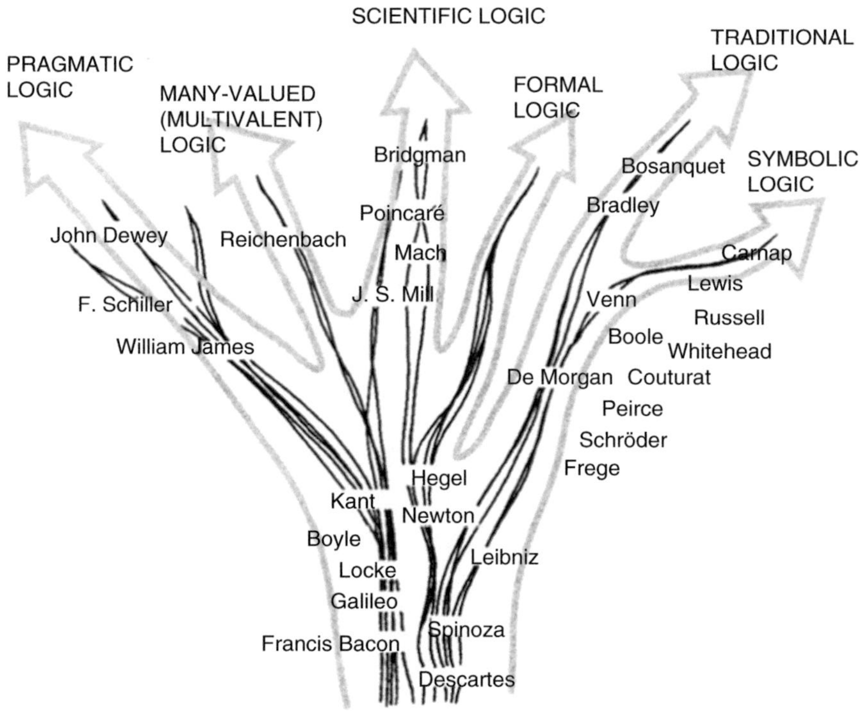

Fig. 4.33 Trends from the history of form transitions in logic. Nearly chronologically labeled with the names of selected main proponents (from Searles 1968, amended)

polymorphic systems—comprising four varying sub-characters—can serve as a simple case study. It turns out that test persons—given the recommendation to proceed by copying, enlarging, cutting and pasting—have some difficulty arranging the items into the requested simple series (diagrammatic representation in Fig. 4.34, top). After ironing out the inconsistencies and establishing a series (Fig. 4.34, bottom), the participants are asked to identify the most distinct trend boundaries.

The result (Fig. 4.34, center) shows a considerable range of opinions. Ultimately, a group effort involving discussion and argumentation yielded the only actual mutual trend shift that encompassed all four sub-characters (namely between objects 10 and 11; details in Riedl 1987a).

This experience represents, in highly simplified form, the process of optimizing a character definition in science. Depending on the type of gestalt perception and 'background assumptions', the process initially intuitively attributes priority to one or the other change in a sub-character. It then takes a convergence of these interpretations to broaden the argumentation and gradually highlight the intuitive nature of the process. In science it often takes decades of research and expert discussions to optimize an interpretation, to find the mutual trend shift as it were.

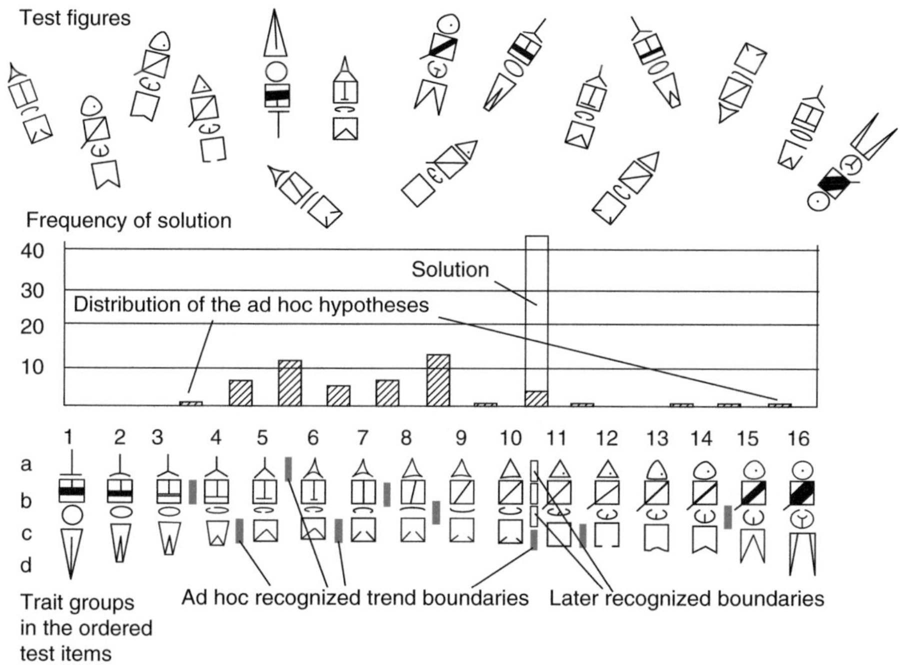

Fig. 4.34 Optimization of a trend boundary: 84 test persons were first tasked with cutting out the above test figures and arranging them in a row. The lower row represents the correct result (after forming a consensus). The second task was to specify the trend boundaries therein both ad hoc as well as after forming a consensus (after the separate negotiation of the trait series a to d). The frequencies are indicated in gray (and in one white) columns (after Riedl 1987a)

Recall in this context (Chap. 3, Sect. 3.3.4 (b)) the restrictions or burdens placed by our definitional form of speech. In polymorphic shifts, this initially forces us to sharply delimit and define terms at our own discretion. This is a common source of semantic misunderstanding.

- (b3) *Weighting a character*, i.e. determining its significance when ordering polymorphic change terminologically, ultimately arises from the overall context.

No character can be assigned a weight from the onset. Nonetheless, our ratiomorphic apparatus—in seeking potential orientation—immediately assumes weightings based on inherent gestalt perception and by automatically recurring on pre-supposed knowledge of some sort. Arriving at an optimum interpretation—achieving the state of the art—then relies on growing experience and on minimizing disharmonies.

This calls for recognizing that every solution that delimits a character also presupposes weighting all other observables. If, as in Fig. 4.34, the boundary with the least contradictions proves to lie between representatives 10 and 11, then that valuation ultimately involved all observations (Fig. 4.35).

The above example involved four characters to help draw the definitive border. They support each other based on the co-occurrence or congruence of their

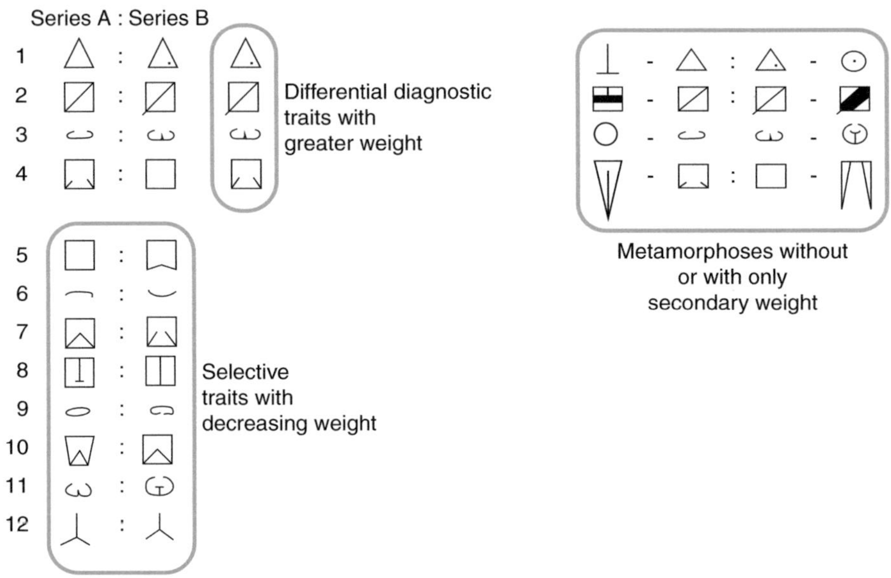

Fig. 4.35 Weighting the traits. When weighting traits with coinciding discontinuities (as in Fig. 4.34, between positions 10 and 11), namely the discontinuities 1–4, higher because they confirm one another, then the weight of the remaining traits (5–12) drops. The former gain differential diagnostic significance, the others are reduced to selective traits (systematics of the trait categories in Figs. 4.39, 4.40, and 4.41; after Riedl 1987a)

boundaries. This gives them the rank of 'differential diagnoses' because each is present in all representatives of one group but absent in all others. All the other characters are then automatically weighted accordingly, receiving successively decreasing rankings. This reflects incrementally lower, so-called 'selective' values, a term borrowed from systematics and treated in more detail in the next Sect. 4.4.

Two summary conclusions: First, no character stands on its own. In a tiered constellation encompassing up to 20 hierarchic levels, each character is embedded in extensive matrices. These range from the broadest framework homologs to the minimum homologs and the full series of homonoms, from the tissues down to the realm of cell organelles and biomolecules. The weightings these positional characters in the lower levels shift based on structural characters such as the homolog 'spinal column'. Conversely, the weightings of the higher-level structures shift due to the positional characters, for example the homonom 'cytochrome-c molecule' (Fig. 4.20). No character can be extracted from the multitude of the others.

Second, we must accept a decisive simplification in such back-and-forth optimizations. This is anchored conceptually, namely in our desire to render complex system contexts understandable.

The entire previous section on 'principles of morphology' was approached as if the systematic entities were already optimized. Moreover, the principles of structural hierarchies were investigated as if the class hierarchies had already been established. These now need to be examined in more detail under the heading 'principles of systematics'.

The above argumentation is derived from experience with organisms. It also proves to be valid for a 'general theory of gestalt' (which, as outlined in Sect. 4.3, Goethe originally had in mind). Similar laws also demonstrably govern how we terminologically delimit human artefacts, be it Old High German from Middle High German, or Romanticism from Gothic.

4.4 The Principles of Systematics

From a cultural history perspective, systematics goes back further than morphology. Tellingly, our grasp of abstract class hierarchies precedes that of physically interrelated structural hierarchies. This again underlines that the 'art of subdividing' requires strong powers of abstraction. Nonetheless, it was a later process both with regard to its goals and its technical prerequisites. Combining things that are not physically coupled is a deeply anchored human faculty.

The languages of all native peoples studied in this respect have developed terms categorizing the organismic realm—no doubt a matter of survival in differentiating between the edible and the inedible, for example. Berlin et al. (1966) reported that the systematics applied by some native peoples rival that of our own. This is clearly based on intuitive competence.

Organisms can, of course, be categorized according to very different perspectives. One purportedly Chinese encyclopedia "grouped the animals as follows: (a) those that belong to the emperor, (b) embalmed animals, (c) lame animals, (d) suckling pigs, (e) sirenians, (f) mythical creatures, (g) ownerless dogs, (h) those belonging to this group that (i) behave erratically, (k) that are drawn with a very fine brush made of camel hairs, (l) and so on, (m) that have broken the water jug, (n) that, from afar, resemble flies" (after Borges, cited from Foucault 1971).

Like many things associated with biology in Western culture, scientific systematics is taken to have begun with Aristotle. Botanical studies, for example, were further addressed by his student Theophrastus (372-287 B.C.). Aristotle differentiated bloodless from blooded animals. While this was incorrect, it did do an admirable job in separating the invertebrates from the vertebrates (although the dolphins in the latter group were still placed among the fishes, etc). Nonetheless, this effort is based on a principle that probably also underlies the achievement of native peoples: contradiction-free order as a pursuable goal. The principle is attributed to nature itself. Accordingly, we ourselves can expect to be programed for this expectation.

Rome made little progress in this matter, and the Middle Ages were marked mainly by a penchant for collecting rarities and books on medicinal herbs. The eighteenth century ushered in a rapid development of comprehensive scientific systems, foremost championed by Buffon (1707–1788) (publications from 1749–1809). More famously, Linné (1707–1778) (main publication 1737, see 1758) then created the binary nomenclature that simultaneously designated both the genus and the species—the approach still in use today.

Adanson (1782) appears to be the first to have reflected upon the conceptual process behind systematic classification. He recommended listing the characters of an organismic group and then step-by-step deleting those that repeat themselves, an approach that would lead from the families to the genera (compare Foucault 1971). Such considerations marked the further history of the process.

Three epistemological issues require consideration: (Sect. 4.4.1) the problem of weighting, (Sect. 4.4.2) optimizing class concepts and (Sect 4.4.3) the nature of the natural system.

4.4.1 The Weighting Problem

It is somewhat ironic that the approach taken by systematists first became a 'problem' at a time (toward the mid-twentieth century) when the product, namely the 'natural system', was universally accepted as being correct. Basically, the problem arose only after attempts were made to explain the process in simple terms.

This brings us to the core issue of this book chapter, namely how the intuitively so successful achievement of the systematists, along with the clarifying controversies, can be understood methodologically.

This calls for examining (a) the attempted simplifications and their prerequisites, followed by (b) seeking clarity about the categories of characters.

(a) *Attempted simplifications*: Two schools of thought arose about possible simplifications in the approaches taken by systematists. These also influenced the taxonomic treatment of molecules. All originate in the desire to navigate around the homology problem, a topic seen to be annoying and/or vexing. The three epistemological issues raised above spawned important schools of thought in the 1950s and 1960s. All three once again intermingled perceptive and explanatory components, requiring discussion both here and in Chap. 6.

I begin with the somewhat older concept of (a1) 'cladistics', followed by (a2) 'phenetics' or 'numerical taxonomy' and, finally, (a3) the discipline of 'molecular systematics'.

- (a1) Cladistics go back to Hennig (1950, compare Ax 1984) and hold that kinship affinities are best represented by reconstructing the branches of the phylogenetic tree. This corresponds to the explanatory principle. That in turn influences the preferences for, or weightings of, what is being perceived.

  If this is the case, then the main issue is the succession of original and of derived characters, termed 'plesio-' and 'apomorphies', respectively. Their respective presence in several systematic units is termed 'symplesio-' and 'synapomorphy'. This terminology itself already clearly involves an explanatory facet.

Examining the process of perception based on the recommended method reveals underlying prerequisites. The hereditary context, the reading direction of the change, the weighting of the characters, and the homologies used must be intuitionally anchored.

Methodologically, the expectation is that commonly held versus special (unique) characters yield a 'tree' with a recognizable main stem and branches. In many cases this will be correct, neglecting convergences, parallel evolution and 'mosaic evolution' (i.e. the fact that characters can change at very different rates).

We expect the resulting 'cladograms' (Fig. 4.36) to depict the reliability of the presumed homology. And that would simply mean, as expressed by Jürgen Remane (1971), putting the cart before the horse.

- (a2) *Phenetics*, along with 'numerical taxonomy', is a more radical approach. This relates to the weighting problem and is based on the assumption that character weighting (and therefore homologizing) is arbitrary. Accordingly, the recommendation is not to weight at all but rather to measure. This method, propagated by Sokal and Sneath (1963) and anchored in fine systematics and bacterial taxonomy, triggered more opposition.

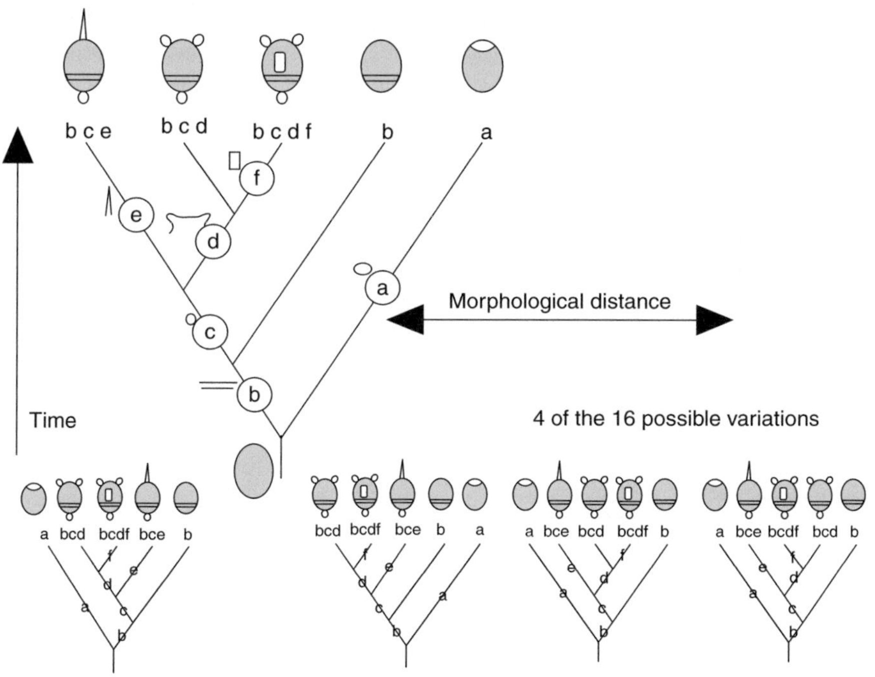

Fig. 4.36 Schematic cladograms. The changes in the traits a to f lead to five different forms (example from Futuyma 1986; illustrations by author). The succession of the changes enables postulating a relative time series of the branchings. This is not the case with 'morphological distance' because, even in this simple case, 16 conflicting alternatives arise (of which only four are illustrated)

That approach clearly reflected the limitations imposed by the paucity of characters in bacteria groups. In complex systems it can be applied successfully only within the closest kinship groups, i.e. in those cases where close kinship has already been recognized. Comparing the segments of the first leg of two ground beetles requires reliable knowledge about all the relevant homologies. Comparing those data with the metrics of the first leg of a spider (Fig. 4.37), an entirely different group, is a senseless undertaking.

It took ten years of debate to determine (Sneath and Sokal 1973) that the process cannot function without accepting the above prerequisites. The term 'operational homology' was introduced and a call made for using common sense. This raised the feared specter of circularity and an inkling that simple definitions were not forthcoming. Overall, from an epistemological (philosophical) perspective, the situation was deemed knotty and embarrassing. The feeling was that the entire process stood on shaky ground (p. 428). At issue was the 'ratiomorphic apparatus' described above, and more closely examining that apparatus helped solidify that ground.

Nonetheless, open questions remain about the metrics themselves. Simply put, 'what is a significant measure?' The method must be adapted to the problem being addressed. This can be illustrated by the question, 'How long is England's coast?' Clearly, the scale of the map and the span of the compass used modify the sense of any measurement.

- (a3) *Molecular systematics* also belongs in this framework of simplifications because relatively simple structures are involved. This issue was already raised in connection with the 'homology of isologies' (Sect. 4.3.1 (e)). The conclusion

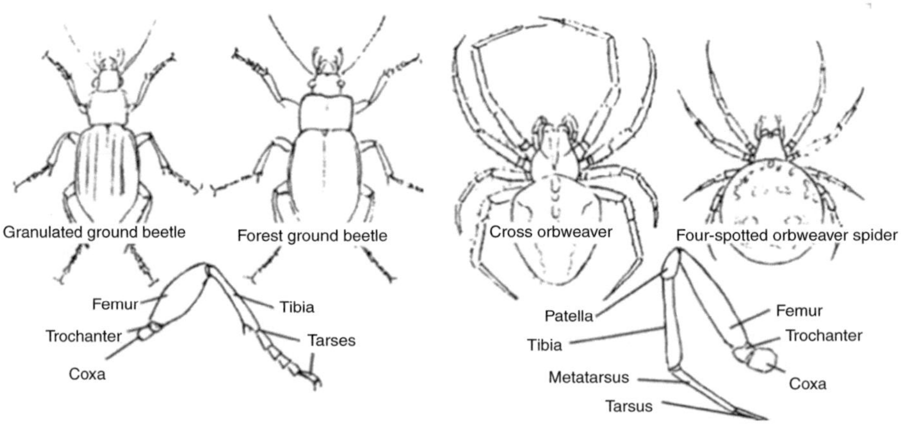

Fig. 4.37 The limits of numerical taxonomy, illustrated based on the homology problem. The type of data that prove to be valuable in metric comparisons of the leg segments of two closely related ground beetles or two closely related orbweaver spiders prove to be worthless when comparing the beetles with the spiders. This is because the identical names of the segments do not reflect any homology; moreover, that terminology stems from human anatomy (figures from Chinery 1976 and Roberts 1985)

there was that positional characters are needed to compensate for insufficient structural characters. This is the only way to sufficiently cleanly homologize a structural element.

The same holds true for using molecules in systematics. A prerequisite for homologizing a base pair in DNA is that, as part of a homologous triplet (a homologous gene), it lead to a homologous large molecule that plays a homologous role in a particular organismic group. These clearly represent homologous or homodynamous structural characters.

Two methodological simplifications are available if one wishes to avoid or cannot comply with this requirement, which cannot always be fulfilled. The first is to remain within in a narrow and reliably determined kinship group when conducting comparisons. This actually describes the typical procedure: the probability of dealing with homologous structures is based on an intuitional cornerstone forwarded by traditional systematics.

The second simplification is to employ one of the many calculations that have been generated to quantify kinship, i.e. that weight and read the direction of changes, even when characters are scanty. All these approaches revolve around trusting maximum parsimony, i.e. the most 'frugal' or simple interpretation. Some are strikingly accurate—for example the minimal probability that characters, once lost, are regained in their original structure (Dollo parsimony).

The discussions on this topic have also led to the insight (and yielded promising methods) that solutions can be optimized through an iterative process of initial weighting and subsequent fine-tuning through recursive experience (Williams and Fitch 1989) (overview in Hillis and Moritz 1990). The trap of hypothetical simplicity is to be avoided. This recognized approach guides every further effort. Our trust in human cognition is strengthened by the wealth of confirmations. Our trust in explanations is strengthened by their simplicity.

(b) *The categories of characters* were recognized early on by systematists. They are eloquently written into the 'definitions' of all system groups. More closely examining such definitions reveals that they already involve weighting. The remaining task is simply to examine the process behind this weighting.

This calls for recognizing that (b1) no clear, reliable guidelines exist. Rather, (b2) the categories of the characters promote (b3) self-organization.

- (b1) *The paucity of clear and reliable guidelines*: No single detail provides real insight into or predictive power about the structure of the world. Our cognitive apparatus (as outlined in Sect. 4.2.2) is, however, broadly capable of operating based on a high redundancy of not entirely identically repeating things. It can also deal with the consistency behind orderly successive events and simultaneities, i.e. the *post hoc* and *simul hoc*.

  ll further expectations about the objects we encounter must be derived from experience, from associative learning. This pertains to the arrangeability of the objects (which improves potential prognoses) and, in the case of polymorphic objects, to the constancy of their character sets. The species concept itself must have been a decisive 'master instructor' (even if breeds or, conversely, species groups are sometimes mistakenly held for species). The gradated diversity of organisms, and every genealogical constellation, has promoted our expectation of orderly structure.

Every classification process is necessarily naive. This only becomes apparent with unusual new discoveries. The discovery of buildings erected by lost cultures or the letters of a lost language are cases in point. Sect. 4.2.1 (b3) illustrated this based on the newly discovered gnathostomulid animal group. That example highlighted the wealth of background assumptions that had to be called up. The same background assumptions no doubt underpinned naive systematics: ever-growing experience and its transmission within a culture, rather than reliable guidelines, drove the process.

An illustrative example: early sailors were greeted with great skepticism when they first offered the skins of platypuses for sale in Europe: prospective buyers had all too often been duped into paying high prices for deceptively patched-together 'rarities'.

We can no longer conclusively reconstruct the development of the above processes because much of the underlying experience is anchored in the pre-linguistic era. This calls for focusing on the present-day situation.

- (b2) In every system, classifying categories of characters initially involves perceiving their hierarchy. The definitions are based on the description of the next higher group, i.e. they presuppose the definition of that higher tier, and each subgroup delimits itself from its equally ranked neighbor groups. The characters used in the definitions are specified with phrases such as 'always with', 'as a rule with', 'generally/commonly with', but also 'generally without', etc. This implied ranking merits further investigation.

  Mammalia, for example, are Amniota (a group that also includes reptiles and birds) and always feature milk glands and a retained fourth left aortic arch. They are as a rule viviparous, have hair and generally exhibit four extremities that are developed into legs.

  Exceptions include the egg-laying platypuses and some whales that lack detectable hairs. Bats, whales and sirens have wings or fins instead of legs, and, once again, whales and sirens have reduced hind-limbs (Fig. 4.38). Note, however, that many reptiles are also quadrupedal.

  For mammals the character 'milk glands' would suffice as a minimum diagnosis. Nonetheless, these are not immediately visible in a bear, for example, and cannot be reliably detected in any male mammal. Moreover, the combination 'with hair' and/or 'viviparous' would also be sufficient because the platypus is at least haired and whales are viviparous. Of course, waiting for a whale to give birth is not an option in the determination process. The diagnoses need to be somewhat richer. This justifies incorporating every feature which, in combination, is more or less characteristic for mammals.

  This yields a tiered range of differentiating characters. The distinction between (M) mammals and (R) reptiles serves as a simple example (Fig. 4.39).

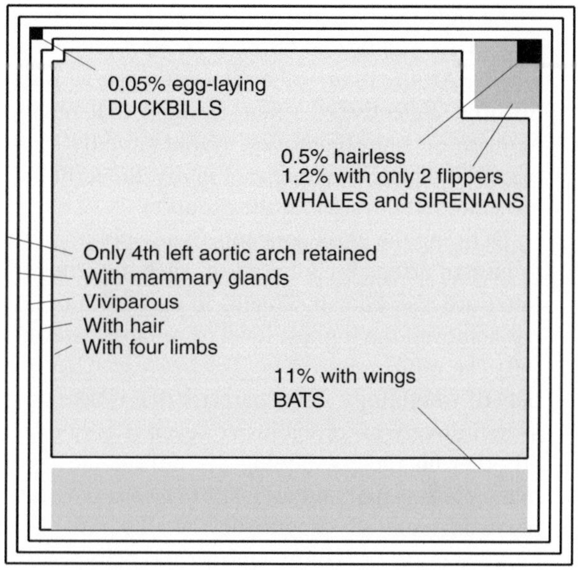

Fig. 4.38 Degrees of coincidence of key mammal traits. The sizes of the deviations are quantitatively approximated by the respective surface areas; based on four species of duckbills, 84 species of whales and sirenians, and about 800 bat species

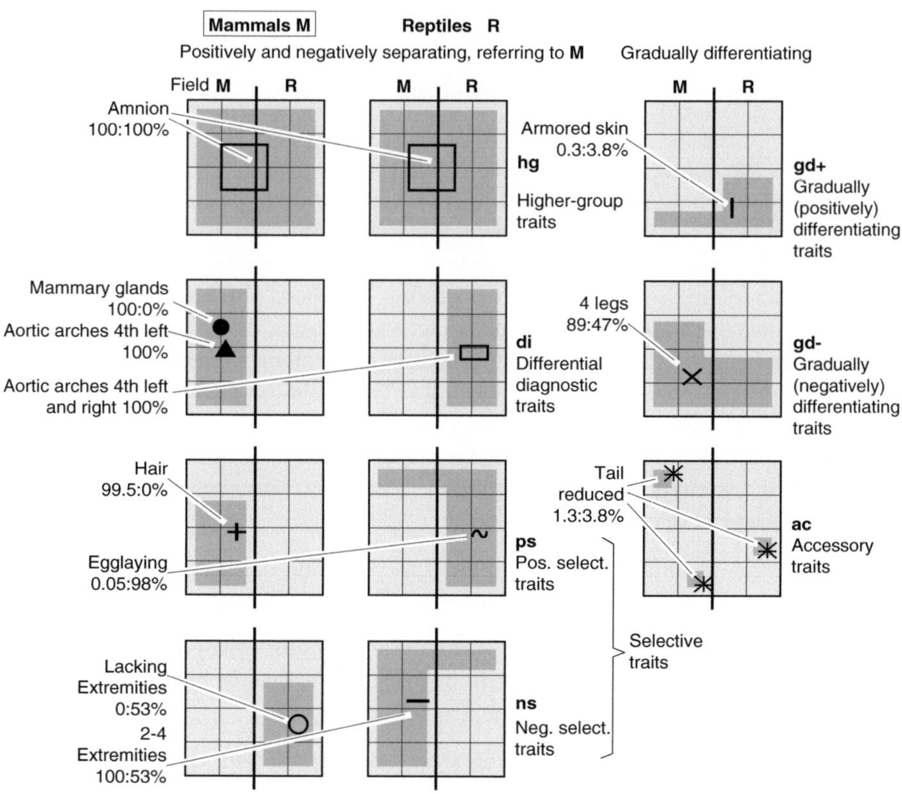

Fig. 4.39 Categories of traits in systematics, illustrated based on the distinction between mammals and reptiles; values in percent, related to 3700 species of mammals and 6300 species of reptiles (compare Fig. 4.38; symbols repeated in following figure)

As noted above, the 'differential diagnostic' characters receive the highest rank. Milk glands and aortic arch Nr 4 left are present in all (M) but missing in all (R). All the many other characters serve selectively or, more weakly, gradationally. 'Selectively' refers to those that function diagnostically for at least a number of representatives. 'Gradationally', in contrast, refers to those that merely indicate frequencies. Finally, the term 'accessory' defines characters that are scattered throughout the group(s).

Defining the class concepts in any polymorphic object, from geomorphology to human artefacts, is based on such specifics of characters. Accordingly, their ranks are also known. Thanks to its wealth of characters, biological systematics has achieved the highest level of reliable determinations.

- (b3) *The self-organization of characters*: Characters are subject to a mutual interplay of weightings. Any character that is taken to be differentially diagnostic, i.e. as definitively in- or exclusive (even if only provisionally), sets the rankings for all other characters. Giving the milk gland of mammals diagnostic rank automatically sets hair and vivipary in a selective rank. If one took hair or vivipary as being differentially diagnostic, then the milk gland would drop to a selective rank because either certain whales or the platypuses would be shifted into neighbor groups.

4.4.2 Optimizing the Class Concepts

Importantly, the degree to which selective characters differentiate organisms, and the degree to which the exceptions compensate one another, play a role. After all, in the mammal example, 98.6% of the representatives with milk glands have hair and 99.8% are viviparous (see species numbers in the legend of Fig. 4.28). Depicting these characters in a field diagram according to their powers of delimitation (Fig. 4.40) puts the higher-level group characters at the top, the differential diagnostic characters at the outer corners, the selective ones along the outer margins, the gradated and accessory ones inside the field. The delimiting power of these four characters in defining the mammals is 99.6%.

Two steps are required to evaluate such a result. The first is to investigate (a) how to interpret the strength of a delimitation, the second (b) how the interplay between field- and character optimization develops.

(a) *The degrees of delimitation strength*: Understanding the range of delimitation strengths and shifts in character ranks calls for considering two aspects.

The first is to examine (a1) how delimitation strengths change when character weightings change, the second to determine (a2) the role that the hierarchic relationships in the fields play for the evaluation.

- (a1) *The relative weighting of characters*: Let's remain with the above example and take the old term 'quadruped' literally. Combining all those Amniota that have developed four walking legs, and juxtaposing them to the 'legless',

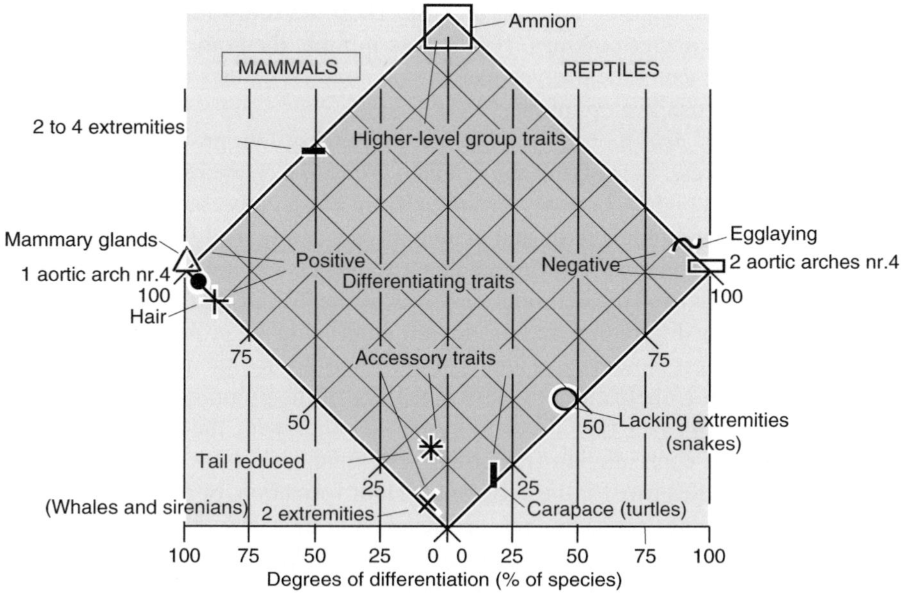

Fig. 4.40 Order of traits according to discriminatory power, exemplified by the separation of mammals and reptiles (after Fig. 4.39). Note the grouping of the differential diagnostic traits along the extreme margins of the diagram. Accessory traits, in contrast, can prove to be subgroup traits (see also Fig. 4.41; symbols as in Fig. 4.39)

fin-limbed' and 'wing-limbed' forms, would yield an entirely different picture. Although this radical view has never been fully implemented, it is an instructive exercise.

Taking quadrupedalism and 'non-quadrupedalism' (of the 'aqua-drupedalists') as the differential diagnostic characters causes all the other characters mentioned so far to lose their rank. They drop from being differentially diagnostic or highly selective to merely enabling a weakly gradated delimitation.

Exhibiting milk glands has a delimitational power of only 39.5%, the aortic arch of 44.4% etc., and the combined median value would be a mere 59%. This is understandable considering that the 'quadrupeds' would now encompass a mixture of four-limbed mammals and reptiles, with the 'non-quadrupeds' containing whales, sirenians, bats and snakes.

The argument is clear. Such a classification, even if unequivocal based on the limbs, contains so many contradictions that no systematist has ever attempted that approach. This means we intuitionally grasp the pitfalls of contradictions and are inherently warned that they should be avoided. It follows that an optimization process is involved both at the level of individual researchers and in the overall discussion in systematic circles. The result in each case is the obtainable optimum.

Characters or traits therefore enjoy advance weighting in the systematics of every scientific discipline. As research advances, however, contradictions crop up, triggering an ongoing revision process. These revisions retain the interplay between the character rankings. If one rises in rank, then others drop. In other words, the characters themselves weight and shift each other reciprocally until attaining the respective optimum.

- (a2) *The role of field hierarchy*: The 'hierarchy of system groups' also plays a role in this process. In studying an organism group, the characters that ultimately prove to have a higher- or lower-group rank are often unforeseeable. For example, the fact that reptiles, birds and mammals can clearly be differentiated based on the position and number of retained aortic arches was not initially apparent. This insight improved the delimitation power of the overall character complex by providing a new differential diagnostic trait, improving the probability of having drawn the correct boundary. Finally, these three groups themselves form the Sauropsida, which helps improve the reliability of the grouping. This is further supported by the fact that the sauropsids, together with the amphibians and fishes, form the contradiction-free group Vertebrata.

  The same holds true for the subgroups. Hoof formation, or retractable claws, are good mammal characters. Yet, extracted from their definitional framework, they would merely be weakly selective. It turns out, however, that the even- and odd-toed animals (ungulates), as well as cat-like predators, are well characterized by these respective traits, elevated them to highly selective and even differential diagnostic rank (schematically represented in Fig. 4.41).

  Such shifts of characters into higher- and lower-level groups also significantly help to optimize the framework. Similar constellations are a mainstay in ethology and comparative linguistics. Reciprocal, back-and-forth weighting is always involved. The process is first completed when the characters in all adjoining higher- and lower-level groups have attained their maximal ranks (Fig. 4.41). Recall that our trust in such expectations and theories grows as contradictions and exceptions are minimized. Again, optimization, not arbitrary processes, is involved. The result is once again a hierarchic system of reciprocal confirmation.

(b) *Reciprocal optimization between field and character*: The final step is to examine the interplay in optimally delimiting the boundaries of fields and characters. In order to more easily convey the processes, we assumed that the field boundaries were already optimized before attempting to optimize the character boundaries. Now we reverse the situation. De facto, both processes rely on each other and, in practice, they run in parallel.

The following example highlights (b1) how a field boundary dominates a character boundary and (b2) the effect of a character boundary on a field boundary.

- (b1) The term 'hair' illustrates the first of the above processes. Beyond mammals, other organisms such as spiders, caterpillars and even plants can be 'haired' or densely furred. The logical field delimitation 'mammals' prompted a critical examination of such hairs. The result: mammals exhibit an entirely unmistakable type of hair.

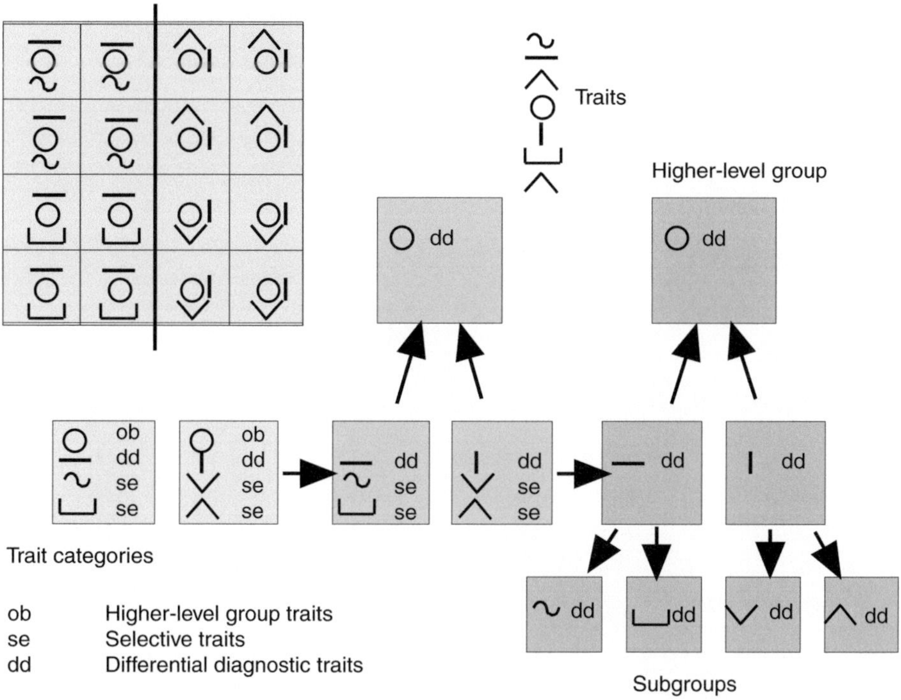

Fig. 4.41 The development of trait hierarchy, exemplified by 16 representantives (species) with a combined 7 defined traits. The degrees of differentiation of the traits (designations after Fig. 4.39) increase recognizably with the hierarchic category (a metrically conceivable solution of the task in Riedl 1987a)

Hair therefore represents a highly selective character. A single such hair suffices to definitively identify a mammal. This 'type' underlined that mammal hair arose independently of feathers and also demonstrably not from scales, but in groups of three from below an individual scale. The type also reveals that spines are derived from hairs, that the current-sensitive thornlets on the upper lip of dolphins stem from whiskers, and that the material making up the horns of rhinos is composed of fused hairs.

The discontinuities that separate the highly polymorphic organ 'hair' from all other hair-like structures are indisputably defined by its bulbus and the rich details of its layering (Fig. 4.42).

- (b2) The aortic arch example illustrates the reversed path of insight. Initially, gill arches were known to be present embryonically in all vertebrates, but retained in entirety only in hagfish and lampreys (and for the most part in fishes). They were then recognized as such and as being present in reduced number in lung-breathing vertebrates as well (Fig. 4.43).

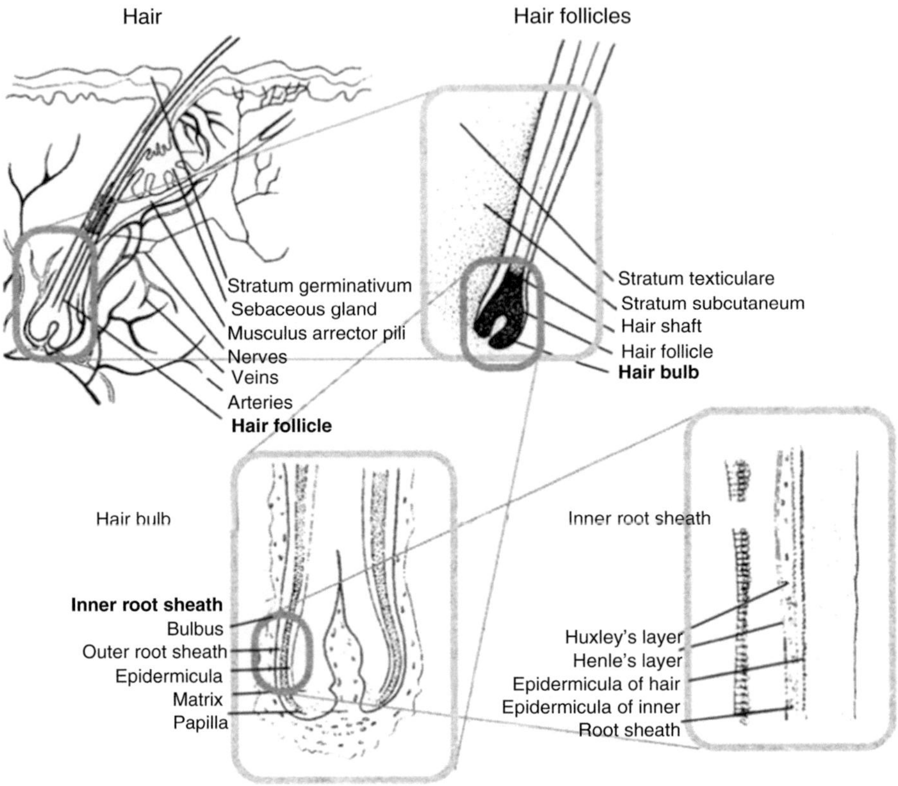

Fig. 4.42 The interplay between field and character in practice. The high systematic weight of hair arises from its coincidence with the 'field of mammals' on the one hand, and with its unmistakeably high differentiation on the other (four levels of differentiation times 4–7 features each are indicated; after Riedl 1975)

This assessment lent further support to the unity of the term 'vertebrates', but also supported their subdivision by yielding differential diagnostic characters for the reptiles, birds and mammals. It also revealed how distantly separated the tailed and tailless amphibians—the urodeles and anurans—are. This interpretation was confirmed with the discovery that the two groups had split early on the evolutionary time scale.

Once again, these characters support discontinuities—in this case between system groups (i.e. class concepts). The distinction is rock solid: recognizing the developed aorta system enables categorizing all vertebrate species into the respective classes.

The same underlying reciprocity is expected to define classes and structures in the systematics of polymorphic systems in every science. In many cases, the impetus to differentiate so deeply may simply have been lacking.

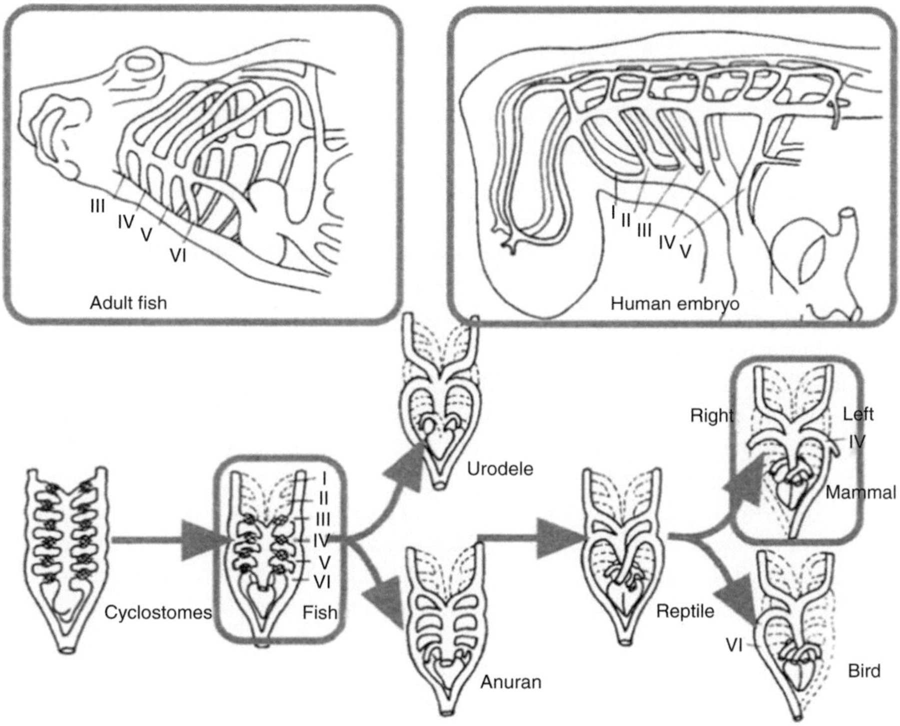

Fig. 4.43 Interplay between character and field in practice, exemplified by the coincidence in the number and arrangement of the aortic arches in vertebrates from fishes to humans (top) with the vertebrate classes (bottom) (after Riedl 1975)

Again, reciprocally optimizing field and character concepts is not arbitrary. In fact, our pre-conscious handling of gestalt perception leads to an astounding degree of differentiation. This is reflected in the rich terminology we need to help grasp the process, even if only schematically.

4.4.3 The Nature of the Natural System

Based on the processes behind the perception of *simul hoc*—orderly simultaneity—2×10^6 species in the organismic realm have now been successfully arranged in 5×10^5 system groups. Assuming an average of only ten differential diagnostic and highly selective characters yields an ordered framework encompassing 25 million homologies.

Even back in the late eighteenth century, five to ten million such insights were probably available. This leads to the final two issues that need to be addressed here.

Using today's terminology, these are (a) the 'reading problem' and (b) the problem of the 'nature of the natural system'.

(a) *The reading problem*: The more dense and complete such a hierarchic, harmoniously diverging system of similarities is, the more likely it yields clear series of similarities. Each such series represents incremental modifications of distinct interdependencies. The question is in what direction such shifts should be read or interpreted.

Once again, this is related tour desire to explain the historical framework, i.e. in a theory of evolution. The reading problem itself was prompted by questions regarding common descent. The theory of evolution itself, however, is not the cause of the reading problem. Rather, the reading direction prompted the call for an evolutionary perspective.

Determining the reading direction is an issue because it proves to be difficult for an isolated series. If a shift in shapes (as in Fig. 4.44, top) is drawn on the blackboard in a classroom experiment, the students tend to remain undecided. Pressured to reach a solution, a weak majority emerges for reading the series from the left, as it was drawn.

Of course, this example was chosen to demonstrate that the reading direction cannot be determined: it is equally possible that the square represents an inflated star and the star a folded square. This interpretation changes only with insight into additional interdependencies, whether they involve bifurcations or a trend-within-a-trend phenomenon. Such an example is illustrated in Fig. 4.29.

Even in this case, however, the reading direction is merely suggested by the symbol type rather than being definitive: the contribution of potential differentiations or reductions cannot be determined. A more solid answer requires some indication of probable and improbable interpretations of the reading direction of individual elements (Fig. 4.44). In fact, the overall constellation—'from unicellular organisms to squids, beetles and mammals'—is what ultimately prompts an interpretation. And such an interpretation, as logical as it may seem, must be founded in theory, namely in that of kinship and anagenesis. This is equally valid for all systems involving genealogy or transmission, up to and including the history of languages or of entire cultures.

Accepting the theory of evolution (which has not yet been necessary up to this point) yields additional evidence supporting a particular reading direction (see Chap. 6 on the structure of things that have been explained).

(b) *The natural system*: One interpretation holds that 'the nature of the natural system' is an artificial product because systems are artificial products of our thought processes. This would make 'natural system' a contradiction in terms. In fact, however, the system represents an astoundingly detailed constellation of structural and class hierarchies that lie far beyond any explanation based on chance or on the combined fantasies of systematists.

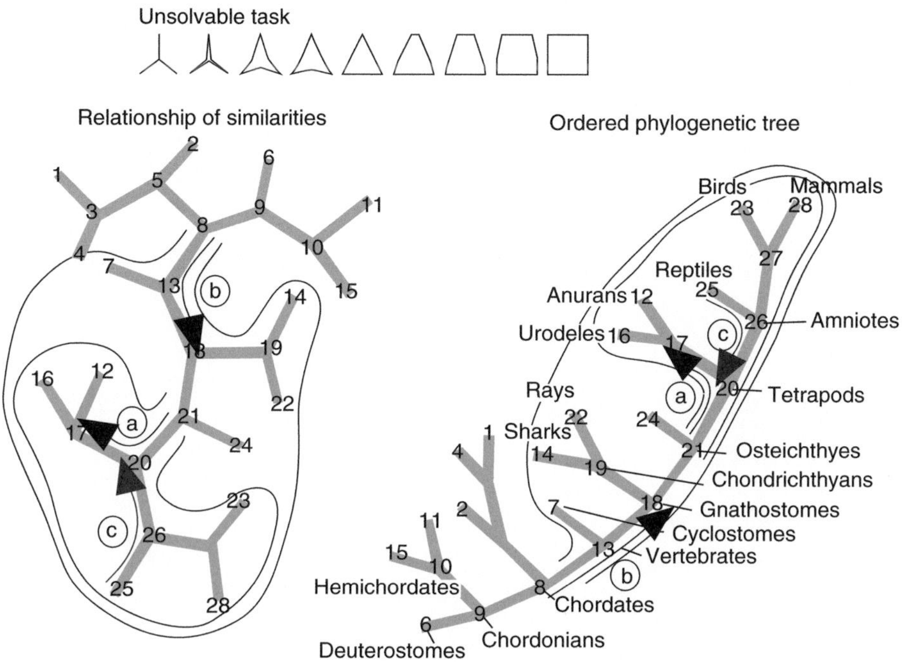

Fig. 4.44. The determination of the reading direction, exemplified by the constellation of similarities in the vertebrates. The single series (top) fails to reveal a reading direction: it can originate from any figure. The same holds true for the vertebrates as a group (left) if solely the spacing of the similarities is considered. If probable (a, b) and improbable (c) individual directions can be assumed, the correct interpretations will confirm themselves in relation to a phylogenetic tree (right) (after Riedl 1975)

This constellation demands an explanation. The clarity with which Lamarck saw the issue is highly instructive. Accordingly, the theory of evolution itself did not trigger our perception of this system. In fact, exactly the opposite is true: the explanation was prompted by the perception of inner order, of the system's internal interdependencies. This theory attempts to explain a broad pattern of order lying far beyond any probability of randomness. Paradoxically, the theory turns out to be founded on two random mechanisms.

The preceding chapters highlight that we have successfully grasped the interrelationships of this world—whether they be shifts in Earth history, in organisms or in cultures—based on our faculty to perceive *post hoc* and *simul hoc* phenomena. The next step (which involves an additional faculty of our innate make-up) is to add the *propter hoc*, the explanation, as a conceptual construct of the world around us.

Chapter 5
The Systems of Explanation and Understanding

As opposed to the 'systems of cognition' discussed above, the issue here is rooted in the more transparent world of the humanities. The humanities, however, are also fraught with theoretical conflicts. This is because cognition is largely controlled ratiomorphically, i.e. it functions almost automatically. This inconspicuous process meant it long remained intransparent, was poorly studied and was simply not considered worthy of serious scientific endeavor. The process of explanation and understanding, however, is much more strongly linked to conscious reflection and has therefore attracted considerable professional scrutiny.

Thus, perceiving and processing the simul hoc is already controlled via gestalt perception. The propter hoc, in contrast, is an addition involving an act of reflection, as already recognized by David Hume (1739/40).

This has yielded a different and considerably richer scientific literature. The data on the underlying cognitive process stem almost entirely from the fields of morphology and perception psychology, consolidated by insights gained through evolutionary epistemology. In contrast, explanation and understanding feature a wealth of standpoints that find their expression in many genres of literature and that are anchored in traditional currents of philosophical systems, epistemology, the theory of science, and in logic.

As might be expected, the respective terminologies differ as well. This calls for critically examining terms that, interestingly enough, have not yet featured prominently in previous chapters. These include 'explaining', 'understanding', 'reasons', 'conditions', 'validations', 'causality', 'finality', 'teleology', 'teleonomy' and 'entelechy'.

Importantly, everything that needs to be explained, understood and substantiated or validated basically stems from the very fields of similarity already treated (and methodologically dealt with based on probabilities) in Chaps. 3 and 4. The further discussion therefore builds on our perception, clarification and justification of those fields. This will confirm our previous insight: changes in an explanation, in an understanding of, or in a validation of a phenomenon have little or no influence on

R. Riedl, *Structures of Complexity*,
https://doi.org/10.1007/978-3-030-13064-0_5

the form and justification of the underlying field of similarity that seeks explanation, understanding or validation.

Introducing the systems of explanation and understanding requires (Sect. 5.1) presenting the conditions behind our faculties and (Sect. 5.2) outlining the difficulties that superimposed reflection has gotten us into over the course of intellectual history. This provides the basis for presenting the rational conditions that we expect from the process of (Sect. 5.3) explaining and (Sect. 5.4) understanding.

5.1 The Conditions and Our Faculties

Perceiving similarities is a matter of survival: correct prognoses are vital and clearly govern the lives of all creatures. Our efforts to explain and understand similarities also reflect that purpose, namely to improve our orientation and prognoses. In today's world that means cramming as many sheer limitless facts as possible into our one-and-a-half liter brain volume.

This pursuit opens up the fascinating prospect of transcending mere perception (a fundamentally passive orientation) to actively making a mark in our world. This ranges from a physicist's ambition to explain matter, up to the politician who seeks to guide society by promoting understanding for its laws. We also need to accept that, beyond any academic considerations, the machinery of science is driven by the pursuit of influence and power. While that perhaps represents a plausible purpose, it introduces an entirely new flavor.

Delving into these issues calls for presenting the underlying faculties and conditions. The first step is to outline the (Sect. 5.1.1) preconditions, then the (Sect. 5.1.2) superimposed level of reflection and (Sect. 5.1.3) the minimally reflective, so-called 'good old common sense' arising from such reflection, concluding with (Sect. 5.1.4) a 'psychology' of the explanatory experience'.

5.1.1 The Preconditions

The hypotheses of the 'apparently true' and of 'the comparable' have already been introduced (Chaps. 2, 3, and 4) as representing the preconditions behind our faculties for postulating reasons and purposes. This is accompanied by certain expectations of time and space. Despite sounding trivial, it is worth repeating that we need to comprehend objects and processes as specified entities, and these must be recognizable when re-encountered in order to then attribute reasons or purposes.

Two additional hypotheses underpin our expectation of causes and purposes. These form the gateway to conscious reflection. In this sense, David Hume's abovementioned thesis is spot on. We add the *propter hoc* to this world. Importantly, no evolutionary step originates without preparation, without constituents. This means that we are pre-programed for these two hypotheses as well.

This can best be illustrated by tool use in animals. This includes organisms other than chimpanzees, which gnaw sticks into shapes that enable them to fish termites out of their tunnels. Certain birds, for example, break off long thorns and use them to spear insects in holes in the bark of trees (overview in Eibl-Eibesfeld 1978). Laboratory experiments have revealed further astounding capabilities in monkeys. These range from assembling tools to multi-step actions—all serving a specific purpose, namely accessing food (e.g. Rensch 1973). The purposeful component in such activities lies in a transitional realm between genetic programs, experience- or imitation-based abilities, and early forms of organized reflection.

Principally, however, all behaviors, like all body functions, are entirely final, namely designed to achieve some purpose. Or, as Lorenz (1978) put it, to enable every manner of activity and every function useful in promoting the survival of the individual or the species. Consciously reflected, this introduces the first hypothesis, that of the purposeful (see below). The second hypothesis, that of the causes, also underpins behavior. In animal behavior it may seem a bit farfetched to seek the 'because' that drives the hypothesis of the causes. Nonetheless, the entire toolbox available to organisms and their behavior can be interpreted, both adaptively and constructively, as 'in order to' or as 'because'. The two merely represent a reversal in perspective.

5.1.2 The Hypotheses of Causes and Purposes

Reflection on causes and purposes is anchored in a predisposition or faculty at the unreflected level. It clearly reached full bloom with the nascence of human consciousness.

Much like the hypotheses of the apparently true and of the comparable (Chap. 3, Sects. 3.2 and 3.3), these two hypotheses correspond with extra-subjective reality. This needs to be demonstrated in order to understand their success and their incorporation into our 'program'. At the same time, we need to examine why they show deficiencies in many of our presumptuous attempts to intervene in the course of nature (Riedl 1980, 1985, 1995a).

The (a) hypothesis of the causes and (b) of the purposeful—as they appear to us upon reflection—are treated here separately. In each case, the blueprint behind the hypothesis must be determined first. The three subsequent steps are to substantiate or validate its success and therefore its incorporation, reveal deficiencies and indicate how these might be overcome.

(a) The hypothesis of the causes leads us to expect that the same things or processes are attributable to the same underlying causes.

- (a1) This expectation is successful and therefore incorporated because it is confirmed in most cases: it proves to be true in everyday life. If we know nothing about a certain situation, then the best assumption seems to be to anticipate that the causal relationship experienced in one causal chain is also behind the novel, comparable situation. The reversal of such an expectation inevitably leads to failure, as would relying on some arbitrary assumption—there are simply too many of them.

In the animal kingdom this behavior is known as the 'win-stay' strategy (Huber 1995), which is pre-supposed in most conditioning experiments. Once a pigeon or rat has recognized the connection between the lever and the food reward, it sticks with that association. This explains the difficulty involved in re-programming a new, contrary connection. The situation is much the same in humans.

Our mistakes often lie in confusing the direction of causal chains. We can confuse a glove expander with a pair of scissors, a water mill with an early paddle steamer, or a generator with a compressor. In the latter pairing, a piston engine drives a generator in one case, and a generator drives the piston pump in the other.

- (a2) The deficiencies of the hypothesis lie less in the expectational stance itself than in the simplicity of the pre-prepared causal concept. We strongly prefer to expect linear, limited, non-interlinked as well as unbranched, direct and non-recursive relationships (see overview in Fig. 3.6). Basically, we expect the simplest possible relationship. This is understandable because organisms, up to and including early hominids (i.e. at an evolutionary stage in which mistakes more often than not meant elimination from the population), were unable to construct more complex causal contexts and had little need to do so. This, of course, has changed in today's complex society and in light of our major interventions into the scheme of things.

  First, we prefer to think in clear-cut chains of causality. We refer to cause and effect, often overlooking that many causes can lead to a particular action and that the initial cause can have many other repercussions as well. In jurisprudence, the so-called 'imputation doctrine' calls for the judge to keep the causal chain leading to a crime as short as possible. This is an attempt to avoid endless regressions: after all, it is conceivable that a shoplifter was ultimately motivated by his deceased grandmother.

  Second, complex networks are unpleasant to unravel mentally. The popular saying that the 'world is a totally interlinked system' proves to be little more than a platitude. Experiments (Dörner 1975) as well as computer games (Vester 1975) show how poorly we anticipate the functions in interlinked relationships.

  The same holds true even for the simpler case of recursive causality. Industry spent the entire last century straightening out causal chains—right up to maximizing the effectiveness of assembly lines. Little thought was given to how the products coming off the assembly lines retroact on the people, materials and energy that enter into the equation at other end of such lines. The approach was much like that in a vending machine: no thought need be given about how the dispensed chocolate bar impacts the coin that was inserted. Within the machine, a feedback can be excluded. Outside the box, however, every product inevitably retroacts on its own value. And this process is so complex that we today still often fail to grasp the full tangle of interrelationships.

- (a3) These deficiencies can be overcome with a recommendation that is both simple yet difficult to follow: whenever experience regularly shows a prognosis to be incorrect, then something must be wrong.

  But make no mistake: we as individuals often succumb to outside pressures to circumvent that wisdom, just as our entire society has become accustomed to failure. For example, society is unable to escape the shear between inflation and unemployment. It seeks to counter the increasing gap only with renewed growth, even though it is well known that any system that maintains itself solely through growth is destined to collapse based on that very growth. This unmasks one of the drives behind the misorientation of industrial society.

(b) The hypothesis of the purposeful leads us to expect that the same things and the same processes will serve the same purposes. This represents a reversal in perspective versus the hypothesis of the causes (see above).

- (b1) Correspondingly, its success parallels that of the hypothesis of the causes. Although the associated expectation may lead us astray from time to time, it still provides the best possible level of anticipation when no background knowledge is available. The simple errors lie in misinterpreting (reversing) the directions of purpose and cause (see the above glove expander, water mill and compressor examples).
- (b2) The deficiencies of this hypothesis, however, differ considerably from those related to causes. This is largely because, in human languages, the concept of causes has clear anthropomorphic roots. The concept is derived from our intentions; it incorporates the functions of both our artefacts and our bodies and, in analogy, all vital functions of living organisms.

  In some languages such as English, this connection is even tighter: 'aim' and 'purpose' more strongly reflect conscious processes (and are therefore closer to 'deliberate intent') than the more neutral German 'Zweck'.

  Attempting to transpose our notion of purposefulness into a more general, non-anthropomorphic concept reveals that we identify substructures and subfunctions as being purposeful when they successfully contribute to establishing and maintaining a higher-level structure or higher-level function. This corresponds to the hierarchic pattern we anticipate behind purposeful interrelationships.

  De facto, bricks seem to have a purpose in building a wall, the wall a purpose in building a house, and the house a purpose in supporting daily life. We can even query the purpose of our own selves. After all, people clearly play a role in maintaining society. While examining such interrelationships may reveal something useful about the 'added value' one contributes to society, this digression tends to muddle the perspective. For example, pursuing the issue further by asking about the purposes of society (which clearly helps determine the status-quo conditions of the biosphere), then the purposeful relationships seem to actually reverse themselves: rather than society being created to fulfil the conditions of the biosphere, the biosphere appears to be created for society, and society, in turn, to serve our purposes.

- (b3) Overcoming this flawed concept evokes yet another simple but difficult-to-follow recommendation: relativize our anthropocentric concept of purpose. This takes some effort because we are unprepared for such a step. Piaget (1978) already reported that children, in their naiveté, believe that lakes were created to enable boats to float on them, and boats created so that children can take a boat ride. We adults also have problems with aligning our demands on society with the added value we contribute to it. If a stroke of luck doubles our fortune, we tend to think this legitimizes doubling our demands (Riedl and Delpos 1996a). The world seems to be here to fulfil our purposes. This misjudgment is intimately linked to the environmental problems we face today.

5.1.3 Common Sense and Intuition

We are all familiar with the phenomenon of 'good old common sense', even though we would be hard-pressed to find an expert definition of the concept. Taking the vernacular meaning, i.e. the sum of our constant unreflected actions and decisions, then these must be driven by the four hypotheses discussed above, namely by the ratiomorphic apparatus.

Another phenomenon—'intuition'—deserves mention here. The term stems from 'to look at', but in the sense of being directly aware of the essential, without conscious reflection having intentionally or immediately led to that insight. One might also speak of inspirations, but intuitions go a step beyond this by including strong emotional and evidence-based experience.

Both phenomena therefore involve a clearly unreflected element of our make-up. This, however, is their only commonality at this point: the guidance they give proves to be almost mutually exclusive, even contradictory.

(a) Common sense and (b) intuition are therefore treated separately here.

(a) Naturally, good old common sense is also guided by experience. It helps us reach decisions in a cybernetic manner with a limited contribution by memory (Chap. 3, Sect. 3.2.1). These same conditions effectively govern the behavior of all higher animals. In humans, reflection will always be part of the mix, but the degree to which this is the case is often difficult to determine.

    The interesting aspect in the present framework of explaining and understanding is how we interpret causes in everyday life. "Fritz broke his leg because he slipped on a banana peel" (Kutschera 1972, p. 103 f). We even accept sentences such as "If he would have slipped on a banana peel, he would have broken his leg" as causal explanations. Kutschera then analyzed the deficiencies of such perspectives from a scientific standpoint.

    Foremost, the explanation is incomplete because slipping can have many different repercussions, for example falling into the water or into the arms of a fair lady. Moreover, the act of slipping is not the sole cause of a broken leg. Rather, it involves an entire chain of special circumstances, conditions and movements made by Fritz.

Such poorly considered speculations reflect a modal argumentation. They are meant to provide us or others with an explanation for the unexpected event 'Fritz showed up with his leg in a cast'. Such argumentation can allude to natural laws but generally does not strive to do so: assumptions based on everyday experience—inductive heuristic speculation—suffices. "Modally", according to Kutschera, "we can causally validate many events that we cannot validate deductively. This is because we are unaware of the applicable natural laws or of all the (initial or) antecedent conditions necessary to arrive at a deductive validation." Deductive control is rarely sought, especially because it is often unattainable.

This type of explanation without reference to laws can be referred to as being 'disponential'. For example: 'The window pane broke … because it was hit by a stone.' Or: 'Fritz pulled Harald from the ice because he is a helpful person.' After all, we are familiar with the dispositions of window panes and of flying stones as well as of ice and helpful people. Some attributes such as 'fragile' can also be considered to be 'manifest', i.e. unequivocally determinable. The distinction is not always clear, but does play a role in testing such assertions.

This issue is also interesting when it comes to differentiating the forms of understanding, explaining and validating (Sects. 5.2.4 and 5.3). The task here was to demonstrate, based on simple examples of causal explanations, that good old common sense first operates inductive-heuristically and that it is initially based on spontaneous appraisals as well as on cumulations of past experience (see also heading 'logical deductive examination').

(b) Intuition, however, must also rely on some element of experience. Creatures that lack any experience lack intuitions as well. Intuition involves experience-based, even revelational vision. This type of inspirational beholding developed differentiated forms as far back as in Antiquity and attracted more attention by philosophers than good old common sense. Descartes himself referred to a mentis intuitus, an intuition of the mind, as a type of thought preceding any method. And Kant distinguished it as the discursive (logical) clarity gained through conceptualization, as opposed to that gained via the senses (empirically).

Improving our understanding of this issue, however, does not require adhering to the tenet of intuitionism, namely that intuition is the principal and most reliable source of knowledge. It suffices to recognize it as a source of knowledge in its own right, albeit one that draws its 'certainty' from other sources. This certainty is based in honed language and involves a legitimation resembling logic. The validation is anchored in an internal, contradiction-free framework of expectations rather than in confirmation through empirically verifiable prognoses.

This perspective recalls the insight (Chap. 2, Sect. 2.1.3 (c); Figs. 2.2, 2.3, 2.4, and 2.5) that communication, language and logic arose under selection pressures that differed from those governing our forms of perception, namely predominantly under coherence rather than correspondence pressure. The initial goal was consistency within the system and only secondarily correspondence with extra-subjective reality.

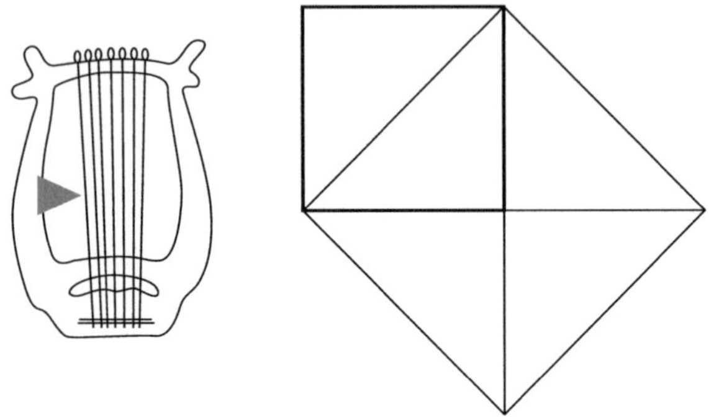

Fig. 5.1 The suggestion of rationality, exemplified by halving the string of a lyre, which represents an octave, as well as a doubling of the square across the diagonal. If you are convinced that the five resulting triangles are identical and interchangeable, then the accuracy of this expectation exceeds any empirical proof (actual measurement)

A case in point: The insight in doubling the surface area of a square across its diagonal (Fig. 5.1) is clearly experience-based. The count yields four versus two equal triangles. For us, that prognosis is convincingly more accurate (by several decimal points) based on intuitive insight than on any empirical measurement. The same no doubt holds true for halving a lyre string and the octave.

The accuracy of prognoses therefore involves a rational rather than an empirical approach. This insight caused irritation as far back as in Antiquity, with the nascence of enlightenment, and has remained a problem to this very day. The validation of mathematics, especially of algebra and analysis, provides an example: why does two plus two yield exactly four? Three explanations are traditionally forwarded. Simply put: because we have always done it this way, because it makes sense, and because this type of symbolism is entirely sufficient. These explanations are referred to as being conventionalistic, intuitionalistic and formalistic, respectively. Accordingly, intuition continues to serve as an explanation.

5.1.4 The Psychology of Explaining and Understanding

Delving further into the realm of reflection reveals how we factor experience into our notions of causes and purposes.

An example: We learn that the form of a parabolic trajectory is determined by two parameters, the 'launch' angle and acceleration. The result is that the parabola becomes longer with increasing initial acceleration and, at constant acceleration, is longest at a 45° launch angle. The two parameters are correlated. In isolation, however, they provide no clue as to why they are correlated. The relationship resembles a description more than an explanation. What, then, do we perceive as being an explanation?

The same thought is expressed in a different terminology in the philosophical definition. There, explanation is defined as "Rendering clear some more or less tangled state of affairs by setting out the cause or the condition or the purposes as to the why and wherefore something is as it is, or presenting the lawfulness of an occurrence that governs a particular case. Some unexplainable fraction almost always remains" (Schischkoff 1991, p. 181). What, then, makes up that unexplained remainder?

To the extent that I can trust my sense of feeling (and supported by the experts), we accept the description of such a correlation as being explained only when—together with additional, comparable correlations—it can be derived from an even more encompassing correlation. In short: "There is no causal explanation for causal laws because these laws form the basis for causal explanations" (Kutschera 1972, p. 118).

Coming back to the above examples: we consider parabolas and free fall to be explained only once they are subsumed under the law of falling bodies, and only when these are further subsumed under Galileo's 'terrestrial mechanics', and these, together with Kepler's 'celestial mechanics', are derivable from Newton's theory of gravity (Fig. 5.2). And admittedly, the laws of gravity also merely involve a description of a correlation between mass and distance—i.e. they themselves require explanation (Riedl 1985).

This raises two points. The first is that the system of explanations remains open in the upward direction and, in fact, in both directions. The second is that we are again dealing with a hierarchic system, one that is both logically conclusive and apparently rooted in our 'instinct' for interrelationships.

5.2 Changes in Cultural History

We cannot fully reconstruct how our ratiomorphic apparatus gradually became accessible to conscious reflection, and we can only speculate about the earliest notions of causality. Nonetheless, comparing prehistorical and ethnological data provides some important clues. Accordingly, the remains of interglacial bear cults correlate well with the belief of recent Arctic peoples that bears are mediators between humans and the Gods.

It is safe to assume that early interpretations of the world were animistic. Nature as a whole was 'animated', soon joined by a pantheon of Gods into which humans projected—in exalted form—their own inner and outer experiences.

In fact, 'self-awareness' may well have arisen late, in a segregation from the 'voice of the Gods' (Jaynes 1993), perhaps even as late as the time of the Iliad. Lévi-Strauss (1968) accurately identified magical thinking and a 'totemistic operator' behind all naive explanations of the world, approaches that helped introduce the first semblance of order into things.

With respect to advanced cultures, I restrict the discussion here to a brief recapitulation of "Western" intellectual history as it pertains to our topic. This history is

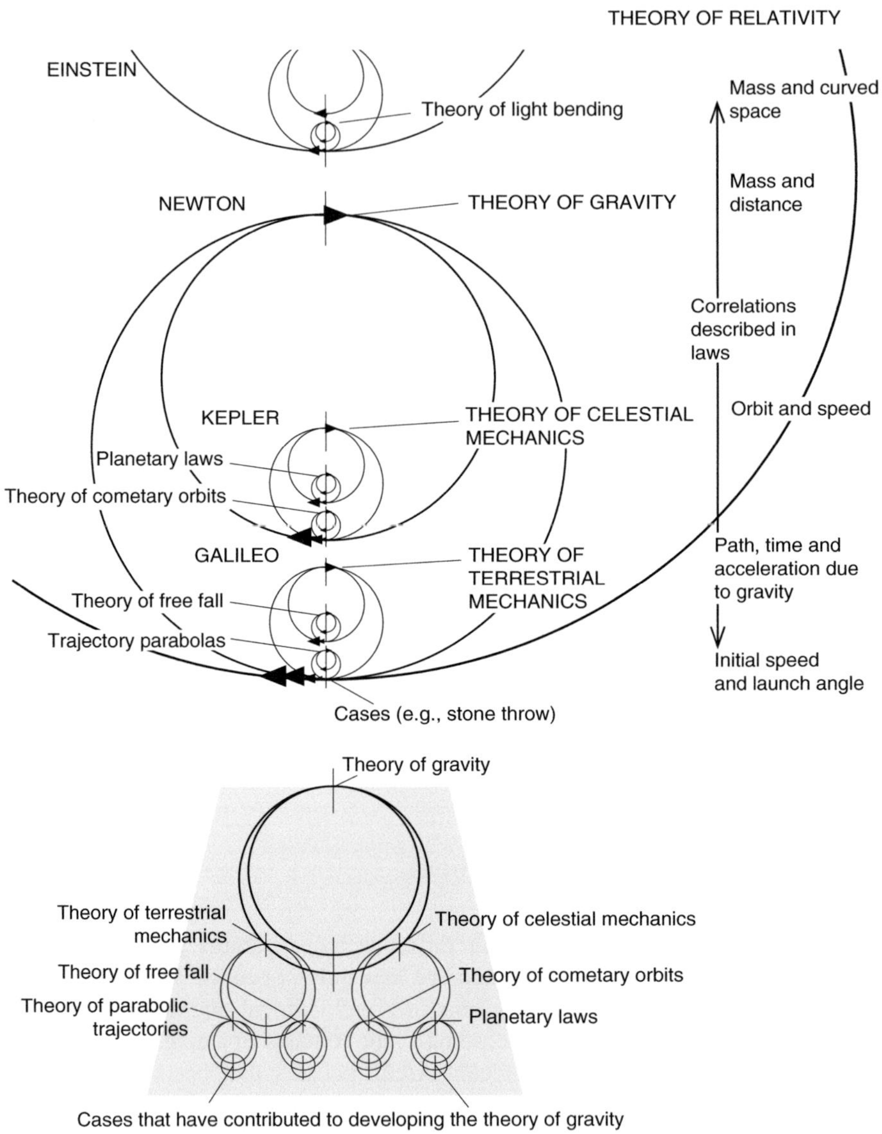

Fig. 5.2 The hierarchic system of explanations, illustrated by the simple example of 'theories in theory' in physics. The correlations involved are listed on the right. Such descriptions are considered to be explained if they fit into the next-higher theory. Bottom: the representational style that is further used in this book

not necessarily any more insightful than that of the Chinese, for example, but it did conquer and transform the world, even if primarily through military and economic might.

This abbreviated approach applies the notions of causes and purposes to illustrate the development and shifts in our notions of explaining and understanding. The three instructive stages are: (Sect. 5.2.1) the beginnings, (Sect. 5.2.2) Antiquity and the Middle Ages and (Sect. 5.2.3) the modern era. Combined, they (Sect. 5.2.4) detail our current understanding of these issues.

5.2.1 The Beginnings in Our Culture

It has often been said that all key cognitive questions were already recognized and formulated by the pre-Socratic thinkers. Despite the meager evidence from this early intellectual era, this assertion is no doubt true (overview in Capelle 1968).

The decisive step and the first documented shift leads from poetic theogonies and cosmogonies, from the mythical thinking of the so-called Orphics, to the beginnings of science as advocated by ancient Ionian natural philosophers. Going further and taking cause and effect to reflect guilt and atonement reveals that three specific causes were already being distinguished. They would later be referred to as causa efficiens, materialis and finalis: drives, materials and aims.

A second cognitive event also stems from this early era, namely a split in our view that has accompanied us throughout cultural history—later referred to as materialistic versus idealistic currents of thought.

The 'ancient atomists', probably as early as Leukipp and well documented in Democritus (460-370 BC), imaged the world as being composed of material particles whose movements were deterministic, without the intervention of any higher instances. That stance was borne of practical, healthy experience and naive interpretations.

In contrast, an intuitive conception of the interrelationships in this world arose in Italic-Sicilian Greece with Pythagoras and, better documented, with Parmenides (515-445 B.C.). Such interrelationships were equally prescribed for the soul and for things. This was conceived as a reasonable and effective drive behind a goal-oriented world order. Only later did it come to encompass the material, blind and irrational drives that could, however, be overcome by applying reason.

This prepared the way for empiricism, i.e. relying on senses and experience, as opposed to rationalism with its trust in reason and thought. These two currents play a defining role in all later developments, specifically in creating the as yet unresolved schism underlying our culture. This schism can be attributed to the different developments of our forms of perception on the one hand and of communication, language and logic on the other (orientation in Fig. 2.5; details in Riedl 1994).

5.2.2 *Antiquity and the Middle Ages*

These two major eras in the history of Western culture are traditionally treated separately. Our focus here on notions of causes and purposes enables a joint presentation: by successively examining (a) the empirical and (b) the rationalistic world view.

(a) A whole range of empirical sciences, as noted above, was founded by Aristotle. His differentiation of four forms of causes was introduced in Chap. 4, Sect. 4.3.2 (b) (overview in Fig. 4.22). This perspective merits fuller treatment here.

We need to (a1) more precisely define the four causes, (a2) justify why exactly there are four of them, and (a3) determine what their intellectual fate was.

- (a1) The four causes. In his 'Metaphysics', Aristotele writes: "We can first say that we understand something only when we believe to know the causes. Causes, however, are referred to in multifarious ways." This is further detailed in 'Book V of Metaphysics'.

  Building a house—today one might equally say: 'producing a chick'—requires (as expressed in Latin) the *causa efficiens*, "in the sense that they give rise to the initial motion or to inertia." This encompasses power, energy and might. 'Power' perhaps best captures the meaning: the term 'efficient cause' as established in English fails to hit the nail on the head and is moreover quite misleading because it suggests that the remaining causae are 'inefficient'.

  Aristotle further notes the need for the *causa materialis*, the materials, "to the extent that they represent that from which something is made, and here the one element is the cause, as the substratum (hypokeimenon), for example of the part". In today's sense, substance, element, component, constituent or compartment and, even back then, "the material for that which is produced thereof."

  Nonetheless, energy and material alone have never formed a house. Another requirement is a form-defining selection and a plan specifying the configuration. This is the *causa formalis*, in Aristotle the tò ti en einai ('essence of the thing'), namely the totality and the composition of the form. This encompasses the selection concept propagated here.

  Ultimately, even the combination of energy, material and a construction plan do not create a house. Someone must insert an intention, some purpose or aim must be pursued, or simply some program be specified: the *causa finalis*. According to Aristotle "as that for which something happens; in this sense, health is the cause behind taking a walk." This encompasses all plans and long-developed programs, whether they involve house building, going on walks, courtship or the ontogeny that ensures that a chicken egg always gives rise to a chick. This purposeful or final cause has experienced a special fate over the course of intellectual history.
- (a2) Why we experience causes in four forms is a legitimate question. One approach to answering it is to use the information presented in Chap. 4, Sect. 4.3.2 (b) and to combine it with an additional synthesis: Recall that Aristotle

recognized a symmetry in these four causes (compare Fig. 4.22). Accordingly, two act from outside, two from within. This corresponds fully to the fact that we experience forces and purposes as functions, but materials and the product of form conditions as shapes and forms. Two different sensory channels are involved. The result: all human languages recognize and distinguish verbs and nouns.

We recognize the second symmetry (Fig. 4.22) based on our knowledge of hierarchically structured complex entities: driving forces and materials act from the lower levels, selection and purposes from the higher levels (Riedl 1998). This represents another cognitive duality, one in which we view the former as constituents (much like 'property'), the latter as being somehow subject to conditions. Linguistically, we express this in the active and passive voice.

De facto, pre- and post-selective processes are involved. These initially decide whether a system can be built or not, and only then, when it is under construction, whether it can endure in the respective surroundings. This insight has often been neglected, with selection being interpreted as restricted to what this book understands as post-selection (see also Sect. 5.3.1 (c) and, for more details, Chap. 6, Sect. 6.2.1).

Aristotle's *causae* merited deeper consideration here because they are simply indispensable for our understanding of complex systems. Nonetheless, their weighting soon shifted in the history of science, and they were then marginalized and finally forgotten altogether.

- (a3) The fate of this concept of causes has two explanations: For one, its subdivision into four entities failed to receive universal acceptance. This is partly because the underlying cognitive dualism went unrecognized. Secondly, many were reluctant to accept that the world could be based on four different causes. Moreover, if four causes were at work after all, wouldn't one of them ultimately be the cause of all the others?

  Kindred spirits of Aristotle such as Epicurus, but even Lucretius (at the dawn of the new age), left this pressing question open. Shortly thereafter, in the third century AD, the rationalistic axis entered the fray, led by the Greek physician Sextus Empiricus: the impression arose that Aristotle himself had already held the *causa finalis* to be the cause of all causes, a view that is incorrect (Kullmann 1979). Nonetheless, the church and the entire Middle Ages followed this interpretation, which came to full fruition at the onset of the modern era.

(b) The rationalistic world view takes its cues from Pythagoras and Parmenides and fully unfolds with Plato. Accordingly, our perception of the world is based on the fact that the souls of human beings, and indeed all things, share in or have a memory of timeless ideas of the good, truth and beauty behind a 'world soul'. This gave rise to Plato's philosophy of ideas 'transcending' experience. A fascinating concept that had immense repercussions.

The question arose as to whether the soul might perhaps partake in the everlasting. Cleanthes, Zenon's successor, already posited that this could be possible—to the extent that the soul of a person itself was good, true and beautiful.

This soon gave rise to the notion of a 'judge of worlds', as subsequently revealed to Paul. This short recapitulation is suffices here.

Plato's transcendental solution to the cognitive problem strongly positioned the causa finalis as the primary cause of all being, attributed thanks to the authority of Thomas of Aquinas to the *causae exemplares*, the foremost intentions of God. This established the philosophy of the church—a 'goal-oriented, purposeful world order'. It acted throughout the Middle Ages into the Renaissance and, as will be shown below, into the stance held by the humanities, which consolidated themselves only much later.

5.2.3 The Modern Era

Starts with the Renaissance, according to cultural historians. This is because it marks a rebirth of the arts coupled with the onset of the scientific thought that characterized the natural sciences in the following centuries: empirical and rationalistic positions came to full fruition, helped constitute the humanities, and then caused their split from the natural sciences (overview sketches in Figs. 2.4 and 2.5).

Again, this calls for treating (a) the empirical and the (b) rationalistic approaches separately.

(a) The empirical approach shaped the natural sciences anew. This initially took place under the dominant influence of the church. A second factor determined the fate of the scientific concept of causality. It has two sides, with Kepler and Galileo being important figures. Both were devout Christians and strove to help humankind partake in the glory of Jesus Christ in a much greater cosmos. They were warned not to expound on the purposes of the world and had already encountered ample resistance in merely studying the forces at work in the cosmos. Secondly, the *causae materialis* and *formalis* played no role in their endeavors. Galileo, for example, was himself the cause behind selecting the materials and forms of his spheres and inclined planes. He could then limit himself to measuring forces and movements. This is one of the restrictions that earned the name 'Galilean revolution' and that ushered in the 'scientific method' of the modern era.

This process was almost concurrent with the rise of 'English empiricism' as expounded by Francis Bacon. Bacon, himself still half anchored in the Middle Ages, postulated the goal of science to be the mastery of nature. This was followed up by Hobbes, Locke and later Hume with realistic theories of the state as well as critical and skeptical theories of cognition. The Age of Enlightenment was heralded in, gaining a foothold in Germany with Christian Wolff Fuß, in France with d'Alembert, Diderot and Holbach. This led—with Laplace and de Lamettrie along with many others—to a materialistic natural science focused on forces, in which humans themselves were interpreted as being mere machines.

This paved the way for the development of 19th century positivism with Comte, J. St. Mill, Spender and Feuerbach. The thrust was anti-philosophical, anti-rationalistic, anti-idealistic. The view was that only positive, tangible facts (whatever that may be) could be the objects of science. This gave rise to what might be called 'physicalism'. Its adherence to the scientific ideal of physics has survived to this very day via the early 'Vienna Circle' with Mach and Boltzmann. This also gave rise to a materialistic 'reductionism' with its unfulfilled expectation of being able to satisfactorily understand even complex systems based on (reducing them to) their parts.

Of the causes, only the forces have remained—now expressed in the framework of four physical interdependencies. This is odd, considering that form- and purpose-giving causes are clearly reflected in the Darwinian principle of selection and the goal of maintaining the species. 'Systems theory' has risen to counter this, but has still failed to make a major breakthrough; 'synergetics' and 'chaos theory', with their 'fractals', have tackled phenomena of complexity anew, but failed to structure the phenomenon of causality (for a treatment of these issues, see Chap. 6, Sect. 6.1.1).

(b) The rationalistic approach is dated in the modern era to Descartes, almost a contemporary of Kepler and Galileo. He differentiated the causes into spatial rules of being and dimensionless rules of thought. And because God created the world, God also remains the first cause and final purpose of all motion and inertia. In light of this dualism, the Cartesians found themselves confronted with the question of how to explain the correspondence of the rules governing being and thought. Towards the eighteenth century, Leibniz outlined a closed system which, following Descartes, must be logical-analytical and must postulate pre-stabilized harmony in the world.

The next cornerstone is Kant, who denied the validity of all the extreme positions: the empiricist one because nothing definitive can be concluded from observations, the rationalistic one because imperatives of thought are not necessarily 'imperatives of being'. Moreover, in the wake of Hume, he established the counter-position that expectations of causes cannot be gained from experience because they themselves are a priori conditions for any gain of experience.

This proof closely approaches the view expounded in this book. Preconditions for our cognitive capacity must exist. In fact, however, with Kant the problem shifts from a proposed transcendental solution (see Plato) to an equally 'transcendental' (even if not meant metaphysically) necessity to go beyond, i.e. transcend, experience—a step not accessible to empiricism.

Then, in 'German Idealism' as represented by Schelling, the transcendental concept regains its speculative-metaphysical meaning. Hegel and others use it in an attempt to overcome Kant. The ensuing developments are less important for the present topic and represent a proliferation of well-known positions—neo-Kantianism, neo-scholasticism, neo-Thomism and anti-metaphysical versus metaphysical Idealism. The metaphysical branch remained finalistic and anchored in the divine, whereas the anti-metaphysical branch differentiated itself either rationalistically or axiologically (theory of value). Windelband, Dilthey and Max Weber cleaved it from the natural science, making it a manner of thinking particular to the humanities—as a finalistically oriented grouping of cultural sciences.

This initially appears to be quite logical because everything that moves people and gives rise to artefacts is difficult to conceive without purposes. At the same time it is also unusual, considering how obvious it is that energy demands—whether they involve power or overextended resources and materials—define the selective framework that constrains human culture.

This returns us to Lord Snow's insight. Our reduction or simplification of causal relationships helps explain why the natural sciences change the world without actually having fully comprehended it, and why the humanities lack an instrument to counter this.

5.2.4 The Concepts of Understanding Today

We have already determined that the terms 'understanding' and 'explaining' seem to be interchangeable colloquially. This is also reflected in epistemological texts. Stegmüller (1983) writes about an 'understanding explaining', Kutschera (1972) about an 'explanatory understanding'. This section takes 'understanding'—as the broader framework term—and outlines the degree of differentiation it has attained for the present topic. The next Sect. 5.3 then more closely examines the forms of explanation because, beyond its forms, the underlying conditions also prove to be crucial.

The first step is to segregate 'practical understanding' from the overall concept of understanding. The phrases 'it is understood that' or 'I understand this to be…' or 'he has a good understanding of...' clearly refer to prerequisites and to matters of course or to the mastery of a particular skill, whether it be a trade, an art or a science. The focus here, however, is on the forms of 'theoretical understanding'. Based on Kutschera's approach to differentiating these forms (1972), seven remain. Examining them from the perspective of our cognitive understanding of causality yields two subsets of forms.

Accordingly, I juxtapose (a) the forms of overarching causal perspectives with the preponderance of (b) final viewpoints.

(a) In the first type, a relationship to all four forms of causes is never explicitly mentioned, but upon closer examination undeniable.
    (1) 'Determinative understanding' refers to the assignability of a process, whether it involve an activity, a device or an instrument.
        This resembles the assignability of a perception by the simul hoc without, however, matching it entirely. The term 'understand' is not applied to perceiving the assignability of an object—at best to being able to follow the

process of assigning. One doesn't say: 'I understand the dolphin as a mammal', but one understands 'that it is counted to the mammals'. This terminological proximity shows that we are poorly equipped to describe the ratiomorphically controlled processes of categorizing perceptions. This is no doubt because these processes operate successfully at the subconscious level and are therefore difficult to fathom reflectively.

(2) 'Causal understanding', in contrast, denotes the common form, which we expect to sufficiently capture the causes behind a process or condition.

(3) 'Genetic understanding' encompasses derivations, insight into situations, conditions, as well as into the intentions behind some development, whether it be an architectural style, an organism, or even the Alps. The focus in these latter two forms is often on forces, yet in some cases more on purposes, although selective material and formal conditions are clearly a consistent component.

(b) In the second type, final relationships dominate.

(4) 'understanding of function' denotes insight into the purpose served by something, whether it be an instrument, a switch or even an organ or a cell. This involves both consciously intended as well as subconsciously established purposes of programs that have passed the test of time. In some sense this is a transitional form. After all, more closely examining the establishment of a function clearly shows that all four forms of causes are required to attain a satisfactory understanding. One approach is to behave as if only the purposes were decisive, a position often taken in cybernetics. The same holds true—if perhaps less transparently—for the following forms.

(5) 'Understanding of meaning' describes the ability to correctly interpret purpose, sense or the message behind symbols, whether they pertain to body language, spoken words, letters of the alphabet or texts.

(6) 'Rational understanding' means comprehending intentions, i.e. putting oneself into a position to empathize with the motivations behind an action. This itself is based on drives but nonetheless pursues goals, for example to ease disappointment or show enthusiasm.

(7) 'Intentional understanding' refers to the same thing but more specifically, namely to recognizing the intention, purpose and aim of an action. These three forms of understanding are rooted, as expressed by psychologists, in 'introspection' and 'projection': (a) 'what do I myself experience from that which I perceive from an acting person', (b) 'I will therefore be able to assume the same for that actor'.

In this second type of understanding, the emphasis on the purposeful or final cause is justifiable. In principle, however, we need to understand all four forms of causes to achieve sufficient understanding—or shall we already say 'for sufficiently explaining' (see below).

5.3 The Conditions of Explaining

The first step is to focus on the use of the term as relevant to our topic. Clearly, explanation is meant in the sense of 'explanatio' rather than of 'declaratio' (for example a tax declaration). Nonetheless, the transition can be fluent, such as statements or declarations before court, which typically border on explanations. The differentiation between 'practical' and 'theoretical explanation' provides a more narrow focus. The practical mediation of knowledge is an issue only in the context of this or other books: the broader issue revolves around the theory that can be derived from the process of knowledge gain.

A second consideration is that explanations can also represent or at least harbor validations. This calls for examining the concept of validations or substantiations (German: Begründungen).

Validations can also be viewed as answers to 'why' questions, whereby two forms can be distinguished. If the question targets causes, then a satisfactory answer can be called a 'causal validation' and one can refer to reasons grounded in beingness or in reality (German: Seinsgründe, Realgründe). In a broader context the question may also target all the factors that validate the situation of something or of some process. The answer can then be called an 'epistemic validation' and one can refer to cognitive and rational reasons. A special form of epistemic validation is 'proof', probando, originating in jurisprudence and later becoming a cornerstone of mathematics: the plausible or logically inferred derivation of a statement, a theorem or an axiom (i.e. of a 'validity', a provision or claim) from a basic assumption, a system of axioms or a theory.

In principle, causal validations are predominantly empirical-inductive, whereas epistemic validations are predominantly logical-deductive. The distinction, however, is not clear-cut. First, not all validations represent an explanation. The oscillation period of a pendulum, for example, is determined by its length, but this still fails to explain the cause behind the relationship. Second, not every explanation represents a validation. Even if an action is rationally explained, this still does not validate its taking place. Rather, it merely substantiates that the action was reasonable based on preferences and assumptions.

We say: pendulum length and oscillation period are correlated, and because we can more easily change the length, we experience it as the cause of the oscillation period. The explanation, however, lies at another level. An action, in contrast, can be explained based on initial conditions, even if the causes behind those conditions remain unknown. This explains the view espoused in Chap. 2, Sect. 2.2.1 that our ability to gain knowledge where none was previously present is based on an iterative and hierarchic system in which inductive and deductive processes interact reciprocally (see Fig. 2.6). No logically deductive validation is deemed satisfactory unless it corresponds to experienceable reality, and no empirical-inductively synthesized theory can shirk logical-deductive validation. For example, we expect the axiom of parallel lines not to contradict our experience in the finite world. Equally, when deriving an action from initial conditions, we expect it not to counter the laws of logic.

Accordingly, this differentiation of the terminology primarily involves differential weighting within the same cognitive program. And we can assume the same mechanism to be at work in the forms of explanation. We can work with the following terms: explanandum, that which is to be explained, explanans, that which is to be explained based on antecedent or initial conditions, for 'necessary' and 'satisfactory' explanations'.

The categorization presented here again follows Kutschera (1972) in order to simplify the comparability of the texts and begins with (Sect. 5.3.1) today's view on causal explanation. Sect. 5.4 (Forms of understanding) then deals with teleological explanation and with the understanding of actions and of artefacts. The causal explanations are appended with (Sect. 5.3.2) the structure of the two hierarchies of explanation and (Sect. 5.3.3) the three paths of recognition, explanation and emergence.

5.3.1 On Causal Explanations

This is the concept of explanation whose definition still includes the word 'causes'. It also encompasses explanations in the narrowest scientific sense. The remaining three concepts, in contrast, can be understood either as special forms or even as having questionable scientific merit—depending on the perspective of the user or the critic. In the present context, however, this is a reasonable starting point.

The approach taken here begins with the concept of explanation (a) in the colloquial sense and then presents (b) today's scientific framework conditions and (c) the modifications recommended for examining complex systems.

(a) In everyday language we tend to accept explanations that are modal (see 'good old common sense' in Sect. 5.1.3 (a)). For example: 'The cat is dead because it fell out of the window'. While this is not a causal explanation, it can be taken as being logically consistent based on 'common knowledge' about the vulnerability of such a creature, about an upper floor, and about the nature of a concrete impact point. The ratiomorphic level already operates on the clear assumption that the fall preceded the death (recall *post hoc*). The *propter hoc* is then appended. And we are willing to generalize the relationship: 'Cats that fall onto concrete from the sixth floor will be dead'. This is because every relevant report that we have heard supports this conclusion.

Importantly, such an explanation never stands entirely on its own, even when common knowledge is involved. Rather, it is embedded in an entire system of higher- and lower-level expectations. We are also convinced that an equally violent collision between an automobile and a cat would lead to the same lethal outcome. Moreover, we also assume that the same cause-and-effect relationship would be valid for any mammal. Such considerations are embedded—whether reflected upon or not—in expectations about the effect of mechanical forces, which in turn are anchored in expectations about the vulnerability of all living animals to injury.

This is reminiscent of the subsumption- or Hempel/Oppenheim scheme (Chap. 2, Sect. 2.3.3 (b)) and of its relationship with hermeneutics (Chap. 4, Sect. 4.2.1). This will come as no surprise considering how closely the processes of cognition and explanation are allied. The re-emergence of the subsumption-hermeneutics connection is interesting in the present context because it demonstrates that the explanatory process, with its strong reflective component, still retains a ratiomorphic component. This turns out to reflect a special connection, namely that between the rational, science-based construction forwarded by Hempel and Oppenheim (1948) and a ratiomorphic, poorly analyzed predisposition that was denied any significant scientific status.

(b) The systematic/technical framework conditions dictated by contemporary philosophy for causal explanations differ in nature. They are more interested in validating—logically or at least rationally—the process of explanation itself. Grasping the emergence of the concepts behind causal explanation is secondary. Accordingly, much of the debate revolves around language.

Well-known authors who drove this discourse for nearly a century include Carnap, Goodmann, Hempel, Oppenheim, Popper, Quine, Rescher, Stegmüller and Suppes. Their positions and perspectives are compiled lucidly and comparatively in Stegmüller (1983). His astute commentary, condensed into seven volumes, yields three conclusions: First, none of the above efforts to validate causal explanations are superfluous. Second, the book is still not closed on any of the key points addressed. Third, it is highly unlikely that any successful researcher following an empirical approach will be familiar with the entire discourse. This has repercussions.

The conclusion can only be that the causal explanations forwarded by most researchers are based on privately held concepts and theories. These involve, much like 'common sense', minimally reflected, supposed self-evident facts. They are founded on the pragmatics of ratiomorphic programs and on the guidance of the respective scientific school.

Separately treating (b1) the determination of causal explanations, (b2) the underlying conditions and (b3) the dogmas provides a better overview, although all three problems are interlinked.

- (b1) Causal explanations are answers to queries about causes. Causes are conditions that yield a causal explanation, best formulated as a law of causality. Causal laws are natural laws. What constitutes a natural law, however, remains an open problem. A logical justification is difficult, and efforts to avoid the suspicion of circularity are controversial. From the intuitionistic perspective, the relationship appears to be almost trivial. After all, this framework has already helped us to explain the real world.
- (b2) It is instructive to examine rule-governed statements based on the conditions set. They must not (1) be restricted to specific places, points in time or objects. They must generalize and be valid at least for classes of situations. They must be (2) inductively structured and (3), according to Popper (1973b), be disprovable based on new experience. All three conditions have limitations, and these have

been reached. Nonetheless, we do know the underlying tenets (Chap. 3, Sect. 3.1.2): 'neuronal memory', the 'conditioned response', and conditioning as the prerequisite for all associative learning. Simply put, the correspondences of isolated experiences are formed into a class of expectations and, if disproved, are once again deleted.

The most straightforward condition was considered to lie in (4) the applicability of the 'unreal conditional clause'. For example in the statement: 'Whenever A were to occur, B would have to follow'. Intuitively, this seems high plausibility. We can better understand this perspective by examining the path that held the promise for Carnap (1961) and Hempel that the world was logically structured (building upon Hume's causality and Mills' natural law). That is, until Hempel himself experienced a turning point, namely the insight that a strict definition of 'explanation' needs to consider the respective 'state of knowledge', which itself is even more difficult to circumscribe. This ushered in the so-called 'pragmatic shift', which even today continues to permeate the theory of cognition as a whole. Its tenet? Focus on those things that prove successful (a familiar principle in the present context).

- (b3) In examining dogmas, a good place to start is with empiricism. These dogmas are still readily apparent when dealing with causal explanations. In a sense, they represent self-imposed conditions. They were already criticized by Quine (1953), which influenced the more recent trends of concepts in the theory of science.

  The first dogma comprises the demanded strict separation of analytical and synthetic statements. This is not always the case, for example when 'dispositional declarations' are involved. 'Fragile', 'water-soluble' or 'magnetic' are some of the 'dispositional monikers' frequently used in explaining actions. Stating that: 'The glass pane broke because it was hit by a stone' is considered to be a causal explanation. In contrast, stating that: 'The glass pane, when it was hit by a stone, broke because it was fragile', merely appears to be rule-based. This differentiation reflects our 'observational language', which can express itself in symptom- or in reduction sentences.

  The second dogma involves the demand that empirical-scientific explanations of all non-logical concepts be derivable from observable phenomena. It turns out, however, that some sciences, in particular physics, very successfully operate with 'theoretical concepts' that are only loosely attributable to the observable.

  The third dogma involves the conviction that the tools of logic must suffice to explicate all relevant terms. This demand was forwarded above all by the modern-day 'logical empiricism' of Russell followed by Carnap. It turned out, however, that the syntactic methods must be supplemented by semantic methods and pragmatic terms. Accordingly, 'confirmation' itself already represents a pragmatic concept because it refers to situations, points in time and persons.

  These dogmas clearly stem from the desire to make empirical explanations contradiction-free—an ambition already torpedoed by the nature of human language. Avoiding such problems calls for taking the step from logical to pragmatic empiricism and further to a 'naturalized' epistemology (compare Callebaut

1993). This is firmly ensconced in EE, 'evolutionary epistemology', of which the present author is a proponent.

(c) In dealing with complex systems, we need to become even more pragmatic by relying on what empirical knowledge gain has to offer. This means determining the reliability of statements based on the frequency of confirmed prognoses. The framework is causal explanation rather than any (contradiction-free) consistency of deductive derivation. Importantly, one particular trap of empiricism (recall the behavior of 'Russell's chicken'; Chap. 3, Sects. 3.2.3 and 3.2.4), namely impermissible extrapolation, must be avoided.

Recall also (Chap. 2, Sect. 2.3.3 (b)) that the controls lie in the degree to which a theory or expectation is integrated within the subsumption scheme. Specifically, they must be embedded in the framework of reciprocal confirmations inherent in a pyramid of tiered theories (compare Fig. 2.12). Chap. 6 then reintroduces the double pyramid concept discussed in dealing with the structure of cognitive processes (Figs. 2.13 and 2.14).

This calls for (c1) newly defining the terms and (c2) testing the four forms of causal explanation on complex systems, especially in relation to (c3) the hierarchic organization of such systems.

- (c1) The definition of the terms can often be approached more loosely. This is because we cannot expect them to be exactly defined from the onset and can trust that they will be optimized based on the context and the respective 'state-of-the-art'.

  'Natural law' therefore encompasses all the detected coincidences of parameters that are not man-made and which, taken as explanans, yield a sufficiently high percentage of experience-based, confirmed prognoses. This basically reflects the constraints that reduce the world's structural chaos and arbitrariness and enable us to experience predictable order. Such expectations can also be understood as rules of thumb. This is because the degree of confirmation necessary to refer to a law in the traditional sense is context dependent and cannot be determined in advance. We can also drop any differentiation between 'eternal' and temporary lawfulness: all coincidences of parameters—even if at highly different times and regularity—can only have arisen after the cosmos gained structure.

  The same is also valid for the term 'cause' (German: Ursache). Every optimization process can involve imprecise and even entirely incorrect explanations. Mutually exclusive solutions can be developed even about whether chance or lawfulness prevails (Chap. 2, Sect. 2.2.1 (b), Figs. 2.8 and 2.9). That depends on whether one and the same problem—depending on our propensity—is viewed ratiomorphically-cybernetically or rationally-probabilistically. It is independent of whether, in a 'Ptolemaic world-view', the planets are held to be embedded in crystal bowls that circle the Earth, the dolphin is placed among the fishes, or 'phlogiston' is taken to be a separate category of igneous material. As soon as theories become interlinked in the subsumption system of higher- and lower-level theories, the contradictions must emerge. While isolated concepts of causality can immunize themselves against refutation, they cannot do so in the overall context.

Circumspection is warranted with regard to the term ‘explanation’, especially because we need to distinguish between necessary and satisfactory explanations. In complex systems, especially regarding their ontogeny, numerous necessary, i.e. indispensable, explanations are often required to approach a satisfactory explanation (see Chap. 6, Sect. 6.2.2; Fig. 6.3 for a more detailed discussion).

- (c2) The four forms of causal explanation can all be postulated to root in the four physical interdependencies. In practice, however, a major restriction is our need to operate with synthetic terms such as ‘drives’/‘driving forces’ and ‘survival conditions’, ‘selection’, ‘elimination’, ‘fit’, ‘growth’, ‘steady states’, ‘program’ and ‘purpose’. These stand for families of complex states or processes. They are indispensable because the case examples can typically be derived only theoretically as a postulate, but not de facto traced back to specific original conditions. Why? Because their historical development is inaccessible to the scientific approach.

  These ‘synthetic terms’ constitute a condition symmetrical to the ‘theoretical terms’ identified above (Sect. 5.3.1 (b)) as being a problem in rationalistic dogmas. Whereas the latter involved disengagement from observable phenomena, here the reverse is true, i.e. a disengagement from derivability. The issue is once again that of emergence (Riedl 1997).

  This calls for examining the four *causae*—as related to complex systems—in more detail.

This analysis is more in-depth than for the other forms of explanation (Sect. 5.4) because background information needs to be introduced first. The issues are (i) the driving cause (*causa efficiens*), (ii) selection, (iii) the material-, (iv) formal- and (v) final (purposeful) cause.

(i) The connection of the ‘efficient or driving causes’ to the four forces is still relatively easy to trace. This encompasses the forces that counter gravity on land, enable body movement and fuel the metabolism that ensures growth, maintenance and survival. These are largely derived from the strong and weak interactions initially present as the energy contained in chemical bonds and that, through various energy conversions, then work as physical forces. This also includes the energy reserves in starch, fat or food depots. In humans this extends to exchange values, precious metals or capital, i.e. energy in all its material guises.

(ii) The issue is more differentiated in the case of ‘selection conditions’, requiring additional compound terms. Even the simplest form of selection, namely sieving, conjures up complex terms such as sieve and sieving- or screening material. Moreover, the results of sieving clearly depend on the structural and dynamic properties of the material to be sieved and of the sieve itself.

It is still relatively simple to describe the forces at work here, i.e. the shaking of the sieve. Nonetheless, tracing this dynamic back through the chain of energy transformations, namely to the source that powers the world in general, proves to be a complex task. It depends on whether the sieve is being shaken by an electric motor operated by a sedimentologist or manually by a farmer's wife. The dynamics of the sieving material are similarly differentiated. Is the material being washed through the meshes of the sieve with a water jet? Or is something more along the lines of a 'Berlese funnel', as employed by soil biologists, involved (where tiny organisms crawl down through a sieve in search of moisture).

The situation differs with regard to the features of the sieving material and the sieve, i.e. of the objects being selected and of selection versus elimination. Clearly, manufacturing a sieve and producing the sieving material requires and uses energy. They represent necessary preconditions. Nonetheless, the forces behind their production provide little causal explanation for what falls through the sieve and what not. This is mostly because the history of sieving material and of sieves has involved all manner of sieving operations that can no longer be reconstructed.

This requires abbreviating the examination of the causal relationship. The entire genesis of sieves and screening material, as a precondition, can be left aside: the new point of departure in examining the sieving outcome is based on the features of the material and the sieve.

(iii) The 'material causes'. In sieving sand, it matters little whether the fraction passing through the screen stems from a stone mill or from river bedload. In soil analyses using the Berlese funnel, it is irrelevant whether evolution produced a mite or a springtail (a primitive, ametabolous insect): the crucial factor is the organism's diameter.

It is an entirely different matter, however, whether sand is available for constructing a building—or whether only snow or palm leaves (or nothing) is at hand. The decisive role of material availability was already discussed in relation to the phenomenon of preselection. Moreover, the type of available material determines the overall possibilities and the form of the unfolding system (Chap. 4, Sect. 4.1.2 (b)).

Material causes could be further elaborated, but to remain with the simple example above, the reasons why only snow or palm leaves would be available to build a shelter are self-explanatory. Although such preconditions can harbor many drives, material causes cannot be simply reduced to driving causes. The difficulty lies in their often unreconstructable genesis. A satisfactory explanation, however, would require such information.

Finally, complex systems are never entirely composed of a single material. Moreover, any one material is always composed of other materials and, in combination with yet other materials, forms even more complex classes of materials.

(iv) The approach to 'formal causes' is much the same.

It is juxtaposed with the pre-selection of materials and involves, after the materials have been assembled, post-selection by an external system.

A simple lock-and-key example suffices: the concept of potential fit is equally valid for the selection acting on an atom in a molecule, a cell in a tissue, an organism in its environment or an idea in society. The configuration of the key's bit represents the features that the lock requires of the key. Keys and locks abound. A particular key will fail in very many locks, but it will fit one lock somewhere. Importantly, many duplicates can be made, for example for the entrance into a hotel or for the doors in train cars. This is applicable in the present context.

Again, turning a key—not to mention its manufacture—clearly requires using and consuming energy. Nonetheless, sufficient knowledge about those expenditures reveals little to nothing about the fit.

The causes behind the particular structural features determining the fit include a historical perspective, which is rarely fully reconstructable: too many alternative strategies have gone unrecorded. Nonetheless, a chain of causes remains an immutable postulate.

Much, of course, can be reconstructed. The phylogenetic tree of organisms, the history of the continents and of habitats, of languages and cultures illustrate many of the causalities required for causal explanations. Nonetheless, they alone remain insufficient to dismiss or forego the concept of formal conditions.

Finally, forms and their causes are not isolated entities in complex systems. Beyond a serial arrangement of formal conditions, almost all forms are themselves composed of subforms which, combined, help define the overall form.

(v) The 'final cause'. Despite the special status the humanities have historically afforded it, the final cause it is amenable to examination in the context of causal explanation. The first step is to avoid the loaded term 'telology' and replace it with 'teleonomy', as suggested by Pittendrigh (1958).

From a natural history perspective, this refers to programs that harbor a goal. For example the program contained in a chicken egg, which as a rule gives rise to a perfect chick, which then becomes a hen, which in turn is programed to lay eggs. Equally, a behavior, an organ or a physiological reaction is programed to ensure the survival and reproduction of its bearer. Such programs should not be misinterpreted as having aims or goals. Nothing actually programed something or set a goal. Rather, among the countless trials, mistakes and malfunctions, only those functional chains that successfully coalesced into such circular programs (through self-preservation and species survival) have remained. All others have been eliminated.

The genesis of these programs involves long chains of dispositions and conditions of material- and formal causes, and requires and consumes energy just as they do. Moreover, as a historical product, they are also resolvable and anchored conceptually.

Accordingly, only things that serve a purpose have been retained. The fitness of the adaptation and internal harmonization at every level of organization is astounding. This ranges from energy generation in the cells to the reflexes and genetically anchored behavioral programs, the 'Instantaneous-Information-Assessment-Mechanisms' (German: AIAMs—Augenblicks-Information-auswertenden-Mechanismen; Lorenz 1978).

It is therefore unsurprising that humans all around the globe experience entirely congruent reactions of aversion, desire or appetence, fear and relaxation, and that we can also interpret the corresponding behaviors of higher animals quite well. The fact that human behavior is controlled by (largely subconscious) emotional states and motivations has only been recognized quite recently (compare Wimmer 1995; Wimmer and Ciompi 1996; Ciompi 1997).

The major shift in our perception of purpose is related to the nascence of human consciousness, which involved intentionally developing programs in imagined space. This reflects an internalization of self-sustaining processes. The (misguided) temptation to project such planning onto all of nature is strong.

Again, much like in material and formal causes, purposes are embedded in a succession of purposes, each one itself being composed of further purposes and, combined, all forming the overall purpose, which is to maintain, improve and set our sights on some conceptual purpose of life. These range—even if in entirely differing functional 'garb' (depending on the respective level of complexity)—from our life plans through appetences and aversions, organ and cell functions, down to the last hydrogen bond that holds a particular molecule in the right place in our genetic material.

(c3) These four forms of causes are embedded in a hierarchic framework and can also be deciphered through that framework. This hierarchy played a role in Chap. 3 and again in Chap. 4, from the cosmic dimension to that of bodyplans (Chap. 4, Sect. 4.1.2 (b); Fig. 4.2 as well as Chap. 4, Sect. 4.3.2 (b)).

What remains to be done at this point is to insert two gained insights into the forms of causal explanation. The first is (i) a further differentiation of the terms 'cause' and 'precondition', the second is (ii) considering how the material and formal causes operate though all levels of the hierarchic framework. Note again that the conceptual and graphical representations of the tiers highly simplify the hierarchic patterns but help provide an uncluttered overview.

(i) Causes and preconditions: As outlined in Fig. 5.3, the functional definitions of the driving and final/purposeful causes extend unchanged through all levels of the systems. In contrast, the material and formal conditions are represented by different terms from tier to tier: they are tailored to the respective gestalt and therefore expressed differently. This explains the different terms encountered in molecular biology, ultrastructure research, cytology, histology, organology and systematics.

Nonetheless, the material and formal causes do operate across the systems, the former from the lower, the latter from the highest levels of the overall system's structural hierarchy. This again calls for differentiating the terms cause, condition and precondition.

We experience a cause as being a relationship that leads us to expect a transition from the immediate effect to the precondition. The strength of chitin, for example, can be interpreted as being one of the material causes for the develop-

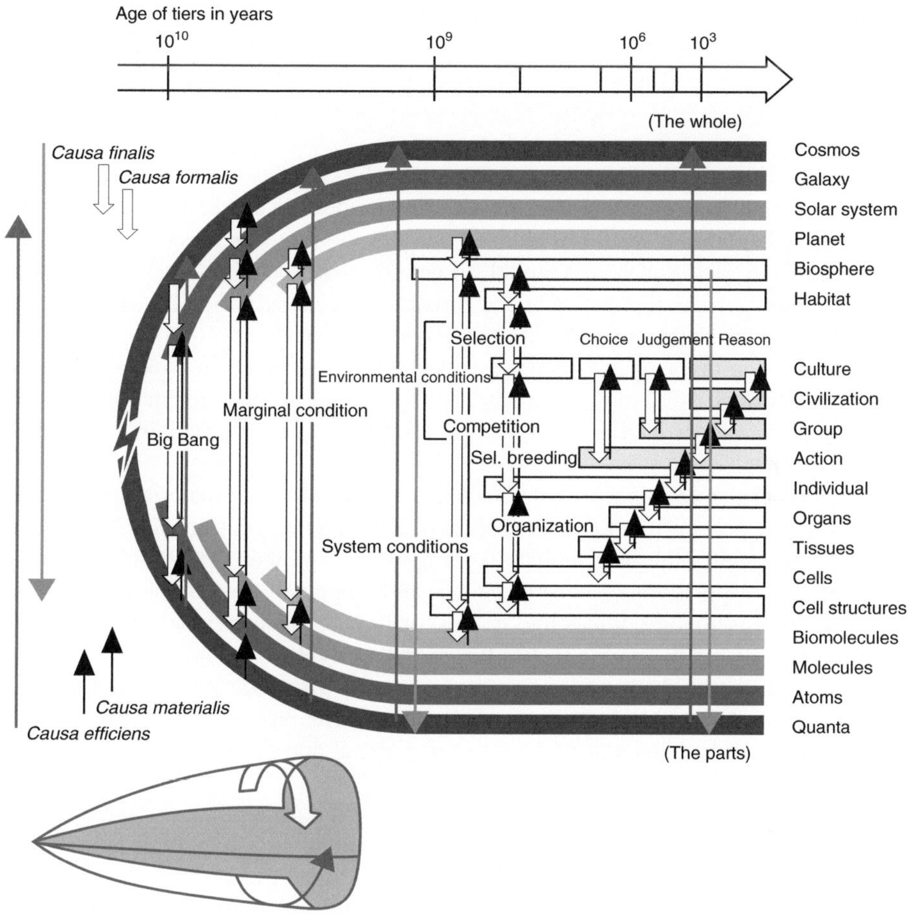

Fig. 5.3 The four forms of causes in the tiered structure of the world. The bars representing the hierarchic tiers (see also Fig. 4.2) are labeled on the right, the age of the tier on top. The material and formal conditions are inserted in all positions, but the driving and final/purposeful conditions only in the key positions to avoid cluttering the scheme (after Riedl 1985, amended). De facto, the scheme should be envisioned rolled up (see insert on bottom) because the physical interrelationships most likely have a common origin

ment of the arthropod exoskeleton. Once an exoskeleton has established itself and become the condition for a particular type of joint structure and muscle insertion (in beetles, for example), then the invention of chitin changes from being their cause to being their precondition. The 'inventions' that trigger further developments take place very slowly (far exceeding observational time periods). Accordingly, we designate most of them as being conditions and preconditions.

This means that the largely established conditions and preconditions and their entire hierarchy were themselves all once 'causes' in the above sense.

They must, as pre- and post-selective material and formal causes, have been directly responsible for the development of a new intermediate level of organization: the structures and functions of new or modified constituents were tested by the higher-level systems. This is treated later in the context of historicity (Chap. 6, Sect. 6.2.2; Figs. 6.1, 6.2, and 6.3).

Clearly, such new interrelationships must be genetically anchored to be preserved in a species. Chap. 6, Sect. 6.3.2 examines the mechanisms that ensure such a fixation. It would be entirely naive to ignore this process and remain mired in a 'string-of-pearls model' of genetic representation, i.e. to expect that complex systems could have evolved via random changes in millions of serially arranged genes that are independent of one another.

(ii) A final step involves examining how to interpret such an 'unbroken effectiveness' through all levels. Again, the action of the driving and final conditions extending across all tiers by no means hampers our understanding of overall function.

The principle initially seems trivial with regard to the material conditions. Everyone accepts that the structural laws of chemical bonds are equally valid in our skin and in the 'metabolism' of a major city. A more interesting question how the dispositions of muscle-, bone- and nerve cells determine those of the overall musculature, bones and nerves, or how the dispositions of organ types and metameres determine that of the overall bodyplan.

The same holds true for complex forms: the disposition of clay determines the brick, and the brick the construction types. The bronze alloy, the tempering of iron, the development of screws, sheet metal and poured concrete changed civilization.

The most interesting aspect in all of this is the astounding conservatism of the construction elements at the lower system tiers (see Chap. 6 for more detail and specific examples).

This unbroken effectiveness is more difficult to convey for formal conditions. It helps to consider functions instead of dispositions, particularly mechanical functions. They help us recognize the influence that the type of locomotion has on the body form of a gazelle, whale or beaver. Or at the next level its effect on the form of the bones in the 'hand' of mammals as diverse as horses, dolphins or bats (Fig. 3.8).

This approach also sheds light on the concurrent influence exerted by the history of the respective structural forms and by the available materials. Such phenomena extend through all levels. Consider the effect that the function of hollow bones had on the arrangement of the bony trabeculae, or the function of the eye on lens shape and on the back of the eye (ocular fundus; Fig. 5.4).

Here, the laws of mechanics and optics still provide a measure of clarity. This clarity is no longer directly perceptible in physiological functions. Nonetheless, the influence of blood pressure on the blood vessel walls remains recognizable, as does the function of the vessel walls on the arrangement of their muscle cells, and, in turn, muscle cell function on the elongated vessel form.

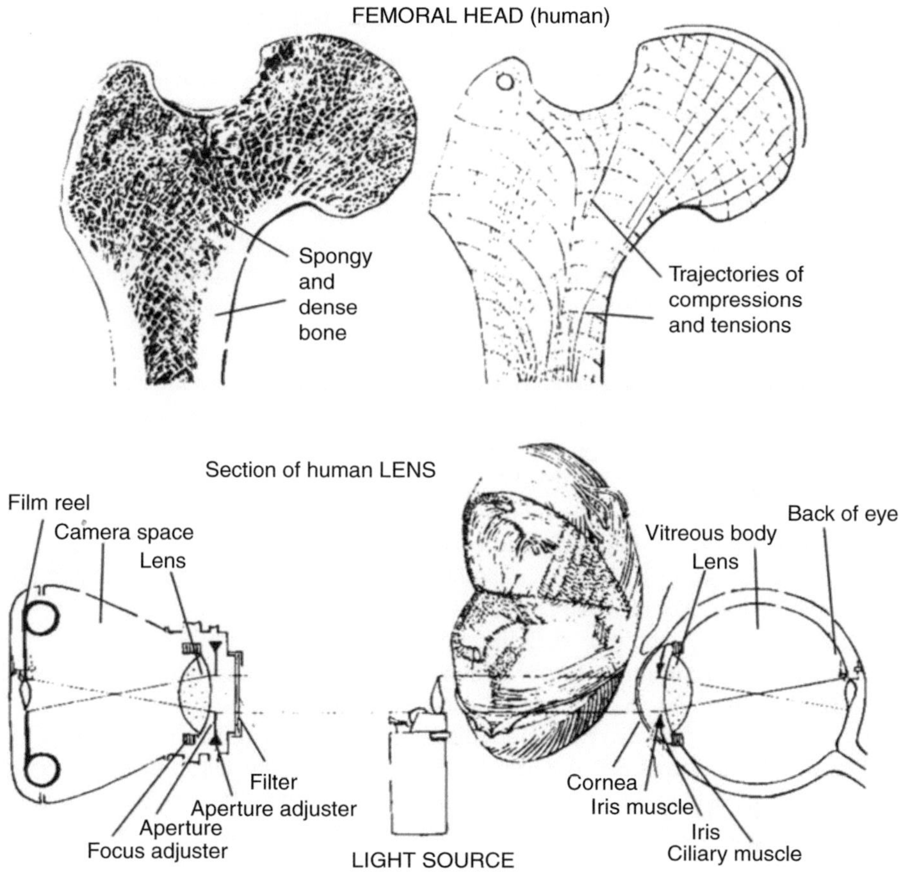

Fig. 5.4 Formal or functional conditions from the higher-level system, exemplified by an optical (below) and a static (above) apparatus. Compare the arrangement of the bone trabeculae with the force trajectories in the model and the analogous correspondence of all key functions of the camera and eye (after Riedl 1980)

5.3.2 *The Double Pyramid of Explanation*

Examining our cognitive systems (Chap. 3) revealed that two hierarchies—structural and class hierarchies—necessarily complement each other (Chap. 3, Sect. 3.4.3 and the schematic representations in Figs. 3.11, 3.12, 3.13, 3.14, and 3.15). Structures form a hierarchy of rule-governed relationships. Their generalizability defines and justifies the obligatory status. That universality builds on the classes of case examples encompassing those structures, just like these classes are themselves again composed of structures.

This relationship is also presupposed in discussing the explanatory process, but is often taken for granted and remains unmentioned. It becomes important in more highly complex systems (such as those used here to establish that very relationship).

The above symmetry is reflected in the symmetry defined by two pyramids. In introducing the issue (Chap. 2) it became clear that two types of pyramids are involved: one representing the structural hierarchies of distinct individualities, the other standard building blocks (Chap. 2, Sect. 2.3.3, schemes in Figs. 2.13 and 2.14). They again manifest themselves as the pyramids explaining complex systems.

Two aspects require validation: (a) the features of this double pyramid and (b) the reciprocity of the prevailing explanatory processes.

(a) The structure of both pyramids exhibits a tip and a base comprising theories. The tips extend to the boundaries of our knowledge, to the respective 'state of the art'. The most widely separated are the theories behind the mega- and microcosmos, for example the theory of relativity at one end and the elements of the quantum theory at the other.

The theories of the bio- and cultural sciences are also embedded in the same framework, except that they end earlier. The biosciences typically gain little from the laws governing the curvature of space and time, or from those pertaining to quarks. Equally, the cultural sciences no longer profit from the laws of geophysics or from those explaining chemical bonds. In all these cases the tips incorporate the ultimate theory which, as discussed earlier, requires no further explanation. Figure 5.5 sketches this constellation.

The bases of the two pyramids adjoin one another, in most cases at the mesocosmic level accessible to direct observation. More precisely: in the range of our reality, where we still feel ourselves to be active players. Astronomy, of course, is an exception. The mesocosmic level encompasses a broad spectrum of disciplines on which this book can focus its attention and address the relevant questions. This is because analyses typically start in a more immediate, familiar sector in which materials and forms are tangibly perceptible—where 'elementary observations' are possible. In genetics and systematics this represents species and individuals. In comparative anatomy, histology and cytology it represents organs, tissues and cells, respectively. In the inorganic and cultural sciences, substances and artefacts are involved.

The proposal to view the relationships of the two pyramids in this manner is a didactic one, designed to offer the simplest model possible. Three additional arguments support the model. In most sciences, the base of the pyramids corresponds to (1) the greatest breadth of their phenomena (the content of the classes), often also to (2) the starting point of the questions being raised and, finally to (3) the approaches to the respective disciplines from a history-of-science perspective.

Recall the insight gained from the 'psychology of explanation' (Sect. 5.1.4; Fig. 5.2): identified coincidences or correspondences are considered to be explained only when they arise along with others from a higher-level coincidence. The ultimate level or authority that all explanations can invoke is positioned at the tips of the pyramids.

Clearly, the laws governing levers and pendulums were described before those explaining quanta and the speed of light, just as the laws of heredity were discovered before those of the genetic code and evolutionary constraints (see also the 'three pathways'; Sect. 5.3.3).

(b) The matter is somewhat more differentiated with regard to the reciprocity of the explanation in both pyramids (compare Fig. 5.5). The discussion on the cognitive process (Chap. 4, Sect. 4.2.1 (b), along with Figs. 4.6 and 4.7) already

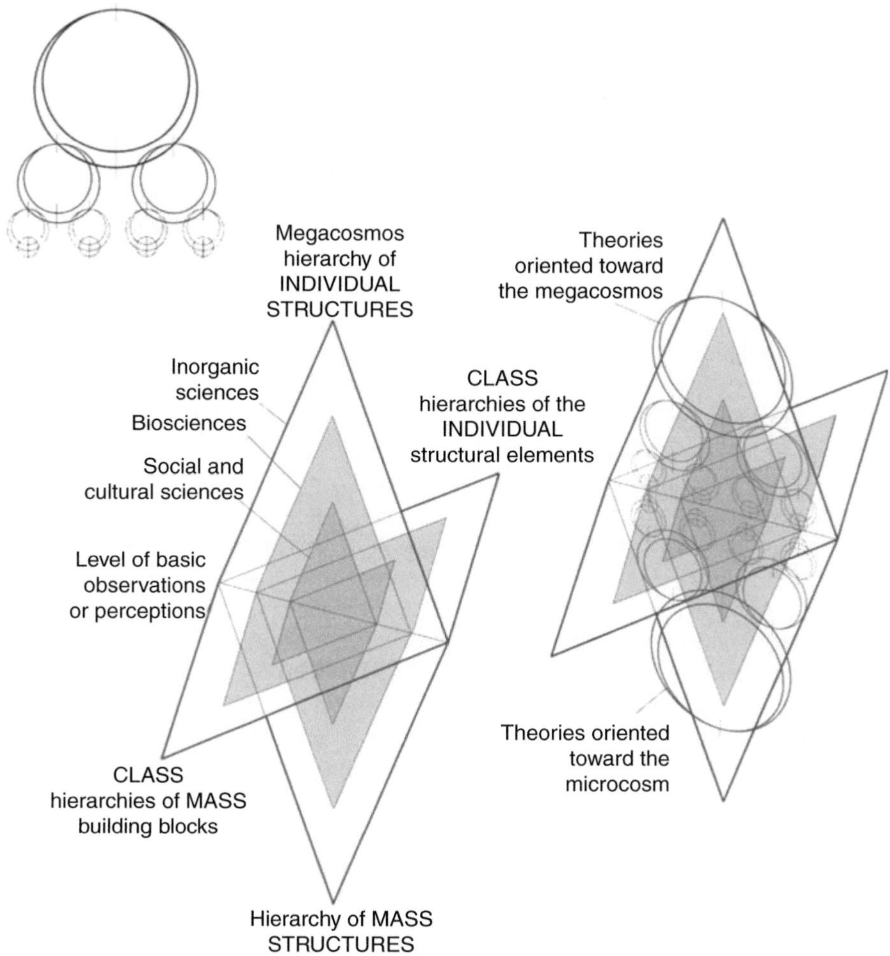

Fig. 5.5 The structure of the pyramids of explanation. Note that the hierarchies of theories of individual structures are juxtaposed with those of mass or standard structures, that the laws of physics encompass those of biology, these in turn the laws of social and cultural sciences, and that all the class hierarchies are aligned with them as depicted in relation to cognition in Figs. 3.11, 3.12, 3.13, 3.14, and 3.15. The sketch (top left) recalls the theory constellation introduced earlier in the context of the cognitive process (in Figs. 4.6 and 4.7)

revealed that resolving a level in a complex system can (and very often must) proceed from two sides, i.e. based on case examples from the next-higher system as well as on cases from the next-lower system.

This requires closer examination. First because it taxes our rational way of thinking and, second, because the process, as a hermeneutic circle, was long misunderstood as untenable circular reasoning.

In the bio- and cultural sciences this reciprocity also repeatedly surfaces in the explanatory process (Chap. 6). The situation is different in the inorganic sciences for historical reasons (see also Chap. 6, Sect. 6.3.1).

5.3.3 The Three Pathways: Assuming, Explaining and Emergence

The discussion on the cognitive process and on generalizing showed that the paths of perception and cognition—along with the reinforcement of generalizations—are related to how the respective objects emerged. Insights are reinforced parallel to the expected developmental pathway. The cognitive path runs in the opposite direction (compare Fig. 3.12). What seems to be a mere peculiarity gains gravitas in the context of explanation below.

It turns out that the same pattern repeats itself both in explaining and understanding. The development of assumptions, such as when case examples from a particular level lead to a hypothesis or theory at the next level, runs from the base of the pyramids to their tips. Conversely, the gained confirmations of the explanations run from the tips to the bases (Fig. 5.6).

The top-down direction once again defines how we conceptualize the development or emergence of entities: this encompasses the most basic constraints and fixed patterns of differentiation and extends to the remaining degrees of freedom in a particular configuration. Accordingly, the explanation follows the emergence of the explained objects.

The theory capping a pyramid's tip corresponds to the oldest law of the respective macro- or microworld. Thus, the laws of gravity and quanta must have preceded any material configurations. Equally, the laws of fitness and mutability must have preceded all organismic differentiation, and all laws of human communication and other faculties preceded any cultural evolution.

The succession of explanations that we apply to things reflects a recapitulation of their genesis.

Although initially surprising, this becomes abundantly clear upon recalling the insight (outlined in Figs. 4.2 and 5.3) that every differentiation must have arisen through insertions between existing constituents and a pre-existing environment. The trajectories of the four *causae* admirably fit into the same model (compare again Fig. 5.6).

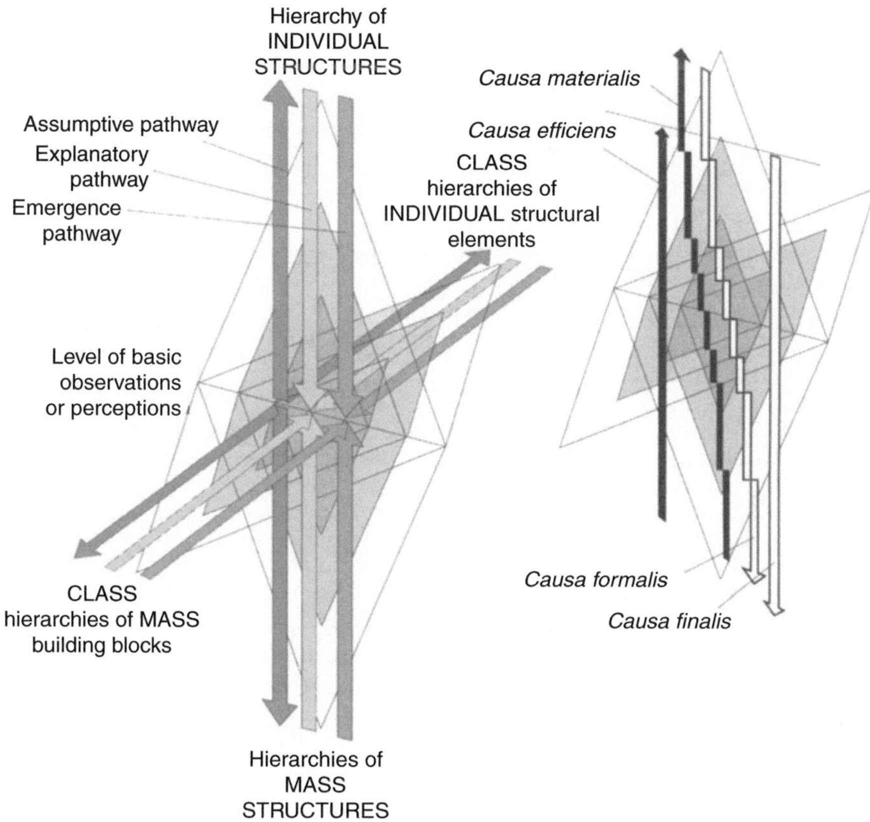

Fig. 5.6 Relationships and symmetries of the 'three pathways'. The development of theories via assumptions broadly leads from specific cases of immediately perceptible phenomena to the opposite end, namely to the theories in the structural and class pyramids. The emergence and explanatory pathways run in the opposite direction, from the most comprehensive theories to the cases. In this sense the explanatory pathways 'recapitulate' the emergence pathways, from the fundamental conditions of the cosmos to the realization of all its structures. The four causae traverse the entire system, the *causa efficiens* and *finalis* only up to the point up still involving life phenomena

5.4 The Forms of Understanding

Among the forms of explanation, understanding must be distinguished from its impostor cousins. This is not because the underlying principles are entirely different, but because a new terminology now governs the discourse. In fact, all forms probably represent special cases of a general method.

Structurally, following Kutschera (1972), the discussion here examines (Sect. 5.4.1) teleological explanation, (Sect. 5.4.2) our understanding of actions and (Sect. 5.4.3) the concept of understanding in the humanities.

5.4.1 On Teleological Explanation

The earlier discussion underlined the necessity of replacing teleology (as used in philosophical systems) with teleonomy in the scientific realm. Much like replacing astrology with astronomy. The long history of the concept, and its special status as a counterpoint to causal explanation, warrants a brief introduction.

Teleological explanations had become a catch-all for everything bearing any semblance to aims, reasons, purposes and functions and pointing to a 'telos', a goal (understood in the previous chapters as causa finalis). Epistemologically, however, there is a differentiation that merits attention.

Again, Kutschera's (1972) structure is followed here, (a) presenting three forms that help delimit the term and (b) juxtaposing teleology with the term entelechy.

(a) Delimiting the term is fundamental to the concepts of explanation, which also need sorting here.

The distinction is between (a1) the functional, (a2) the intentional and (a3) the determinative-teleological explanation.

- (a1) Functions need not necessarily harbor intentions: an organ, or the component of a machine, serves a clear purpose without itself being conscious of doing so. Functional teleonomic explanations have their limitations, namely at the point where the effect of a function can be identified but not its purpose. This can reflect ignorance, hidden or lost purposes, as well as luxuriating developments (e.g. hybrid vigor in biology).

  The tiny canine tooth in the jaw of elk (which some hunters like to stick on their hats) was long held to be a useless rudiment—until researchers discovered that elk still use them in threat displays (despite the modesty of these erstwhile weapons). A mill wheel upscaled in a rustic-style dining room has also changed its function, namely for the edification of the guests. A tinkerer's 'useless machine' merely serves for the tinkerer's own pleasure.

  The English language unfortunately has no perfect equivalent for the German 'funktionale Zwecke' (literal translation: functional purposes). This has led to considerable misunderstandings. Although one can say 'the purpose of life', that already reflects a strong literary bent. 'Aim' and 'purpose' refer more to 'intention' (German: 'Absicht'), to conscious purposes.
- (a2) The term 'intentional explanation' refers to something we experience or interpret as intention, i.e. by knowing or supposing that a purpose is being consciously pursued. This has its limitations as well. On the one hand because intentions may be subconsciously motivated, on the other because the intention may never attain its goal. This can reflect 'desiring to' or 'desiring but being unable to', or sometimes even 'desiring to but not actually attempting'.

  Intentions are already rooted in animal behavior. In primates they gradually merge with states of consciousness and, in humans, they liberate themselves (but never entirely) from the subconscious drives. The transition from functions to intentions is gradual.

- (a3) Finally, determinative-teleological refers to explanations based on more underlying motivations. Such 'determinative' causes can go back as far as the reasons that gave rise to (or probably gave rise to) some stance or a certain action. The message 'mother called' does not reveal the purpose of the call. In many cases, however, that purpose can be surmised based on knowledge of the situation or on background information.

  Returning to the above-mentioned limitations: the boundaries lie less between the three concepts themselves than in defining the type of scientific explanation accessible to us. Importantly, whenever causes can be sufficiently and reliably identified, the strength and singularity of teleological explanations wane. Their particularities are mostly language-based—they ultimately blend into a family of causal explanations.

(b) The term entelechy is singled out here because it was long held that no causal explanation could be provided for purposefulness, prompting the entelechy concept. In the sense of Aristotle, the word designates 'that which bears its goal within itself'. The concept was born out of amazement at the purposefulness of organisms and of animal behavior. And this clearly explains why, for centuries, the creationistic explanation of the world interpreted this as reflecting the purposes of God rather than involving mechanistic causes. The theory of evolution and genetics marked the turning point. Nonetheless, the natural philosophy of the modern era (e.g. Driesch 1908) still held room for a 'process of organized wholeness' (German: 'ganzheitsstiftenden Prozess'), which operated only in the organic world.

    This perspective could not be brushed off lightly: it was by no means clear how random mutation, combined with adaptation to the environment, could create the internal and external order of organisms. Chap. 6 of this book shows that the causal explanation for this required expanding Neodarwinism to incorporate major recursive causal relationships in the framework of a 'systems theory of evolution'.

    Finally, it is self-evident that organisms do carry their 'goal' within. From the very onset, only those operating on successful, targeted programs survived. The concept of entelechy superbly confirms our expectation of complex, systemic causal relationships.

5.4.2 *Understanding Actions*

'Motivations' are again the issue here, both one's own but especially those of others: how can they be recognized and do they represent causal explanations? This topic is especially interesting in the social sciences and law. It is further complicated by the methodological problem of introspection and projection (touched upon in Sect. 5.2.4 (b)) and by the general issue of free will.

The methodological problem kept mostly the psychologists busy, but clearly had deeper roots. Back in 1963, Konrad Lorenz already published an article addressing the question: “Do animals experience subjectively?” (‘Haben Tiere ein subjektives Erleben?’). On the first pages he states that, were he to know the answer, he would actually have solved the mind-body problem. He then concedes that he had to assume such subjective experience in order to able to understand his animals. This hits the nail on the head.

The psychologists who struggled to turn their discipline into an exact science were informed by behaviorists that introspection was unacceptable as a method. Projection, in turn, was relegated to the depth psychologists. How, then, can we know why another person has acted in a particular manner? This is a vital assessment in everyday life, one that is being continuously made and in most cases proves to be correct. How, for example, can I recognize that a person is panicking? Clearly only introspectively by asking how I would experience myself if I acted like that person, and then projecting that feeling back.

This may not be the most exact of methods. After all, I could misinterpret that action, falsify my introspection and then cumulate those errors into the projection. Nonetheless, this is how life works, and successfully at that. It simply involves inductive heuristic trials tempered and optimized by controls against supposed background information. In other words life experience.

Rationalistic approaches, in turn, additionally conjure up the issue of free will. Determining whether someone is acting rationally requires presupposing that that person considers his or her motivations to be rational. Simply put, that person must be perceived as being rational. This assumption does not always hold true. And even if it is made: did that person make the decision freely or did he or she act out of compulsions, phobias or suppression? In such cases, can motivations still be recognized as causes?

From this perspective, not all causes of an action are based on motivations. The causes of actions, in turn, yield motivations. From the present perspective, rationally experienced motivations represent only the outer layer, concealing a variety of entirely causal (even if concealed), ultimately detectable causes of actions.

The concept ‘theory of mind’ has become established in the more recent English-language literature on the behavior of primates and humans. It posits that the capacity for belief can be inferred from the behavior of another creature (overview e.g. in Baron-Cohen 1995). An effort is made to avoid the term ‘consciousness’ because, from a behavioristic perspective, it is hard to grasp. This book, in turn, argues that ‘experimenting in imagined space’ can be more easily demonstrated.

A customary German term in ethology is ‘Du-Evidenz’ (direct translation: ‘thou-evidence’, describing the fact that understanding can be achieved between humans and higher animals). Discussions with Konrad Lorenz even raised the thought that such evidence-based experience could be brought into the domain of the *a priori*. Methodologically, this approach differs from that of the ‘theory of mind’ in accepting introspection. It is difficult to see how we can understand another creature without introspection and projection on our part.

5.4.3 *Understanding in the Humanities*

The humanities have received mention three times in this book. First (Chap. 2, Sect. 2.3.3 (a)) in the problematic distinction between explaining and understanding. Recall that the natural sciences were thought to be characterized by their explanatory method, as differentiated from the humanities with their understanding method. The terms applied included 'nomothetic' versus 'idiographic', i.e. establishing laws versus describing historical facts. Nonetheless, biology also describes historical facts, and, epistemologically, describing and explaining turned out not to be acceptable alternatives.

Chap. 4, Sect. 4.2.1 (d) examined these methods and determined that the subsumption method and hermeneutics, understood methodologically, are structurally equivalent. This makes it useful for science as a whole to explore this methodological agreement (Riedl 1985).

Of course, the natural sciences and the humanities are differentiated largely based on their topics, which in the latter are mostly restricted to society, culture and cultural history. Nonetheless, physics and biology also differentiate themselves according to their topics, and anthropology along with psychology and ethnology can be pursued both with natural science and humanities approaches. All such trans-boundary disciplines have tended to strive toward a natural science approach. This reflects the increasing demand for more precise treatments. Moreover, the natural sciences have increasingly tended to embrace topics in the humanities, such as human and cultural ethology (Eibl-Eibesfeld 1978; Koenig 1970).

These attempts at a topic-based segregation of disciplines are all good and well. More important is recognizing that all laws governing deeper levels extend up through to the higher ones as well. Although the lower tiers are themselves insufficient to explain the higher levels, they are definitely necessary to explain the forces and materials. At the same time, recall the reverse direction: the conditions of the upper tiers work down through the lower ones as well, for example a political system down through to society, or the structure of a psyche triggering damage to the nervous system (e.g. Ringel 1997).

At the same time, as human beings, our make-up enables a more intimate, 'empathetic' relationship with the actions and artefacts of our conspecifics: we experience them more vividly than, let's say, chemical processes. At the same time, gradients resembling human empathy have been reported from bacteria to chimpanzees. The humanities, of course, deal with higher realms of complexity. There, the gradient extends from microphysics to cultural history. This makes causal explanations all the more difficult.

In philosophical parlance this has been termed 'absolute understanding' as opposed to explaining the 'blind causality of nature'. This, however, introduces pathos, which is less informative than the insight that our make-up is obviously tailored to dealing with humans. The difference lies more in our intellectual history, in the rationalistic-idealistic versus empiricist-materialistic axis. This insight was our starting point (Chap. 2, Sect. 2.1.3 (c), Figs. 2.4 and 2.5) and we used it to more closely examine the cultural history of the explanatory modi (Sect. 5.2.3 (b)).

In my opinion, the complexity, the tangible proximity and the experiencable relevance of topics in the humanities argue for neglecting the physical 'lowlands' of the lower tiers in the hierarchy. At the same time, we have all too often settled down in comfort into something resembling conjectural empathy (somewhere in the realm between theory and literary endeavor). Should we disapprove of such poetry? After all, it may even have given rise to certain sciences, for example like Dostoevsky's novels helped pave the way for psychoanalysis. There is also nothing wrong with the cognitive methods, which are described here in detail as the ratiomorphic, superbly prepared precondition of all explanation. The tenacious mix-ups are what has protracted the confusion.

The following chapter deals with the structures of the explained and of the understood and introduces these very same issues on yet another, third level.

Chapter 6
The Structure of the Explained and Understood

The next step involves going from systems to structures, i.e. from the process to the product. The organization follows that of Chap. 4 in order to simplify comparisons with the treatment of the 'recognized' and the 'explained' there. That chapter juxtaposed the theory of the structure of the world (Chap. 4, Sect. 4.1) with a theory of its explainability. The order of things (in Sect. 4.2) was juxtaposed with the order of the causes. The separation into the principles of morphology (Sect. 4.3) and of systematics (Sect. 4.4) coalesces again here in the explanatory context.

The discussion on the path to a (Sect. 6.1) dynamic explanation of the world is followed by (Sect. 6.2) ordering the causes. These are categorized based on the objects into (Sect. 6.3) patterns of the explained and (Sect. 6.4) patterns of the understood. This approach parallels the division into the natural sciences and the humanities.

6.1 The Path to a Dynamic Explanation of the World

The paradigm of evolution has successfully established itself as an explanatory principle in virtually all disciplines. Beyond the biologists' theory of evolution, chemical evolution as well as an evolution of the cosmos and of the Earth are accepted concepts. Moreover, the evolutionary perspective is also being applied to the development of 'cognitive apparatuses' up to and including that of humans, to the development of social systems, of languages and of the sciences. This reflects the realization that, at all these levels, the conditions of the preceding levels of organization provide key elements of explanation. Nonetheless, the issue is multi-faceted and continues to spawn controversies (treated later in this chapter). The task here is to present the broadly accepted 'state of the art'.

R. Riedl, *Structures of Complexity*,
https://doi.org/10.1007/978-3-030-13064-0_6

This development is useful from two perspectives. For one, it provides a framework to approach an overarching theory of evolution and helps to identify generally valid principles. Secondly, it provides an opportunity to once again test the subtheories based on the respective higher-level theory (Riedl 1976, 1982).

Importantly, philosophy takes a back seat to the empirical sciences here, simply because many philosophical questions have entered the realm of empirical research and yielded testable hypotheses (Stegmüller 1987). The result is a convergence of scientific practice and epistemology—key to the discussion here. Stegmüller refers to a 'pragmatic revolution' because practical experience is prioritized whenever the rational concept conflicts with experience. This also reflects the current trend toward a 'naturalized epistemology' (overview in Callebaut 1993), a framework which encompasses the present topic.

The next step is to further pursue the history of explanations (in continuation of 'cultural history' as outlined in Chap. 4, Sect. 4.2).

The history of the explanatory attempts is presented first, followed by the paradigms (Sect. 6.1.1) of the inorganic, (Sect. 6.1.2) of organismic evolution and (Sect. 6.1.3) of conscious processes. The account is initially restricted to the respective conceptional type because the principles of explanation (Sects. 6.3 and 6.4) can be fully depicted only after appending the notion of causality (in Sect. 6.2).

6.1.1 Origin of the Inorganic

The key issue in presenting each paradigm is to describe and validate the development from the classical to the modern world view.

Accordingly, the (a) static and the (b) dynamic evolutive interpretations are juxtaposed even in the inorganic realm.

(a) The cosmos, above all the firmament, was long thought of as being 'eternal and immutable' by almost all cosmogonies. In Western culture this is exemplified by the cosmogony of the Greeks, which held that predeternined primeval forces, along with the fantastic stories of the 'Castration of Uranos' up to the birth of Zeus, gave rise to the Olympian realm of gods (Schwabl 1958). First transitions from mythos to cosmology can be identified in the Ionian natural philosophers, but the cosmos itself remained frozen in unchanging order—at least in the 'translunar' realm (beyond the moon). The astronomy in the Ptolemaic worldview was equally rigid: crystal bowls in which the stars are embedded circle the Earth. The story of creation in the First Book of Moses also depicts a world that is predefined for all time.

    The shift from the geo- to the heliocentric worldview changed little in this respect—up until the discovery of the formation of new stars. This prompted a fundamental controversy in which Kant and Laplace posited a dynamic cosmos. Nonetheless, even the question whether Earth owed its structure to an act of creation or rather to a developmental history was first decided in favor of development around 1830 by Charles Lyell.

Characteristically, this transition to the heliocentric worldview, designated as a ‘Copernican revolution’, triggered perilous controversies (for the proponents). Interestingly, when the sun was later relegated to a marginal position along the outer arm of a galaxy, this was accepted without much ado. Dethroning the ‘crown of creation’ had been the problem; once that was digested, the physical distance was no longer such an issue.

(b) Today, this concept of the development of the world has led—via the discovery of the so-called ‘Hubble effect’—to the notion of an expanding cosmos and with it the Big Bang theory. This requires transcending the imaginable because the source of the explosive expansion of energy eludes scientific study, and the origin of space and time from nothingness is simply unimaginable. Much the same-sized step, by the way, was required of us by Einstein’s concept of curved space, by the wave-particle duality in microphysics, as well as by the uncertainty principle. This brings us front and center into the present century.

The notion of the origin of matter, in turn, presupposes a cooling cosmos whose expansion is decelerating. This would enable the development of successively lighter quanta, i.e. the origin of matter (the demonstrable ‘capturing’ of an electron by a proton to form hydrogen).

Regarding the planets, a plausible theory postulates that their heavy elements, stemming from cosmic catastrophes, condensed in the marginal vortices of the proto-suns. Equally, theory holds that the crust of our Earth solidified from a magmatic state, and that the atmosphere initially consisted of a thick layer of hydrogen that, in combination with the crust formation, gradually became composed of the most hydrogen-rich gases: methane, hydrogen suldife and water vapor. These are the conditions under which life is posited to have arisen about 3.5 billion years ago. And the consensus, supported by several pieces of evidence (Urey 1952; Eigen and Winkler-Oswatitsch 1975), is to accept that life originated in accordance with the laws of chemistry (see below).

The laws of physic continued to be held as being ‘perpetually’ valid. The explanation is that there were simply no grounds for devising alternative laws of nature. The dominating role of mathematics in physics may also have played a role because many mathematicians leaned towards a platonistic validation of their discipline. This position has shifted only recently (Thirring and Stöltzner 1994).

Important in this respect is the insight that the cosmos, counter to the laws of entropy, can experience a differentiation into complex order (as outlined in Chap. 2, Sect. 2.3.2). Semi-stable systems, far removed from equilibrium, can arise in the environment when they develop into higher forms of organization at the expense of their surroundings (Prigogine and Stengens 1990). This notion was introduced for the organismic realm as anagenesis and is recognizable even in the cultural realm.

Three issues need to be addressed: (i) our recognition of the historicity of the cosmos, (ii) a pervasive, mechanistic explanatory optimism, and (iii) the acceptance of limits to our powers of imagination—a worldview as poor in contradictions as it is full of empirically untestable preconditions.

6.1.2 The Origin of Organisms

At issue here is the transition from the static concept of creation to the dynamics of an evolutive worldview. This closely resembles and is almost concurrent with the above 'inorganic' shift. The transition was basically one from creationism to evolutionism.

This calls for juxtaposing the static (a) with the dynamic (b) interpretation.

(a) The early notions of the origin of life were also static, attributed to the intervention of demiurges, in Christianity to a Creator God. Interestingly, the same sequence is recognizable both in Greek mythology and in the biblical story of creation. For Aristotle, for example, stones were completed before the differentiation of plants, these before the mobility of animals, and the latter before the human soul. The days of creation in the bible also clearly describe genesis as proceeding from the inorganic realm over to vegetation and creatures and ultimately to humans. This may well have introduced a notion of kinship or comparability (the controversies surrounding our relationship to apes and monkeys were rife even back in Antiquity). The overall picture, however, remained static.

Reflections on development are documented as far back as the last century BC. In his didactic poem 'De rerum natura', Lucretius appears to have envisioned extinction, transformation and even the process of selection. Thereafter, however, Christian doctrine stifled the explanatory principle well beyond the Renaissance, attributing creative powers to rock formations and the oddness of fossils to the Flood.

(b) The transition to a dynamic worldview then took place within less than half a century, inspired by the Age of Enlightenment as advanced by Maupertuis' 'Venus physique' (1745) and Lamarck's 'Philosophie zoologique' (1809). Initially this involved free thinkers such as Pierre Louis Moreau de Maupertuis (1698–1759), diplomat, physicist and mathematician. As president of the Prussian Academy of Frederick II, he already published under the pseudonym Dr. Baumann because, back then, mechanistic views of human beings put their authors in peril.

Independent of Maupertuis, other pioneers followed, including George Louis Leclerc de Buffon in Paris and Erasmus Darwin, Charles Darwin's grandfather, in southern England. Buffon, Grand Seigneur of the Imperial 'Garden of Plants', was the skilled author of many tomes (as of 1749) on comparative anatomy and had a clear concept of the relationships among species. Importantly, he was Lamarck's teacher. Erasmus Darwin, country physician, a poet well versed in nature, published didactic poems such as in 'Zoonomia, or the Laws of Organic Life' (1794). He was influenced by David Hume, embued with a vision of changing species and once again with first inklings of the principle of selection.

Finally, Jean Baptiste de Monet de Lamarck—monastery school pupil, decorated officer, taken ill, accepted as a student by Buffon—produced the first clear draft of a theory of descent as well as a first concept of causality. This dichotomy is emphasized here to draw attention to the amalgamation of both efforts that characterized the subsequent history of theories.

The key problem that the natural sciences of the time were called upon to solve was the age of life on Earth. The chronology of the bible was still binding:

based on the genealogy of the family tree after Adam, it prescribed a time period of somewhat over four thousand years. Documents covering one to two thousand years were already available back then. Since the species had not changed over that period, it was clearly necessary to extend the time axis. Many clues were available. Lamarck therefore also devoted his work on 'Hydrology' (1805) to this issue, advancing the argument for a need to entirely discard the time dimensions based on human lifespans. This suggestion was the first in science to reject the imaginable as a measure.

The decisive point in Lamarck's concept is a clear recognition of the transition from perceiving natural order to explaining it, specifically the need to add an explanation (an insight also underlined at the end of Chap. 4, Sect. 4.4.3).

In his 'Philosophie Zoologique' (1809) he wrote: "How could I have perceived the peculiar gradation in the organization of animals, from the most incomplete to the most complete, without asking about the cause (sic!) of such a positive (unequivocal) and important fact, a fact that seems supported by so many pieces of evidence". He continues by noting: "Would I not have to assume that nature has given rise to the different organisms successively, progressively from the simplest to the most complex; … It developed the organization stepwise; in that these animals generally spread to all habitable parts of the Earth, each species attained the habits (behaviors) and the modifications of their parts that we observe today simply by the influence of the conditions in which they found themselves." (taken from the translation into German by Heinrich Schmidt 1909, with parenthetic matter added by the author).

The types of causes that he names—along with the controversies surrounding them—are discussed below. Again, three insights are highlighted: (i) the theory of descendence was founded and was merely further solidified over the following two centuries; (ii) a firm conviction about scientific explainability had been added, and (iii) warnings were issued about the powers of human imagination.

A second 'Copernican revolution' was in the making, but Lamarck himself had not yet reached that crossroads. When Charles Darwin's teachings become popular fifty years later (slipping human beings into the animal kingdom), the controversy enflamed anew. Ernst Haeckel, who loudly supported this revolution, was even proclaimed as the anti-pope by the free thinkers of the 'Monist League' in front of the memorial to Giordano Bruno in Rome.

It would be incorrect to state that the controversy between the new evolutionists and the creationists has been fully resolved today. Although our place in the cosmos has been accepted, our place in the animal kingdom has not yet gained full acceptance. The skirmishes, however, have become rarer and often drift into the bizarre.

6.1.3 Paradigms of the Origin of Reason

The phenomenon pursued here revolves around the terms 'consciousness', 'rationality', 'reason', 'soul' and 'spirit'. All are interrelated and all are rather indeterminate. Of these five, four are treated here in more detail (omitting spirit). The Greeks initially applied the latter to a waft of air, then to breath, to the bearer of life, a breath of life that can temporarily or permanently leave the body. Romanticism then juxtaposed it with nature.

Somewhat simplified, consciousness is taken to be the ability to purposefully reflect upon memory content and upon oneself (Riedl 1987a). The matter becomes more complicated with soul: not only did the definition change within the Greek period itself, it takes on different meanings in everyday language, in philosophical discourse and in religion. Colloquially it might be taken as the embodiment of all conscious stirrings. In religion the soul—segregated from the material world—is considered to be immortal. The various philosophical schools take on both positions as well as all those in between.

The terms reason and rationality have even interchanged their meanings. Based on Kant, reason today is considered to be the ability to establish terms, make judgments and create rules a priori based on perception—as an activity providing material for rationality. Rationality, in turn, is to be understood as an intellectual capacity of a culture, its ability to generate wholeness and values. This makes it a predetermined principle in the world and for the soul—a god-given faculty.

This spawns the recognition that rationality and consciousness must have an evolutionary history because animals form concepts and can exhibit simple states of consciousness. The term soul is too broad. Scientifically, it can be understood as merely being the embodiment of all conscious stirrings, whereby reason is culture-dependent and superimposed upon rationality, perhaps already helping to promote species survival.

This definition of reason is central to the third shot at the transition from (a) a static to (b) an evolutionary, dynamic status.

(a) The paradigm of reason as a given principle was already raised by the early Rationalists. It is attributed to a world soul, one that is eternally beautiful and good and which the human soul can partake in. The transition of this concept through Plato, Cleanthes and Paul into Christianity was outlined above, as was the dualistic worldview (Descartes) as the necessary consequence (see Chap. 5, Sects. 5.2.2 and 5.2.3 as well as Fig. 2.5).

    Briefly summarizing the history of 'Christian philosophy' helps to better understand this position in its modern-day form. It begins with the church fathers, gains shape with Augustine, differentiates itself in Scholasticism, weakens in the Renaissance and regains influence in the Reformation (as Neuthomism) and with the papal decree as the Philosophia perennis. Many prominent personalities who are open to the world yet religious lean toward that position.

(b) The paradigm of reason as an evolutionary principle was first alluded to in Darwin (1871). The concept was bolstered by the theory of descendency as advocated by Ernst Haeckel along with Mach and Boltzmann, who were interested in cognitive issues.

    Haeckel wrote (1868, p. 88), explicitly regarding Kant: "He views the soul of humans, with their innate faculties of reason, as a complete, given being… He gave no thought to the notion that this soul could have developed phylogenetically from the soul of the most closely related mammals. The wonderful ability to arrive at insights a priori, however, originally arose via inherited brain structures that were gained gradually and stepwise by the vertebrate ancestors of humans through adaptation to synthetic linkages of experience, of knowledge a posteriori."

Lorenz was unaware of this passage in Haeckel, but his work of 1941 presents this idea more explicitly. This insight (Chap. 2, Sect. 2.1.3) marks the starting point of the present discussion. In fact, Lorenz already full grasped all the material on the evolution of 'cognitive apparatuses'—from simple organisms to humans—at the time of his professorship in Königsberg, still under the aura of Kant as it were. His 'Russian manuscript', which was rediscovered only after his death and published by his daughter (1992), clearly presages his monograph (first published in 1973).

It is revealing that Lorenz initially referred not to a 'theory' of cognition but rather to a 'natural history' of humankind and later to a 'teaching' or 'discipline' of cognition (German: 'Erkenntnislehre'), i.e. to a didactic construct involving a particular constellation of objects, concepts and their application. Donald Campbell (1974) was then the first to speak of 'evolutionary epistemology', and Vollmer (1979), who built on this, sought to validate its elevated rank. The term has been retained until today (German: 'Evolutionären Erkenntnistheorie'). Firstly, because it goes beyond mere doctrine to enable causally anchored prognoses underpinned by theory. Secondly, because it validates why and how human reason fits into this world, something we expect a cognitive theory to deliver.

Looking back, all our attempts to explain the world suddenly became dynamicized. In the wake of this development, cognition as a discipline also took a 'pragmatic turn' and became 'naturalized'—a perspective advocated in this book. In all the significant theoretical frameworks, the static (often even deterministic-creationistic) theory defining our nineteenth and twentieth century understanding of the world transformed into an evolutive one with all its attendant changes (early overviews in Riedl 1976; Jantsch 1979).

Such upheavals that encompass our overall worldview are rare in cultural history. Karl Jaspers (1957) identified one such shift that marked the transition from a mythic to a critically reflective concept of the world in all advanced civilizations between the seventh and second centuries BC. History will judge the import of the present shift.

6.2 The Order of Causes

If the shift to evolutive explanations is interpreted as an expansion of the old, static perceptions, then, oddly enough, the shift in the causal concepts turns out to represent a constriction or reduction. This has repercussions in the history of science. Adequately explaining complex and therefore usually historical systems requires further differentiation.

This calls for (Sect. 6.2.1) separately examining the reduction of the causal concepts, then (Sect. 6.2.2) juxtaposing them with the growth of historicity and finally (Sect. 6.2.3) examining the relationship between the four cognitive forms of causality and the four physical interactions.

6.2.1 The Reduction of the Concepts of Causality

Again, the discussion benefits from briefly touching upon early cultural history, for example on pre-Socratic philosophy, which yielded the first documented frameworks for explaining the world: Thales, Anaximander. The sentence "Just as our soul, being air, holds us together, so do breath and air surround the whole world" is attributed to Anaximenes from the sixth century BC. These thinkers were referred to as 'hylozoists'—from 'hyle' (wood or forest). Later, Aristotle sought the roots of the designation in 'as yet unformed primordial matter' and 'zoe' (life). In their interpretation everything was animated by breath and soul. This notion of the inseparability of soul and matter was further spun by a broad spectrum of thinkers including Giordano Bruno, Diderot and Goethe. Today, this would be referred to as a 'systemic relationship', which expresses itself in the inseparability of a cognitive dualism, namely of information and energy.

Chap. 5, Sect. 5.2 outlined the subsequent history of philosophy, with its split into an empirical and a rationalistic axis. The task here is to examine the history behind the reduction in our notion of causes with its split into materialism and idealism. This development is tied to the two cognitive dualisms or symmetries behind Aristotle's distinction of four causes. At issue is their reduction, specifically into two mutually exclusive solutions.

The axis to the materialistic explanation of the world (a) is centered around the so-called 'Galilean revolution', addressed in a bit more detail here because of its relevance for modern science. The axis to an idealistic explanation of the world is (b) centered around Christianity, in the philosophical-cultural history tradition. The (c) materialistic and (d) idealistic manifestations of these reductions are then summarized.

(a) The Galilean revolution is at the center of the reduction process but also accompanies the whole history of philosophy and of the sciences.

Shortly after Aristotle, the opinion arose that the master himself had considered the causa finalis to be the most fundamental cause behind all other forms of causality. This notion, consolidated in Scholasticism, extends up to the present day despite having been repeatedly refuted by competent workers (for example Kullmann 1979).

The subsequent interpretation, known as 'Aristotelianism', was chiefly cultivated and promoted in early Arabian and Jewish spheres, for example by Averros and Maimonides. The Christians, however, among them Albertus Magnus and Thomas Aquinas, adapted and strongly modified the doctrines. This was the Aristotelianism that determined the thinking at the time of Galileo and Kepler. Although other interpretations also reached southern Europe (via Byzantine scholars), Galileo considered the conventional Aristotelianism of the day to be blather and a hindrance for research.

This prompted him to liberate himself of the entire baggage. The goal was to measure and make measureable. This forms the second nucleus of the 'Galilean revolution' (see also Chap. 5, Sect. 5.2.3). Material and formal causes played no role in Galileo's and Kepler's efforts. Not only did the causa finalis need not be addressed, but, due to religious dogmas, it was actually off limits.

In fact, the writings of Galileo and Kepler make no mention of the causa efficiens either, although both were aware of Aristotle's texts. Only one form of explanation was involved and there was no cause to differentiate it more precisely. This position undisputably initiated the development of a new chapter in the natural sciences and—along with the Reformation, the printing press and the discovery of America—ushered in what is commonly known as the 'modern era'. It also coincided with a constriction of the explanatory concept.

This process paved the way for the subsequent currents—the enlightenment and positivism, along with scientism—and elevated physics to its status as the current scientific ideal.

Enlightenment, founded on reason and anti-metaphysical in nature, advanced rationalism and trust in science. In England this took place with religious and political undertones since the sixteenth century, in France on a societal level flavored with moral critique beginning in the seventeenth century, and in Germany as of the eighteenth century. The overall sense was a quest for the direct causes explaining the world rather than for past conditions (in which human interventions are no longer possible).

This current continued in the positivism advanced by d'Alembert and Turgot. It was anti-philosophical and even anti-rationalistic. Comte, for example, held that only those insights verifiable by experience were positive, true facts. Today, we concur with Popper that experience and verification invoke different demands. Experience represents 'assumption knowledge' backed by sufficient degrees of confirmation. Verification, in contrast, appears possible only in the sense of a 'proof' in an axiomatic-deductive framework, for example in mathematics.

The optimism of that time nurtured great expectations. A case in point is scientism, which held everything to be solvable scientifically and that encroached on the cultural sciences by applying methodologies of the natural sciences. Physicalism, in turn, advocated that everything that could not be derived based on physical methods and concepts was unscientific. These are, of course, extreme, untenable positions. Physics as a scientific ideal, however, has remained. In all natural sciences, causes are now ultimately thought of as being forms of 'energy conversion' (see Sect. 6.3.1 for more details).

(b) The second axis is driven by Christianity but, as noted above, was already set out by pre-Socratics such as Pythagoras and Parmenides and expanded upon by Plato. This paved the way, via Cleanthes and Paul, for the idealism underpinning church teachings.

This community of faith initially spread across the Roman realm, then across all of Europe, defining our culture. Islam and Judaism left only few traces therein. Recall that Scholasticism differentiated (via the causa finalis) the most fundamental cause behind all things as being the causae exemplares of the Creator.

Naturally, this notion of the world's primary purposes took on many forms, including secular ones. Examples include the 'subjective' idealism advanced by

Descartes and Malebranche, the 'objective' German Idealism of Schelling and Hegel, the 'magical' idealism in Romanticism advocated by Rudolf Eucken up into the twentieth century, and finally the 'New Idealism', in which the concept of God simply referred to absolute spiritual life.

The present topic is only minimally affected by such differentiations, but the message is that this notion of primary purposes and therefore of a 'purposeful, goal-directed world order' stretched across two millennia. Even today it continues to underpin the worldview of certain currents in philosophy and in the humanities.

This two-sided reduction in the concept of causality is driven in both of its alternative forms by the expectation of a cause of all causes. This might seem natural if we were unaware of the double cognitive dualism that formed our starting point. And it also underpins the carefree interventions of the natural sciences into the world and the inability of the humanities to muster much opposition.

Both approaches to the world around us tender necessary but insufficient explanations. Any attempt to examine the deficiencies of either position should begin with the two forms of 'reductionism' (a term introduced in Chap. 2, Sect. 2.3.3 (c), Sect. 5.2.3 (a)).

The first step is therefore to differentiate (c) 'materialistic' from (d) 'idealistic' reductionism.

(c) Materialistic reductionism encompasses a well-known controversy. Its three forms merit closer examination in the present context:

These are its (c1) theoretical, (c2) pragmatic and (c3) ontological manifestations.

- (c1) Its theoretical form is characterized by the legitimate (and necessary) manner in which we condense similar entities into concepts, hypotheses and theories. It also involves the manner in which we make theories into cases for higher-level theories. This enables us, despite the limited amount of information we can store in our one-and-a-half liter brain capacity, to quite correctly predict a growing constellation of case examples (or at least to predict them well enough to survive).
- (c2) Pragmatic reductionism can be described as an analytical process, one that drives very successful research for example. Nonetheless, such analyses raise the expectation that the parts can then be combined to form a whole or, put differently, the whole can be sufficiently understood based on its parts, i.e. it can be 'reduced' to those parts. To a limited extent this is actually possible, namely when so-called 'reversible processes' are involved and certain preconditions fulfiled, particularly in the inorganic realm.

This calls for delving more deeply and separately examining three perspectives (i–iii).

(i) The limits of reversibility bring the 'hysteresis effect' into mind, as convincingly represented in the phenomenon of 're-magnetization'. Everyday-life examples also shed light on this: a wire can be bent back and forth repeatedly but, after a certain number of times, it becomes 'fatigued' and breaks. The positions of the molecules change with every distortion.

'Historicity' also appears early on in the molecular realm. Lasers are a familiar example (Haken 1978). Charging a rubidium crystal with energy, for example with light, pushes electron shells into more excited trajectories. If they fall back on other trajectories, the emitted energy excites neighboring trajectories. The constituents of the system mutually disturb each other until, in a 'parliament of molecules', one random direction of disturbance dominates and the laser emits light in that direction. The emission of energy remains a (self-)repeating process, but the direction of the light is already a historical one: the direction of the next emission cannot be predicted. The ideal billiard ball is another case in point (Chap. 2, Sect. 2.3.1 (b)): after only a few deterministic steps (collisions), microphysical indeterminism already introduces a dimension equivalent to the surface area of our macroworld.

(ii) In principle, reversibility is a three-pronged conceptualization.

First, it holds true as long as the exchangeable elements of a system are claimed to be identical or their differences negligible. This abstraction is probably valid even for quanta (Prigogine and Stengers 1990). And it certainly holds true into the levels of molecular genetics, population dynamics and public opinion research, whose respective suppositions are that alleles, individuals and opinions are largely identical and exchangeable.

Second, this conceptual reversibility is valid even when non-exchangeable components/elements are involved, namely whenever the disassembled parts are reconstructed abstractly. This is the case in dissecting an organism, successively removing fossil-bearing layers or conducting an archeological dig at a burial site.

Third, some disassemblages are done entirely abstractly. Examples include the analysis of a painting, of a handwriting specimen or of an era. The re-assembly is no different from an analysis, namely residue-free.

(iii) Where, then, do the deficiencies lie? In the organismic realm, physical disassembly almost never allows re-structuring the whole. An orange cannot be reconstituted by pouring the squeezed juice back in, scoffed Hans Mohr (1981).

Subliminally, pragmatic reductionism represents an idealogical schism between organic and inorganic scientists. Medawar and Medawar (1986) addressed this issue, which is rare enough for reductionists. As justification, they posited that the reductive analysis, "of all the conceivable methods to understand the world is the one with whose help we can best recognize how the world might be changed should the need to do so arise" (p. 267). While that does ring true, it also reveals the underlying current: the issue is human meddling in the world, often prior to any sufficient understanding of that world.

The phenomenon of emergence excludes reduction. The two Medawars acknowledge that this phenomenon is amply evident in biology—and we can add to that 'in all complex, historical systems'. They also argue that the utility of the concept simply lies in "giving a name to that which escapes reductive analysis" (p. 270). While that may be correct definition-wise, it misses the mark. This is because it is an invitation to view historical development as being unimportant. That development, however, harbors all of the many decisions and realized alternatives that every historical system has gone through in order to have become what it has become. Intervening at a particular point may well change some part of the world, but triggers uncontrollable chaos when applied more broadly to historically evolved conditions.

The material and formal conditions, in particular, are destroyed and no synthesis can reassemble them. The situation differs somewhat for a system's driving forces and purposes. Although these both also appear in the guise of the destroyed material and formal conditions, they can at least be reassembled abstractly thanks to the conceptual consistency extending though the levels.

- (c3) The truly damaging variant is ontological reductionism. It encompasses those reductionistic positions which assume or even presuppose that only those things that can be analyzed are in fact worthy of consideration.

  Clearly, any such expectation is absurd. Although explicit support for this stance is rare, it subliminally continues to broadly underlie the development of causalistically oriented, explanatory natural sciences. This phenomenon was already introduced in Chap. 2, Sect. 2.3.3 (c). At issue was the weight or significance we attach to the explanatory, experimental (and even more so to the mathematically treatable) sciences as opposed to the so-called purely descriptive ones.

  Recall the hidden agenda, namely the mantra of the achievable coupled with the pursuit of influence and power. Clearly, ontological reductionism is driven by interests that go beyond the pursuit of truth, even beyond any attempt to understand the world around us. Simply put, it can deceive and mislead.

  Of course, the *causa efficiens*, for example the function of our lungs, can be attributed to the body's energy and oxygen demands, just as breathing can be attributed to the musculature, the alveolae to the effect of large surface areas, cell aeration to the laws of gas exchange, and these in turn to the laws of chemistry and further to quantum mechanics. Conversely, the *causa finalis*, the purpose of the system, can be attributed in the same sequence back upward, namely to the energy demand.

  This, however, yields no knowledge about the materials and forms involved, for example why such lungs are present only in tetrapods. Nothing is learned about the cardiovascular system, about the overlap of the respiratory and digestive tracts, nothing about why this lung is not tube-, book- or bush-shaped like in insects, spiders or certain molluscs. We equally gain no insight into why our lung tolerates small quantities of DDT but no water and why we can choke to death on our food. If we were to put our faith in ontological reductionism, i.e. solely in the reductive analysis (where practical intervention and active change is possible), we would be unable to treat any lung disease because we would not know what a lung is.

That example was chosen because its triviality alone makes it convincing. In the cases remaining to be examined below, the reductionistic restriction becames dangerously complex and even life-threatening in the literal sense of the word—even if it merely involves thoughtless, pragmatic reductionism. This becomes amply evident in how we (mis)treat our biosphere.

(d) Although idealistic reductionism is diametrically opposed to the above, it promotes a comparable misconception, albeit in the opposite direction. This reversal tries to explain the world based on its ultimate purposes, the *causa finalis*, or in the sense of church theory the causae exemplares. Expressed in comparison to materialistic reductionism: a top-down explanation.

The origins of idealistic reductionism (as opposed to the materialistic variant) call for starting with (d1) its ontological form, continuing with (d2) the pragmatic form and closing with (d3) its theoretical form. This issue has been less explicitly articulated and differentiated in cultural history, but is no less effective.

- (d1) In its ontological form it is anchored in the philosophy of Christianity. If a Creator God gave this world its sense and, moreover, did so for the good of humanity and well as to test humankind, then this yields an 'ontological finalism' based on that 'purposeful world order'.

  This stymies research (Lorenz 1976, 1983) because it would effectively close the book on many of the remaining open questions we face today. The dispute between evolutionism and creationism, which has still not ebbed entirely, is also rooted in this issue. Although this concept is not pursued further here, it should not be overlooked. It has exerted its influence more from the background, namely on its apparently secular, pragmatic forms.
- (d2) Among the pragmatic forms of idealistic reductionism, the sciences prove to be instructive in the present context. They were clearly influenced by the various currents in idealistic philosophy, although this has often gone unrecognized.

  The social and cultural sciences, quite pragmatically, tend to neglect the effect of subsystems—in a reversal of the approach in the natural sciences. "Accordingly, art sciences and law consider the conditions of a culture along with its economic and social systems to be a given. The same holds true for the psyche and consciousness for a sociologist, or for the biological make-up of test persons for a psychologist" (Riedl 1985, p. 123).

  From this perspective the physical driving forces and material conditions of the subsystems (and the selectively operating formal causes) play no role. The focus is consistently on the final causes operating from the higher-level systems. Thus, actions are justifiably attributed to the conditions of groups of people and these, in turn, to the conditions of a culture and ultimately to the zeitgeist. The proponents—explicitly or not, due either to ignorance or based on principle—still feel rooted in the ontological form. This is accompanied by the clearly erroneous assumption of having sufficiently understood the interrelationships.

  The result is 'top-down' cultural, historical and social theories that exclude the multitudes of 'ordinary folks' along with their talents and needs. This ignores the fact that there would be no cultural or social systems—and no history—without these people. Bertold Brecht sardonically commented on this situation by stating that, "when one learns that Alexander the Great conquered India, one must assume that he was accompanied at least by his cook."

- (d3) Finally, the theoretical form brings us firmly back to the reductionism problem. Although articulatable laws play a minor role in the cultural and social sciences, researchers in those disciplines are no less strongly motivated by the desire to attribute phenomena to final, higher-level causes. It took a new-found opposition to draw attention to the need to differentiate more precisely.

  Accordingly, the traditional 'history from above' was countered with a 'history from below', beginning with Le Roy Ladurie. Later, that moniker was even incorporated into the titles of the respective publications (Ehalt 1984). The behavior of ordinary people became the explanatory principle behind the behavior of the master. Even esthetics, which had hitherto been ascribed to the major stylistic epochs, became countered by an 'esthetics from below' (Wygotski 1976).

  Clearly, extreme positions never yield sufficient understanding of complex systems. The process always involves the reciprocal clarification and confirmation of prognoses. This principle is introduced here as the subsumption scheme (based on the tradition of the natural sciences) and as hermeneutics (as anchored in the social sciences). This is valid for all complex systems with a historical narrative, calling for more closely examining historicity.

6.2.2 *The Causes and Growth of Historicity*

This issue builds on two insights rooted in the forms of causality differentiated above: all differentiations arise as 'insertions' between constituents and a milieu, and all laws that have arisen in hierarchically organized systems 'operate through' the subsequent levels (in relation to 'cognition' see Chaps. 4, Sect. 4.1.2 (b), Sect. 4.2.1 (c), Sect. 4.3.2, (b), Figs. 4.2 and 4.22; in relation to 'explaining' see Chap. 5, Sect. 5.3.1 (c), Fig. 5.3).

In the present context, this recognition helps take the next three steps: (a) differentiating more closely between cause, condition and precondition, and examining how causes beget preconditions, (b) depicting the divergence of conditions and consequences, and (c) presenting several examples of bifurcation and alternative decisions.

(a) The transition from cause to precondition is related to whether we interpret a condition to be the ultimate cause of a very specific result. Does its role as a precondition lie in the distant past or somewhere in between?

This calls for separately examining (a1) causes, (a2) preconditions and (a3) their intermediate forms.

- (a1) Examine the causes of an inorganic system, for example a geomorphological structure such as the formation of the Lido in Venice, provides insight. The conclusion is that the development of this beach required a supply of sediment from rivers, accompanied by wave action to sort those sediments. These initial considerations do not incorporate rock formation and tectonics, photons or the gravitational pull of the sun. Nonetheless, closer investigation reveals that rocks must have been formed and uplifted for rivers to produce bedload and sand, and that photons must be produced and irradiate Earth's atmosphere to create the temperature and pressure gradients necessary to produce winds, which in turn generate the waves that hit the shores as surf (Fig. 6.1).
- (a2) The most universal preconditions in the megacosmic realm are attributable to the gravitational fields in the cosmos, which contract clouds of matter into suns. In the microcosmic realm they are attributable to the strong and weak interactions that enable the genesis of matter (and therefore also of minerals). This is broadly acknowledged but, in studying the Lido, relegated to the status of a background premise (and equally so when validating the existence of organisms and artefacts).

  Examining the cause for the structure of a chicken's flight muscle (see again Fig. 6.1) has a macro- and microscopic component. The macroscopic direction proceeds from the formal and final or purposeful conditions via the preconditions of the breast and wings up to the evolution of the fowl family, birds as a whole, vertebrates and animals in general. In contrast, the direction taken in the microscopic realm proceeds from the material- and driving conditions via the preconditions of muscle fibers, cell organelles and the contractile myosin molecule down to the chemophysical laws of metabolism and energetics.

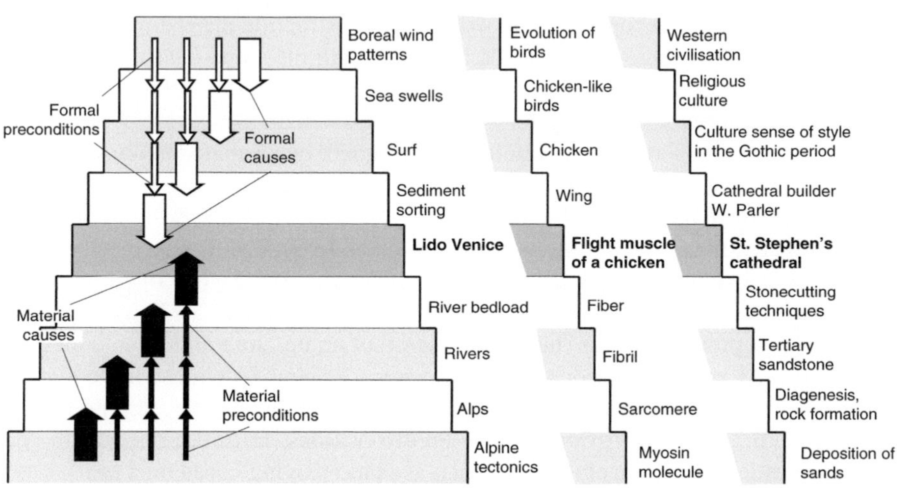

Fig. 6.1 The transition from causes to preconditions with regards to material and formal conditions; illustrated with one example each from (1) geomorphology, (2) anatomy and (3) culture. These examples are further used as a basis for Figs. 6.6, 6.14, and 6.27

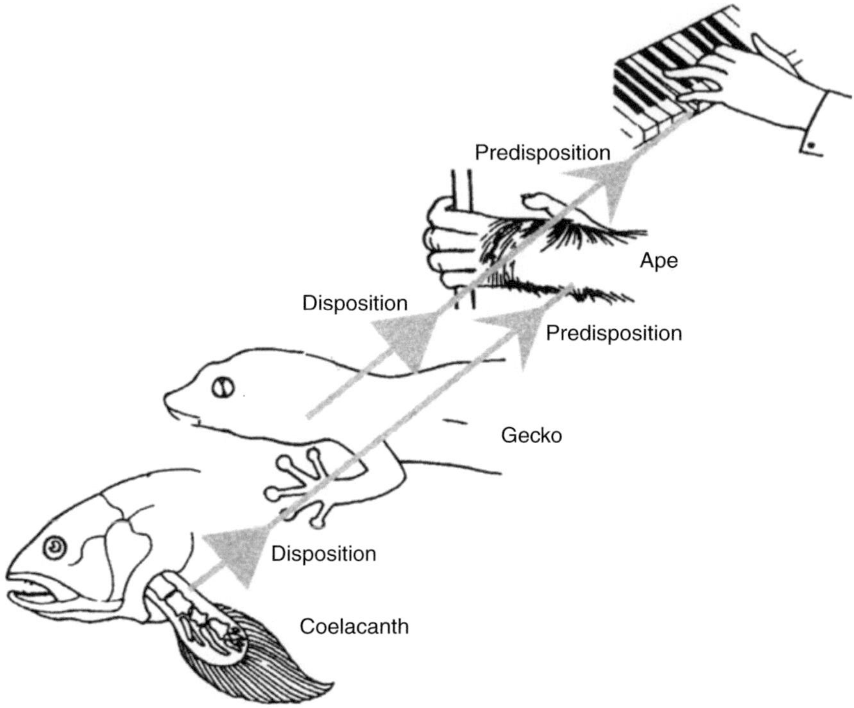

Fig. 6.2 Dispositions and predispositions, differentiated based on possible and no longer possible anticipation of future functions (after Riedl 1994). Symbols as in Fig. 2.5

In the case of artefacts, for example the cause behind St. Stephen's Cathedral in Vienna, the initial consideration will be the construction plans drawn up by the master cathedral builder Wenzel Parler along with the stone-cutting techniques of the time. Ultimately, however, the process will need to touch upon Parler's interpretation of the Gothic style, which reflects back on the Occident, on culture and on human society. The technology, in turn, will be attributed to the soft sandstone available near Vienna and the processes involved in the formation of that stone, i.e. its diagenesis.

- (a3) Finally, the term conditions in the narrower sense, e.g. as intermediate conditions, can be placed between the immediate causes and the preconditions. This parallels the cognitive perspective, where it helps to differentiate between dispositions and predispositions (here in the sense of an up-close and remote diagnosis). In this context, dispositions would refer to new functions that could be anticipated based on what is present. Examples include a leg developing from a fin (Fig. 6.2) or a hand from a paw. Predispositions, in turn, represent the no longer anticipatable developments, such as piano playing based on a fin.

This border will no doubt shift as the 'state of the art' improves. Bear in mind, however, just how few reliable dispositions can be predicted when complex systems are involved. The most surprising developments are clearly subsumed under predispositions. At any rate, the cosmos is predisposed—but certainly not predestined—to give rise to the Lido, to a flight muscle and to St. Stephen's Cathedral.

It is helpful to recognize the wealth of preconditions we can expect behind the development of complex structures and, conversely, the full range of additional phenomena for which these structures themselves then serve as preconditions.

(b) The divergence of conditions and consequences is depicted by one example because presenting even a single one already requires an astoundingly rich vocabulary. This again underlines how poorly suited language is to convey such phenomena.

The selected example is that of the conditions enabling the construction of the Parthenon on the Acropolis. The sculptors Ictinus and Callicrates, under the supervision of Phidias, are immediately responsible for or the direct cause of its form. The material cause is the traditional method of working pentelic marble in the fifth century BC (Fig. 6.3).

The first step is to tease apart the four conditions: the cause and consequences of the (b1) material- and (b2) formal conditions (as in Fig. 6.3) followed by remarks on the (b3) driving- and (b4) final or purposeful causes.

- (b1) The spectrum of causes of the materials begins when the planet condensed due to gravity, the oceans arose, marine organisms produced calcareous shells, these shells accumulated and petrified into marble, and the resulting formations uplifted above sea level. It continues in the physical emergence of terrestrial animals, of mammals, primates and ultimately humans, who then used stone. And it ends with tool production, the use of metal, building construction and ever more differentiated stone working techniques.

  The consequences of these material conditions, however, clearly go beyond the construction of the Parthenon. A case in point is the oldest and youngest of these conditions. Gravity is a factor in all of the above-mentioned causes. Gravity was not only responsible for the planets and for making marble from the shells of unicellular organisms, it also held terrestrial animals on terra firma, enabled the Greek sculptors to sleep in their beds, kept the tools at their designated spots and the worked stones on top of each other, much like it holds together all worldly objects. The differentiated ways of working stone gave rise not only to the Parthenon, but also to the hand axe and every stone edifice in human history, from the Egyptian pyramids to St. Stephen's Cathedral in Vienna.
- (b2) The spectrum of causes of the formal conditions, in turn, begins with the selective formation of the continents, the shells of unicellular organisms and the diagenetic stages of rock formation that give pentelic marble its structure. It continues with the shaping of terrestrial animals and of humans along with the structuring of human society and spoken language—and ends with the birth of consciousness, the realm of the gods as envisioned by the ancient Greeks, and the sense of style in the fifth century BC.

  Once again, the consequences of these formal conditions encompass far more than the Parthenon. Millions of species arose, many human races and even other

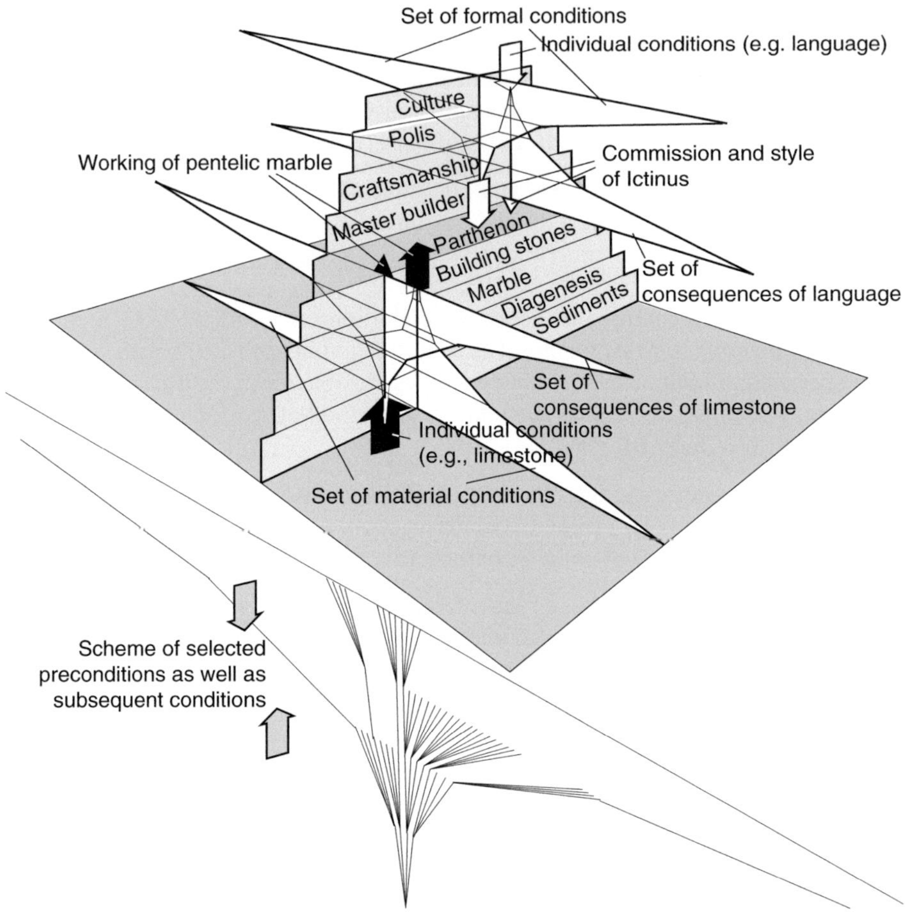

Fig. 6.3 The broad spectrum of conditions and consequences, exemplified by the origin of the Parthenon, in each case between three higher and three lower tiers (after Riedl 1985)

Greek tribes developed—all without constructing a Parthenon. The use of language itself, which was no doubt a prerequisite for this construction project, ranges from the roar of the military phalanx to squabbling Greek fishmongers.

All the above, even if only sketchily outlined, represent necessary conditions, none of which on its own is sufficient to explain the Parthenon as an output. Moreover, all these necessary conditions gave rise to entire worlds of objects, far beyond the Parthenon itself.

- (b3) The driving or operational causes (*causa efficiens*) are recognized as being an energetic phenomenon. They begin with quantum forces, represented in the various forms of matter as the energy of chemical bonds. They underlie the metabolism of unicellular organisms, their calcium secretion and the fact that this calcium can be solidified to form marble. They also drove the metabolism of the animals that conquered land, up to and including humans, fueled civilization, underlay the action of tyrants, the missives of the Oracle of Delphi and powered the construction of the Parthenon with slave labor.

- (b4) The purposeful or final causes (*causa finalis*) then come into play once processes enter cycles that lift the respective systems beyond physical steady states into the dynamic equilibrium characterizing higher forms of order. Electromagnetic effects (the lightning storms in the ancient secondary atmosphere of our planet) already played a causal role here: they gave rise to the first protospecies by distilling energy-rich molecules and autocatalytic processes in the primordial sea (for further discussion see also Sect. 6.3.1). At the species level such causes are represented by the programs that, corresponding to our concept of purpose, promote species survival, whether they be marine unicells or terrestrial organisms including mammals and humans. All these causes—just like consciousness, our notion of the gods and our intent to be on good terms with them—underly the Parthenon.

  All these driving and purposeful causes, and only in concert with the material and formal causes, are necessary to provide an adequate explanation. Importantly, such drives and purposes created not only the Parthenon but in fact everything else in the world (that is, everything that required drives or represents a form of order distinctly removed from physical equilibrium).

(c) All these conditions have undergone bifurcations and alternatives. They beget historicity and define uniqueness, demonstrate the non-repeatability and irreversibility of processes, and underline the finality and irretrievability when something is destroyed.

  In complex inorganic systems these considerations may not seem very poignant: their potential destruction lacks true drama. This changes quickly in the organic realm—all the more dramatically when we are personally involved, when our environment, our culture, or even our survival as a species are at stake.

The way forward, after this introduction, is to discuss the (c1) historically unique conditions and (c2) the irreversibility of the processes.

- (c1) The historical uniqueness is based on the randomness behind the conditions of pre- and post-selection. This is equally true for the material compositions and for the surrounding milieu conditions determining the emergence (or survival/destruction) of a new system. Why? Because, as briefly outlined based on the ideal billiard ball example (Chap. 2, Sect. 2.3.1 (b)) and on reductionism (Sect. 6.2.1 (c2)), physical randomness can influence macro-phenomena even when short causal chains are involved.

  Accordingly, the chances of arriving at the same constellation again are highly unlikely, and when a chain of such constellations is involved, they approach cosmic impossibility. The same holds true for any effort to experimentally recreate such constellations: they turn out to be unreconstructable and remain forever clouded.
- (c2) Another peculiarity of historical systems is the irreversibility of their developmental pathway. This law was first formulated by Dollo (1893) for evolutionary processes in organisms (see Dollo 1922). It is valid for all the systems

introduced here, even for the apparent revival of cultures and their elements: the Renaissance never matched classical antiquity, 'naive painting' poorly mimicked early human art, and 'organic farming' failed to emulate ancient peasantry.

Such irreversibility lies in the complex linkage of functions, in the coherent conditions within the systems themselves. If a system is to operate smoothly, these conditions cannot arbitrarily untwine or be arbitrarily unraveled experimentally.

A short retrospective: Most developmental conditions are unreconstructable and this is disconcerting because we do seek the causes behind even the most complex systems. We have every expectation that (Chap. 5, Sect. 5.3.3; Fig. 5.6) the explanatory pathway of such systems corresponds to a repetition of their developmental pathway. The problem is that we can reconstruct only a small window into those causes. This refers to the 'invariants' characterizing the system, i.e. causes that emanate from both the mega- and micro-realm and then range through all the levels inserted between them, effecting further differentiations. All other causes have become extinguished, have vanished.

We need to recognize this if we wish to avoid severely disrupting the complex world around us. This is all the more urgent considering the increasing gap between knowledge and action: science's simplification of our notion of causality has gone hand in hand with increasingly presumptuous interventions into ever more complex frameworks of causality.

6.2.3 Four Interactions, Four Forms of Causality

The previous chapters have repeatedly addressed the four forms in which we perceive causes in complex systems. These forms served as a foundation for differentiating between cognition and validation (Chap. 2, Sect. 2.3.3 (c2)). Pursuing their fate through cultural history (Chap. 5, Sects. 5.2.1 and 5.2.2), however, revealed that they have lost their defining role in modern science (Sect. 6.2.1).

In contrast, the four physical interactions—the most elementary conditions in the world—were clearly at play in the earliest cosmos (just as they are today). In this book they cropped up in connection with the conditions of transformation (Chap. 4, Sect. 4.1.2 (b)), the explanation of complex systems (Chap. 5, Sect. 5.3.1 (c)) as well as in relation to the development of historicity (Sect. 6.2.2 (a1); Fig. 6.1).

The following discussion looks back in retrospect on the transformations these eight terms (four interactions, four forms of causality) have experienced (compare Riedl 1997).

I separately present (a) the causes of the transformations and (b) the function of the substituted terms.

(a) Four terms related to the causes of the transformation from the physical to the Aristotelian concepts were already introduced in this book: 'indeterminism', 'phase transition', 'emergence' and 'historicity'.

Recall that indetermination manifests itself in the realm of macro-phenomena even via short chains of causality, based on the behavior of quanta (see the 'ideal billiard ball' example; Chap. 2, Sect. 2.3.1). This gives rise to conditions that are not predictable (the emission direction of a laser; Sect. 6.2.1 (c)) and that therefore represent 'historical events'.

The four physical interactions are indisputably involved in such events, yet even the most thorough knowledge about them is insufficient to adequately describe the event. Rather, we need to address a higher level of complexity in the constellation of materials, for example the positional relationships of billiard balls after the first shot.

These are the circumstances behind any phase transition. Even in the simplest physical examples, for example the freezing of water, it proves impossible to predict which group of water molecules will form the first crystal and the direction of that crystal, which then determines the orientation of the subsequent crystals.

This unpredictability is self-evident in the case of complex systems. In the cosmic dimension, knowledge about underlying interactions is clearly insufficient to answer the question of how many water molecules needed to coalesce (preselectively) and in what gravitational field (post-selectively) to form a celestial body. Equally, the pre- and post-selective effects of available compounds and the conditions in their environment are insufficient to explain the origin of life, etc.

The term emergence was coined to describe such a process of coalescing into new system properties. The characteristic feature of emergence is that either the preselective material conditions or the postselective formal conditions, more often both, are insufficiently reconstructable.

Finally, the term historicity refers to all those systems that have experienced emergences and gone through phase transitions. It describes all special, unique structures in the cosmos, all describable geographic and geomorphological constellations, all life and most certainly all artefacts.

(b) The functions of each set of four terms are always twofold. The four physical interactions are clearly active in every process of even the most complex systems. Applying the four forms of causality proves to be equally indispensable.

Whatever materials are involved (the bricks making up a building, to stick with Aristotle's example), all are subject to the interactions of matter and to the pull of gravity. The selective conditions behind a construction plan encompass many aspects. For example, even the planners themselves, as dissipative systems, will have consumed and degraded matter and personally been subject to selection processes. Their own development and even their intentions are attributable to programs rooted in mammalian evolution and then matured in the respective culture. Importantly, both programs will ultimately have been melded by the strong and weak forces, supplied with energy by radiating photons and held on Earth by gravity.

It would be fully justified, if demanded by the explanatory principle of the physical natural sciences, to ultimately attribute all processes in the cosmos to 'energy transformation' (as in Sect. 6.2.1 (a)). This is a recognized postulate but becomes a platitude in the framework of historical systems: applying it yields little gain in knowledge.

Note, in turn, that the four forms of causality are substitutes in the theoretical sense: they represent broader concepts embracing the effects that the above physical parameters continue to exert and, as the theory claims, also led to said forms of causality.

In the practical sense, however, they are also substitutes for the above systems of physical interactions (beyond the reconstructable history of cause and effect). At the same time, after the entire course of history, they represent the cognitive scheme that structures the remaining, currently operative conditions based on their origin, activation and often antagonistic effects. They structure the current mechanisms in the framework of the cognitive symmetries discussed earlier, helping us to take the step from the necessary to the sufficient explanations.

6.3 The Principles of Explanation

Treating the principles of cognition required examining more highly complex systems, namely the principles of morphology (Chap. 4, Sect. 4.3) and of systematics (Sect. 4.4). This is because gestalt perception and our ability to recognize underlying principles focuses on their phenomena.

In contrast, the explanatory process today is oriented in the other direction, namely on physics as the ideal and on pragmatic reductionism. We proceeded as if the complex world was composed of preformed building blocks from a construction kit, neglecting that differentiation arises via insertions. Bidirectional explanations are required.

Accordingly, the causa efficiens was prioritized, and the material causes, represented by the causa materialis, were relegated to a background prerequisite. Physics once again headed the traditional arrangement of the disciplines, and most explanatory models were based on a bottom-up approach.

This sequence partially corresponds with the chronological order of evolutionary processes, making it logical to start here with (Sect. 6.3.1) the inorganic followed by the organic realm, and then to present (Sect. 6.3.2) the development of evolutionary theories before discussing (Sect. 6.3.3) the explanatory models.

6.3.1 Explanatory Models in the Inorganic Realm

Didactically proven examples, one from each key level of complexity in the inorganic disciplines (Riedl 1985), are presented here. The focus is on the ‘state of the art’: no historical overview is necessary.

The complexity of inorganic phenomena is illustrated by one example each from (a) physics, (b) chemistry and (c) geomorphology.

In every case note the (x1) the pre-existence and dominance of cognition as well as (x2) the relationship between structural and class hierarchies. The (x3) bidirectionality of the necessary explanatory models is presented and (x4) the arising questions commented upon.

(a) Cosmic evolution was touched upon earlier (beginning with Chap. 4, Sect. 4.1.2 (b)) in connection with transition and nascence. This involved the explanatory model based on the Big Bang coupled with the creation of space and time. The model holds that (i) the four physical interactions or forces probably derived from one another, that (ii) with cooling, the lighter quanta gave rise to matter, and that (iii) the rapid expansion of the clouds of matter led to gravity fields, which then drew matter together to form the structures of the cosmos.

- (a1) It is instructive to again emphasize the relationship between cognition and explanation: the former must precede and take precedence over the latter. After all, how long did humans live with the tides and were cannons fired before gravity and parabolic trajectories were identified as causes.

  The movements of the planets were known even longer and, despite being an immutable phenomenon, the explanation swung from a heliocentric to a geocentric worldview. Taking a new experience seriously (such as the discovery of the Jupiter moons), however, requires dropping the entire preceding explanation.
- (a2) The differentiation between structures and classes continues to be an issue here. Both can be presupposed—bearing in mind the relationship between their two hierarchies. Consider, for example, how few cases of 'wandering stars' (planets) were necessary before the class concept of planets had to be juxtaposed with that of fixed stars (the suns). The following diagrams omit the class hierarchies to avoid complicating the picture, but their presence is understood.
- (a3) On the bidirectionality of explanations: The cause of light on our planet is presented here as an example for an explanation of a phenomenon in physics (in Fig. 6.4). The diagram shows this to involve a subsumption scheme of theories in a symmetrical configuration. The observations initially involved perceptions gained with the naked eye, then expanded to phenomena resolvable using instruments: the 'bubble chamber' in the direction of the micro-realm, telescopes and spectral analyses for the macro-realm. In both directions, the theories of one level (for purposes of simplification, only two are illustrated) themselves represent case studies for the theories of the next higher level. In both directions they culminate in the (still incomplete) theories of gravity and of quarks.
- (a4) On the three pathways: The two cognitive pathways lead from directly observable phenomena to the more distant theories, whereas the explanatory pathways lead from those theories back to the full range of more tangible phenomena. Regarding the pathway of emergence, we expect that gravitational forces existed before gravity fields, and that these existed before the galaxies as well as before the suns and planets. Equally, quantum forces must have existed before matter, and matter before the reactions of matter in the suns. The explanatory pathway parallels both sides of the emergence pathway.

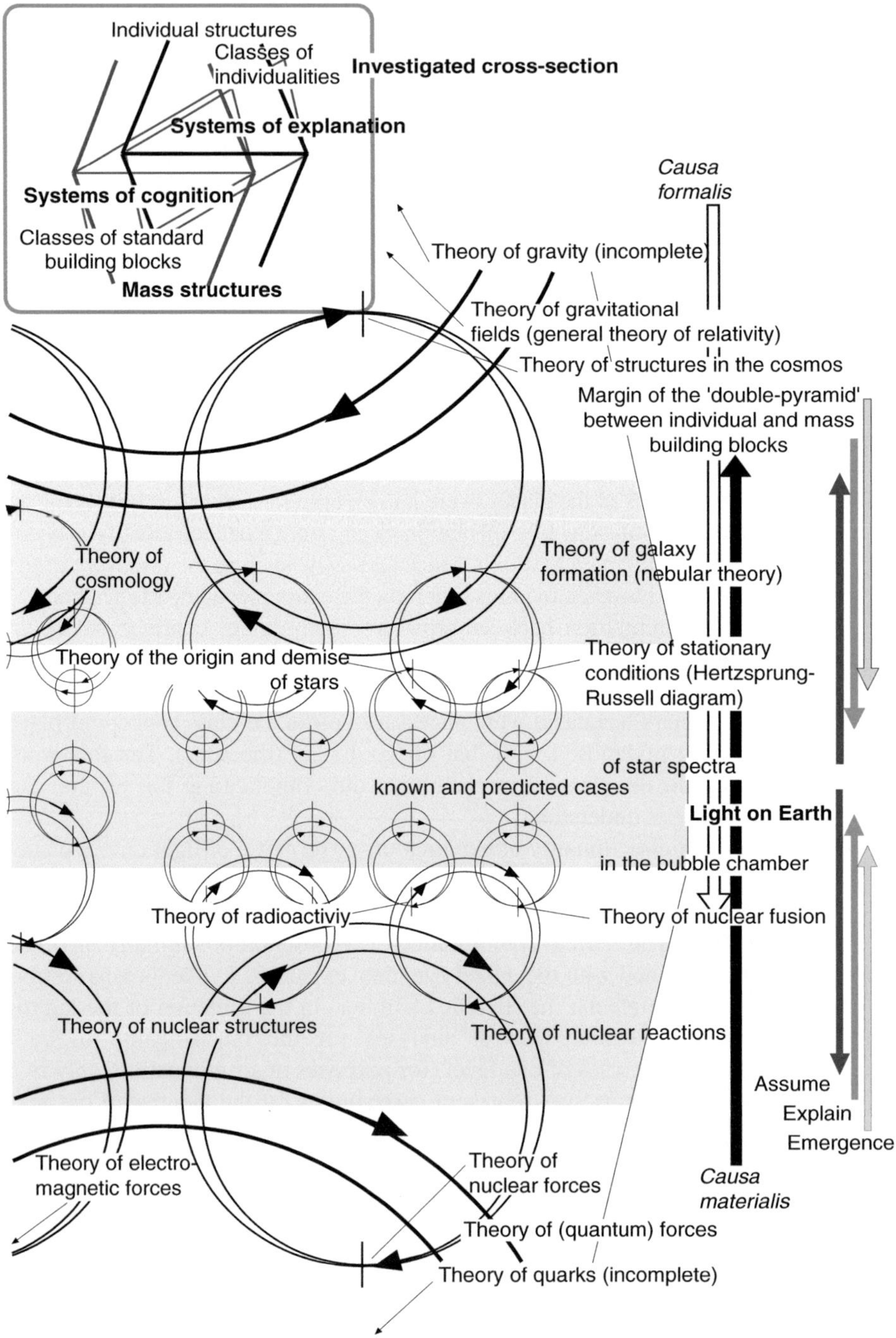

Fig. 6.4 Theory constellations in physics, exemplified by the cause of light on Earth, starting from observed cases in the central field and proceeding stepwise to the higher-level theories encompassing the micro- and mega-realm; the final theories once again meet each other (compare Fig. 5.3). Insert on top left indicates the investigated cross-section (as already explained in Fig. 2.14)

This relationship harbors a prototype of reciprocal causes, a phenomenon termed 'coevolution' by Erich Jantsch (back in 1979) to describe the interplay between the part and the whole, between the constituents and their milieu conditions. Due to the extensive reach of physical interactions, this principle acted even in the early cosmos and continues to play a role in determining all further structuration.

This is well detailed for suns, where gravity and electromagnetic interactions in the overall system (versus the strong and weak interactions of the parts) have established billion-year-long dynamic equilibria (Riedl 1985). It turns out "that coevolution means neither the assembly of basic building blocks nor the permanent differentiation of an initially homogeneous cosmos, but rather the development of hierarchically ordered complexity yielding full structuration at all levels" (Jantsch 1979, p. 141).

Such efforts to underline the birectionality of causal interrelationships may appear trivial to physicists, but only because this example has been kept simple. At higher levels of complexity, this becomes less transparent, loses its triviality and can easily be overlooked.

Many questions in cosmology remain open, only one of them being what triggered the Big Bang and the different extents of the reciprocal interactions. Much remains to be discovered about the origin of the planets as well, especially the origin of the heavy elements in the 'terrestrial' planets (those with Earth-like features), which must have arisen in the active phases of other suns (Oberhummer 1993; Völk 1993).

(b) Chemistry views itself as a science dealing with the properties and transformations of substances and as the extended arm of physics. This prompted the desire to derive its laws from those of physics, an endeavor that can never succeed entirely (see earlier discussion on the laws governing tiered hierarchies; also Bunge 1982). This warrants separate treatment here.

- (b1) The precedence and dominance of cognition is reflected in shifting explanations, for example those involving 'phlogiston' to explain combustion or the discussion surrounding the existence of the atom. Although the substances themselves clearly did not change, new experience required reshuffling the deck.
- (b2) Structures are a decisive issue in chemistry (think complex organic compounds). At the same time, class concepts such as acids, bases, catalysts and many others also play a role. These class concepts serve as the cornerstone for all learning but have at the same time paled into self-evident phrases. This reflects the divisibility and untold molecular variants of virtually all substances.
- (b3) The cause of water on our planet is presented here as a classical example of chemical evolution (Fig. 6.5). Even this apparently simple topic involves an asymmetrical view of the subsumption scheme: chemists are interested less in the laws governing water's origin than in the stability conditions of the underlying multi-particle systems. These conditions range from the theories of spectra and of stable energy levels to the axioms of quantum mechanics.

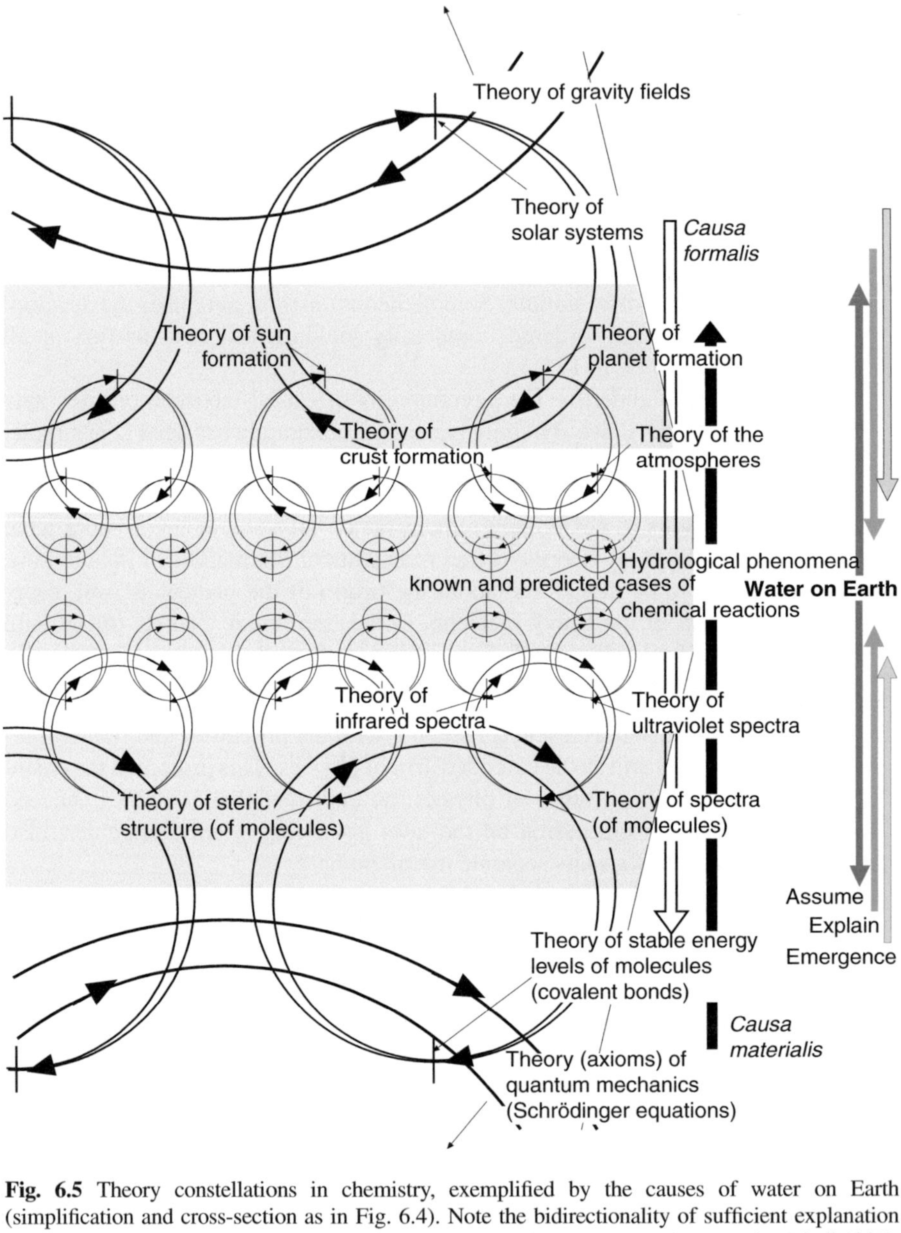

Fig. 6.5 Theory constellations in chemistry, exemplified by the causes of water on Earth (simplification and cross-section as in Fig. 6.4). Note the bidirectionality of sufficient explanation and the relationship between the cognitive, explanatory and emergence pathways (after Riedl 1985)

This helps explain why the compound H2O forms stable molecules—in all three of its aggregate forms. Nothing is learned about why, on this very planet, this particular compound is present—and in such enormous amounts. This requires theories from higher-level systems, i.e. those treating atmospheres and the Earth's crust. These, in turn, lead to the theory of gravitational fields via theories governing the planetary and solar systems. All are very individual forms (compare Fig. 2.13), making them phenomena that prove to be difficult to grasp and non-repeatable.

- (b4) As a rule, chemists themselves represent the acting higher-level system, deciding on the pressures, temperatures and concentrations under which substances are mixed. The effects of these higher-level systems become readily apparent based on the constituents every time researchers successfully approximate evolutive processes, whether it be in a test tube or on the computer.

  The classical experiments by Urey and Miller (compare Urey 1952) deserve mention here: they demonstrated the creation of organic molecules in a reconstructed secondary atmosphere of our planet—an aqueous condensate of hot gases highly enriched in hydrogen and energized by lightning.

  Equally, the models describing how nucleic acids are linked and transcribed into chains of amino acids, the proteins, require the high-level systems of transfer molecules in order to form 'proto-species' in the environment ('smudged' in the sediments of the primordial seas). The inputs for models describing the origin of the first true species must go beyond theories on suitable environments. They must also include theories on the encapsulation of autocatalytic processes in lipid membranes.

(c) The evolution of geomorphological structures is illustrated here by returning to the Venice Lido example used earlier to discuss historicity, causes and preconditions (Sect. 6.2.2 (a); Fig. 6.1). General principles of coastal morphology govern the formation of this beach, creating equivalent shapes along all the world's oceans. Common terms in the Mediterranean include lido and lagoon, in the Baltic Sea spit and haff, along the East Coast of the US 'outer bank' and 'sound'.

- (c1) The pre-existence and dominance of cognition becomes particularly evident in the complex inorganic realm. This holds equally true for mineralogy as its does for geology and physical geography, particularly for geomorphology. All the relevant structures were known for centuries before explanations (that once again shifted over time!) were forwarded.
- (c2) Similarly, the class concepts encompassing these structures predate all scientific explanations, although they were often accompanied by a potpourri of assumptions based on 'good old common sense'.
- (c3) The symmetry of the subsumption model is especially apparent here (Fig. 6.6) because the theories—both in the direction of higher- and lower-level systems—are already differentiated according to the individual disciplines. In the upward direction they range from sedimentology to geo- and astrophysics, in the downward direction from geomorphology to mineralogy, chemistry and quantum physics.
- (c4) This division of an explanatory relationship into so many disciplines along with their different languages calls for promoting theoretical reductionism, for further legitimizing pragmatic reductionism and for excluding ontological reductionism.

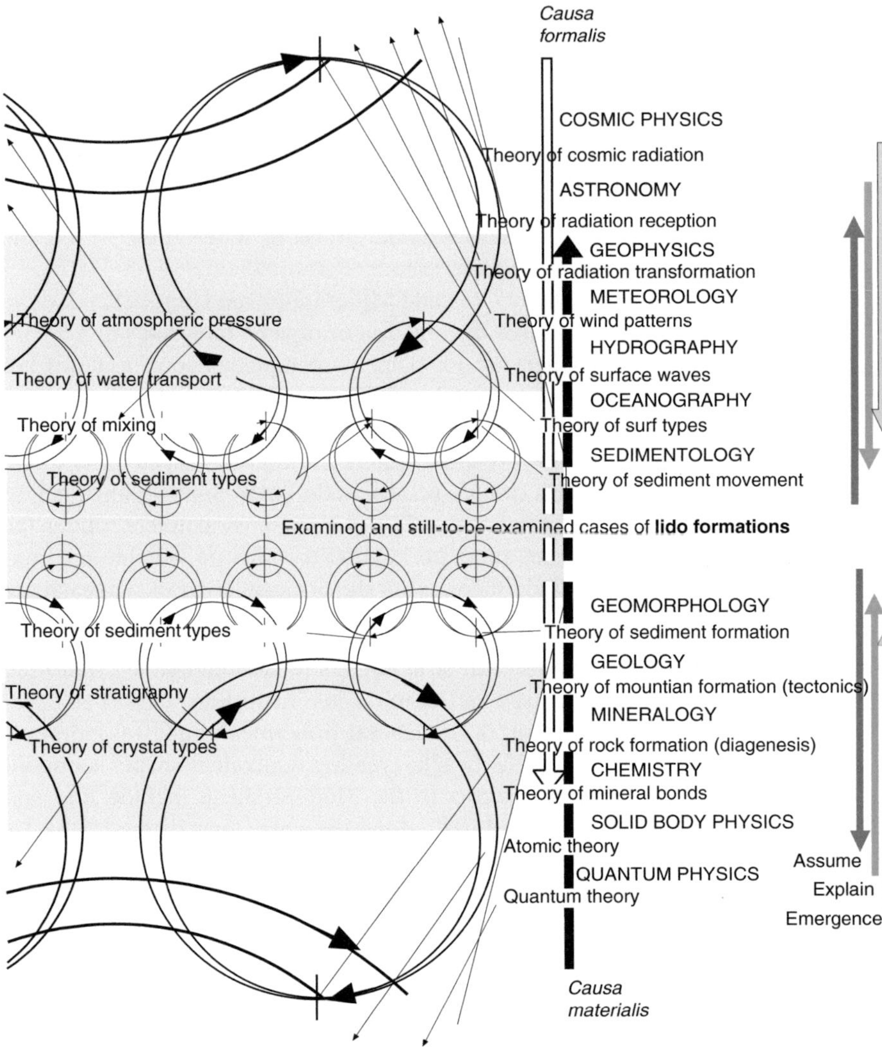

Fig. 6.6 Theory constellations in geomorphology, exemplified by the causes of the 'Venice Lido' (scheme and topic cross-section as in Fig. 6.4). In this case the theories of different tiers also involve different disciplines

Oceanography, for example, does not hesitate to invoke the laws of meteorology and has even coined the term maritime meteorology. This helps hone theories describing how water bodies move and how they can erode, transport and deposit sediments. The laws governing the sediments themselves, however, are then left to the sedimentologists. Equally, the mineralogists justifiably focus on the laws governing chemical bonds and on stereochemistry, explaining the types of rocks. They leave explanations on the types of movements that the rocks undergo, whether they involve tectonics or erosion, to the geologists.

The examples of inorganic evolution presented here underline an important fact: the essential bidirectionality of explanation correlates with the three 'families' of features comprising the foundation of this entire book.

The first family encompasses the four interactions: the strong and weak forces that determine the structure of matter from within, along with the electromagnetic and gravitational forces that act on matter from the outside. The electromagnetic interactions are exemplified by the electrical storms in the above-mentioned Urey-Miller experiment (Sect. 6.3.1 (b)), i.e. the 'energy pump' responsible for sustaining all life processes and that remains—in the form of photons—the foundation of life.

The second family encompasses the four causae, of which the material causes and internal forces are responsible for structuring the constituents. The formal causes function post-selectively from the higher-level systems, whereby the final causes in the present sense occur as soon as processes become part of self-sustaining systemic programs (something already present in proto-species and based on stereospecific autocatalysis).

The third family comprises the six pathways, of which both cognitive paths lead from tangible cases to the higher-level theoretical frameworks. The respective explanatory and developmental pathways, in contrast, lead from the most universal principles, theories and invariants to the differentiation of the cases.

6.3.2 *Evolutionary Theories in the Organic Realm*

As opposed to inorganic evolution, evolution in the organic realm features a much more highly tiered complexity and more numerous phase transitions. In the inorganic realm, such transitions are known from the quanta to the atoms and further to the molecules; we even know some of the principles behind large molecules, e.g. their helices and folding. An additional twelve to eighteen such transitions can be expected in the organic realm.

Initially self evident, matters become less trivial quickly. The full range of complexity clearly taxes our powers of imagination and dampens our willingness to engage. This is a serious hurdle to adequately explaining organic phenomena.

Of course, there has been no lack of effort to deal with complexity. Popular theories include 'synergetics' (Haken 1978), the 'catastrophe' and 'chaos theory' (Thom 1975; Gleick 1988), along with the 'complexity theory' (Kaufmann 1993). Theses theories, however, are designed to avoid transitions across successive phases or to discover those phenomena that remain the same or prove to be analogous across such transitions. This has been elevated to a fundamental principle in the theory of 'fractality' (Mandelbrot 1983; Peitgen and Richter 1986).

All these developments have taken place against the backdrop of pragmatic reductionism (the scientific ideal of dealing only with graspable entities, still advocated today). Considering the competition for research funds, reviewers and reputations at conferences, this attitude may well also harbor ontological reductionism. This calls for approaching the issue from another angle.

Explanatory theories in biology abound, each with its own flavor depending on the subdiscipline. The typical forms of explanation are discussed in more detail in Sect. 6.3.3. All allude to the theory of evolution in one way or another. This is because all facticity in the organic realm can be adequately understood only in the historical context.

The first step is to examine the phases that the theory of evolution has gone through, each of which has a historical background. Six successive conceptual stages are distinguished:

(a) the early notions, (b) the positions of Lamarck and Darwin, (c) Darwinism, (d) Neodarwinism, (e) the synthetic and (f) the systems theoretical concept.

(a) The early notions were already split into two schools: One encompassed the creationistic concepts of creator demiurges and the world of the Greek gods. At the same time, materialistic, functional interpretations were already held by pre-Socratic philosophers, in particular by the so-called 'Ionian physiologists'. They, for example, sought to explain the broad soles of human feet by life in swamps (thinking that the clean separation between land and water occurred only later).

    Empedocles (about 483-424 BC)—physician, philosopher and miracle worker—taught that, initially, plants sprouted from the soil. This was followed by animals that still lacked eyes, of which only those that reproduced and were viable survived. A notion of developmental had already gained hold in Aristotle's day. The idea was that stones had not attained the level of plants, plants not yet the level of animals, and animals still had not attained reason. Even the notion of a kinship between humans and monkeys was raised but was contentious.

    Recall that the First Book of Moses also envisioned such a succession, albeit as independent acts of creation—a concept that has survived up to the creationism still expounded today. The most astounding document originates from the first century BC, namely Lucretius' didactic poem 'De rerum natura'. It anticipates adaptation and selection:

    "These things, then, which are invented to suit the needs of life, might well be thought to have been discovered for the purpose of using them. But all those other things lie apart, which were first born themselves, and thereafter revealed the concept of their usefulness."

    or

    "But those to whom nature granted none of these things, neither that they might live on by themselves of their own might, nor do us any useful service … you may know that these fell a prey and spoil to others."

    Development and selection are presaged. These thoughts lay fallow during the later Roman Empire and throughout the Middle Ages, even in the Renaissance.

(b) The theory of evolution in the modern era arose within a span of little more than a century. Maupertuis and Buffon paved the way in the enlightened second half of the eighteenth century, and the theory took shape in the first half of the 19th century (Glass et al. 1968).

A key obstacle for the concept of descent was the notion—anchored in church tradition—that the Earth was little more than 6000 years old. This, coupled with the information on the time span during which the species had not changed (more than one thousand years), made the theory of evolution improbable. David Hume, Erasmus Darwin and quite specifically Lamarck (1805) undertook efforts to extend this time period, but Lyell was the first (1830) to bring these dating efforts up to modern standards. This contributed significantly to making the theory plausible.

The systematics and comparative anatomy of the day (see Sect. 6.1.2 (b)) had reached a level that enabled Lamarck to take the decisive step: to call for an explanation for the convincing relationship behind the modified similarities of features. I quote: “How, indeed, could I understand that singular gradation found in the organisation of animals, from the most perfect to the most imperfect, without enquiring as to the bearings of so positive (certain) and so important a fact?”

The duality in the concept of causality—namely change in the internal system and adaptation to the environment—is already raised in Maupertuis and Erasmus Darwin as well as in Lamarck. They made a connection between the mechanism of selection and the notion of utility and success in life. This is quite similar to our modern understanding of selective pressures exerted by the environment. What, however, causes change?

The answer requires going into a bit more detail: the foundations of the explanatory theory help better understand its development.

The first step is to discuss (b1) the envisioned causes of change, followed by (b2) selection, and then to return to (b3) the issue of change and introduce (b4) Darwin’s ‘Pangenesis theory’ and (b5) ‘internal mechanisms’.

- (b1) In this respect, Lamarck is known to have advocated a direct form of change. Accordingly, organs would undergo change simply out of necessity, based on the demand for their use, whereas unused organs would degenerate and disappear. This was logical in several respects. We ourselves can observe that ‘body building’ strengthens muscles and that taking them out of action, for example by putting them in a cast, leads to atrophy. A lost kidney can be compensated for by enlargement of the remaining organ. And we can all agree with the view that behavior is a motor of evolution: in modern phrasing, a competitively superior behavioral trait can direct the selective pressure and the success of mutational changes in the necessary direction.

  What failed to be confirmed, however, was the tacit assumption that changes made during an individual’s life would to some degree be hereditary. Such interpretations had been bolstered by the observation that the family members of tailors tended to remain narrow chested, those of blacksmiths broad shouldered. No differentiation between given predispositions and individual adaptations was yet possible.

  Two other considerations are embedded in the concept of active adaptation. Lamarck may not have pinpointed them, but the ‘Lamarckists’ and ‘Neo-Lamarckists’ thereafter did. The turning point came when hereditary shifts were shown to involve small, chance changes rather than a response to any plight an organism might be facing. Doesn't the profound order behind organismic organization such as the phyla themselves point to some form of directed control? This argument, in its latest iterations, will continue to accompany the further discussion.

In contrast, a second argument had to be discarded: Wouldn't we downright expect the utility of active change in nature? What a massive moral dimension that would have unleashed: an individual's progress in honing his or her body and mind would stand a chance of being passed on to the offspring! Today we must acknowledge that allowing our bodies and minds to degenerate is nigh to irrelevant—none of it is transmitted to the heirs.

- (b2) The principle of selection by the environment was discovered by Alfred Russel Wallace, a modest, self-sacrificing tropical researcher, and by the better known Charles Darwin, a reclusive member of English society—independently of one another as one would hope (researched by Brackman 1980).

  More correctly put, they extended the discoveries of their compatriots—Malthus' studies on human populations (1817) and Spencer's 'Synthetic Philosophy' (1850)—to the animal kingdom. Wallace was the first to explicitly state (1858) that he suddenly recalled Malthus' 'Principle of population' during his malaria attacks in the tropics. Accordingly, in a human population that produces surplus offspring yet remains stable, some form of selection or—as formulated by Malthus—a 'survival of the fittest' can be expected (cited from Brackman 1980). Imagine the success of this principle when applied to the considerably greater reproductive surplus in the remaining organismic realm!

  Interestingly, Malthus' insight made few waves and Wallace was largely ignored. The publication of Darwin's (1859) 'Origin of Species', however, roused all sides of the political spectrum and became a resounding success (Desmond and Moore 1991). Beyond the support of his influential friends, this can be attributed to two factors: First, in contrast to Malthus' interpretation, a natural law was invoked. Second, the guilty conscience of the English Puritans (about having just created the industrial proletariat) was eased by a legitimation anchored in that very law.

- (b3) Well-educated persons such as Darwin were naturally aware that the principle of selection could work only if sufficient and suitable variation was available, and if that variation was amenable to becoming hereditary. After all, if the 'improved' individuals reproduced more often but their 'improvement' failed to be passed on to the offspring, then no shift would take place within the population.

  This heritability of variability seemed to be a foregone conclusion. Darwin was well acquainted with Lamarck's work: the books owned by his father, a Lamarckist and contemporary of Lamarck, had a place of honor in his library. In short, Charles Darwin was himself a Lamarckist. It is a very unfortunate corruption of historical fact that his biographers play down this aspect and that textbooks tend to skip over Lamarck entirely. Darwin was in fact even more a Lamarckist than Lamarck himself. After all, he believed the reports in travel journals that the male foreskin had already become shorter within generations in peoples who regularly practiced circumcision.

Be that as it may, the concept was founded on the notion of a heritable variation that served the purposes of the organisms. No one had in mind a mechanism involving random changes.

- (b4) In order to provide a theoretical foundation for Lamarck's concept (the inheritance of actively acquired traits), Darwin developed his Pangenesis theory, which was published in a volume on the 'variability of species' in 1873. Two generations had passed since Lamarck's 'Philosophie Zoologique' of 1809: new material had emerged and biology faced new challenges.

  Presupposing active adaptation, Darwin wrote (p. 405): "Every one would wish to explain to himself, even in an imperfect manner, how it is possible for a character possessed by some remote ancestor suddenly to reappear in the offspring; how the effects of increased or decreased use of a limb can be transmitted to the child."

  Darwin's theory envisions a flow of information from the phenes to the genes. He assumes that all cells produce tiny particles that convey information to the germ cells via body fluids. If the number of cells making up an organ is increased due to use, then the message—bolstered by the respective increase in the amount of particles—would be transmitted to the following generation.

  History dashed this expectation. As we know today, no chemically coded message can reach the genes via the phenes. Nonetheless, Sect. 6.3.2 (f) introduces a theory that envisions such a return flow via stochastic processes: this might be mistaken for Lamarckism, but in fact involves a transmission of functional relationships within the organism itself rather than of changes induced by the environment. Accordingly, the principle promotes internal organization rather than adaptation.

  This is again related to the Darwinian notion because his Pangenesis theory encompasses another class of phenomena, one involving 'inner organization' (treated separately here based on contemporary interpretation).

- (b5) Darwin recognized that selection by the environment alone could by no means explain all the phenomena. Rather, an inner organizational principle had to be assumed—a principle he assigned to the Pangenesis concept (Fig. 6.7).

  He asked himself how the regeneration bud that appears at the site of a salamander's lost limb can 'know' that a leg must be rebuilt there (and know all the details of the process). He also sought the explanation behind the double limbs that sometimes arise during such regenerations.

  Darwin also asked how traits that have long disappeared in the ancestors of an individual can all of a sudden reappear ('spontaneous atavism'; compare Fig. 4.13).

  He sought an explanation for the phenomenon that, during regenerations, an in itself perfectly formed appendage—whether it be a leg or an antenna—could arise at an entirely incorrect position (Fig. 6.8). Today, these phenomena are all subsumed under the category of so-called 'heteromorphosis'. Moreover, he asked for the explanation why ovarian cysts can exhibit fully developed hair and teeth.

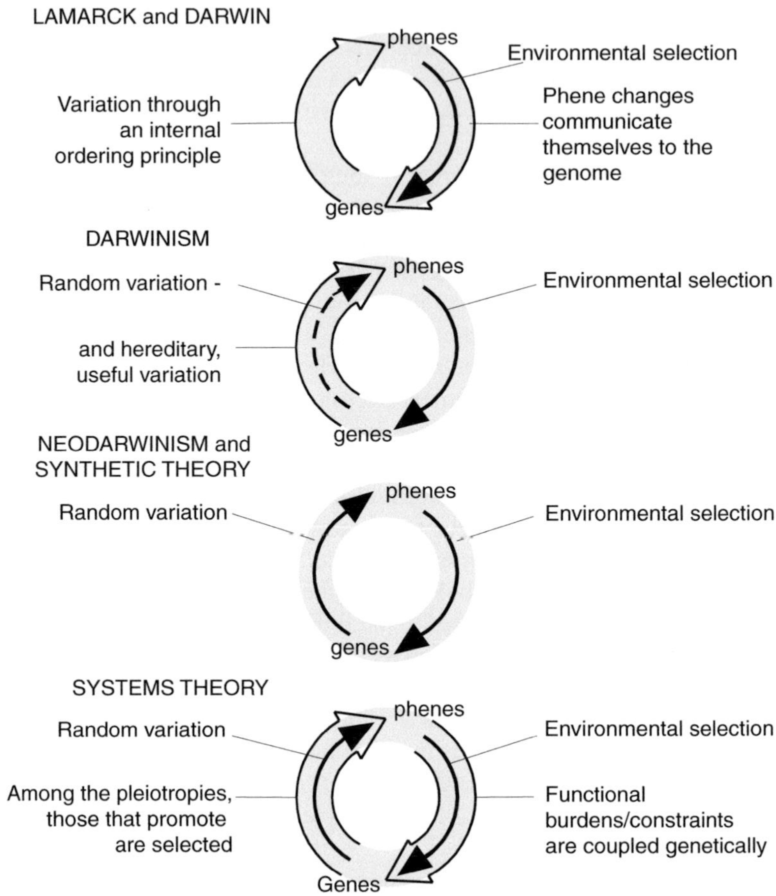

Fig. 6.7 Reduction and restoration of the theory of evolution. Four phases of the theory are distinguished. Left: the expected effects of genes on the phenes; right: those of phenes on the genes

Many more such phenomena have been described today, above all from mutation processes, underlining the role of internal system contexts (Riedl 1975; for more details see also Sect. 6.3.2 (e2)).

(c) Darwinism as we know it today actually first arose through the efforts of Alfred Russel Wallace based on the Lamarckian-Darwinian doctrine. One of his books achieved great renown. It was published in 1889, seven years after Darwin's death, and bears the title 'Darwinism' in honor of the great man he so revered. By and large, it focuses on those aspects that Wallace felt were sufficient to explain the evolutionary process. The Pangenesis concept was already omitted.

He held active adaptation in the sense of Lamarck to be nonsense. For example, he considered the giraffe's neck to be adequately explained by those individuals with somewhat longer necks simply eating away all the tree-top food inaccessible

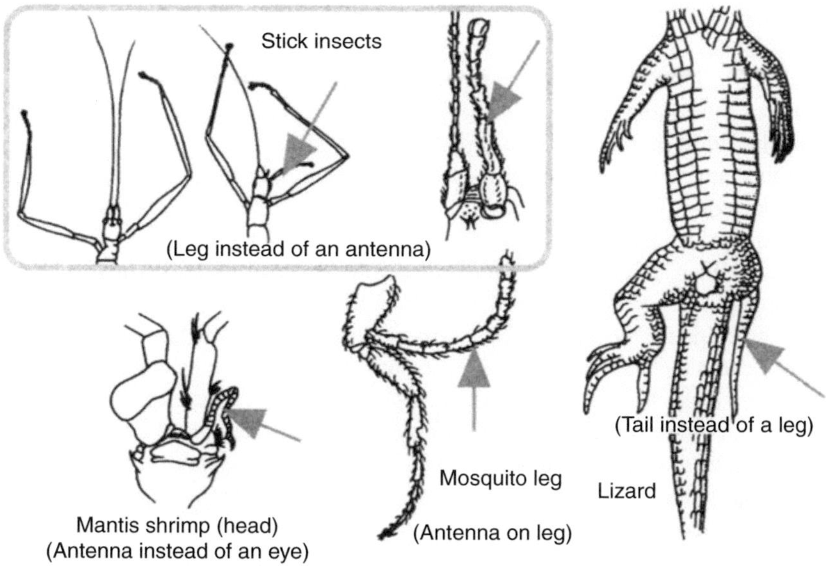

Fig. 6.8 Examples of heteromorphosis, the regeneration of organized body appendages at an incorrect position: a leg replacing an antenna, an antenna replacing an eye, an antenna on a leg, or a tail where a leg should be (after several authors, from Riedl 1975)

to the others. The question how the longer necks arose and how this elongation was inherited remained open. This meant that the phenomena prompting Darwin's call for incorporating internal principles were left unaddressed.

The following discussion juxtaposes (cl) the concept itself with (c2) the emerging opposition to it.

- (c1) This defines the process behind the first reduction of the Lamarckian-Darwinian theory (Fig. 6.7). The expectation of 'internal mechanisms' was discarded, the question how variation arises and is inherited was left open, and the whole process of species transformation was attributed to selection through the environment.

  Environmental conditions, however, shift unpredictably. The factors that drove certain arthropods and fish to leave the water, but prompted insects, spiders and whales to return there, are a product of chance events. Clearly, the prospects of an even more favorable environment are not in the cards for all species. What is necessary is a wide range of dispositions for all manner of potential transformations. This, however, is anchored in organismic organization itself: the prospects provided by the environment remain a game of chance (i.e. the environment is blind to the dispositions of the species).

  Not all the biologists of the day were on board for this reduction. Contemporaries such as Carl Ernst von Baer refused to relinquish the notion of internal principles. In the following generation, Ernst Haeckel, his students and his successor Ludwig Plate (1925), for example, referred to themselves as 'Old Darwinists' to underline that they did not adhere to the abbreviation of the theory, yet still honored the master. Nonetheless, it was the theory's destiny to solidify the context of a simple, catchy Darwinism that immortalized its founder and should in fact have been termed 'Wallacism'.

One genetic intermezzo also failed to trigger any change. Although Gregor Mendel's discoveries coincided with Darwin's late phase, the Academy in Brno sent the experimenting clergyman, head shaking in disbelief, back to his cloister garden empty-handed. The famous botanists he had submissively confided in, like Nägeli in Munich, dismissed him: he 'had the wrong plants'. His laws were first found again or rediscovered at the turn of the century and made their way to England via America. Wallace and the Darwin friend Hooker, both elderly gentlemen at the time, felt they were unsuited for the mathematical perspective and were convinced that the Americans were simply suckers for every new fad anyways.

- (c2) The birth of Darwinism triggered that of Lamarckism, a hitherto unnecessary opposition. That development is not further pursued here because its influence was short-lived. Nonetheless, Lamarkism did reject the one-sidedness of solely environmental selection and insisted on an internal principle that governed the organization of organisms.

  At the same time, additional phenomena were discovered that conflicted with the Darwinian solution (summarized in connection with a critique of today's conventional wisdom in Sect. 6.3.2 (e2)).

(d) The shift to Neodarwinism took place at the turn of the twentieth century. Mendel's laws were widely confirmed, the process of mutation discovered. Both triggered major advances in biology, albeit hand in hand with a second reduction of the Lamarckian-Darwinian perspective.

The 'radicalization' of (dl) the standard position and of (d2) the opposition…

- (d1) A second reduction: Darwinism restricted itself to selection via the environment and excluded internal mechanisms, but one question remained open: how does variation arise and become heritable? The expectation was raised, whether declared or not, that the variation of the traits develops in a 'reasonable' manner that promotes the overall organizational framework.

  This second postulate collapsed as well (Fig. 6.7). The genetics of the time was naturally surprised that the instructions for producing the traits were strung out like beads on a thread and that their sequence showed no recognizable correlation with their function. This suggested some sort of 'string-of-pearls' model of genes, further distracting the research efforts.

  It was also discovered that changes in the genes—mutations—were rare, entirely random, and had absolutely nothing to do with the utility of a modification for the life and survival of a species. This was surprising in the sense that Darwinism (which had already accepted the notion of selection acting blindly) left open the question whether heritable variations might not show some useful connection with the possibilities and necessities of adaptation after all.

In the end effect, two random mechanisms were responsible for controlling evolution: blind selection and the roulette of applied changes (Fig. 6.7).

Again, the theory's fate was to embrace the seemingly simpler solution. Consider the suggestiveness behind having cleanly distinguished between the phenes (which are visible) and the genes (which were not yet visible). And of having recognized in one the effect and in the other the cause of all modalities of change. This prompted August Weismann's 'germ line concept', which his successors molded into a 'Weismann doctrine'. It postulated that the genes persevered across the generations, uninfluenced by the phenes (see Sect. 6.3.2 (e2) for the modern iteration).

This interpretation remained unchanged even after the discovery of pleiotropism and polygenism because it became apparent that genes could influence numerous phenes and, conversely, several genes were at work in expressing a single phene. The one-gene, one-phene concept eroded, yet without significantly changing the string-of-pearls model.

- (d2) Once again, not everyone jumped aboard the train. Biology became differentiated, splitting apart into many sub-disciplines. Genetics became an independent field: comparative anatomists and paleontologists didn't quite know what to make of it and, conversely, geneticists were somewhat bewildered by the complexity of life.

  The emergence of Neodarwinism triggered a parallel shift to Neolamarckism, albeit under an entirely different guise. Experimental approaches were applied to Lamarckian perspectives, politically inspired witch hunts were instigated, and Paul Kammerer's career, for example, ended in suicide (researched by Koestler 1972). Nonetheless, the discussion turned to 'internal selection', to perfectioning principles. Driesch referred to entelechy, Francé to an 'inner intelligence' of living organisms. Little remains of these notions today. The principle of internal selection, however, merits further attention.

  A series of additional issues arose. They had less to do with Lamarckism than with the expectation of 'inner order' or 'inner principles' (treated together in Sect. 6.3.2 (e2) below).

(e) Modern textbooks teach what is generally called the synthetic theory of evolution. It is enriched by many modern insights, above all from molecular genetics, and successfully incorporates and synthesizes a range of disciplines focusing on 'microevolution'. Its theoretical foundation, however, distinguishes itself little from Neodarwinism.

This calls for juxtaposing (el) the various positions held today and (e2) the remaining open questions.

- (e1) We owe the new dynamic primarily to the interaction between the systematist Ernst Mayr (1942), the paleontologist G.G. Simpson (1952) and the geneticist Dobzhansky (1951). Interestingly, their successes can again be attributed to a renewed reduction. This one reduced the object being observed rather than reducing the theoretical concept itself (such as in the previous '-isms').

The efforts focused on microevolution, i.e. phenomena at the species level, primarily those that were open to experimentation, where interventions were possible. This highlights a shift toward pragmatism in the overall trend toward pragmatic reductionism (along with all the above-mentioned pitfalls of ontological reductionism).

The initial progress was based on new insights into the basic principles of heredity. In the 1940s, Avery recognized the function of nucleic acids, Watson and Crick (Watson 1968) their structure. This was followed by deciphering the transcription process: groups of three, the triplets, were 'read' and translated into chains of 20 amino acids, much like the 24 letters of our alphabet, into long 'words' (in this case into the polypeptides or proteins). Variability and heredity were shown to have a molecular basis.

This yielded a key insight for all the theoretical considerations, namely the definitive exclusion of any chemically coded return flow of information from the phenes (via the proteins) and any translation back into the sequences of the nucleic acids. In laboratory jargon this was referred to as the 'central dogma' of molecular genetics, on the tailcoats of the Weismann doctrine. The discussions lasted for decades and the dogma came packaged in a soft and in a hard version. The latter opined: 'no information transfer whatsoever is possible', the former: 'technically, no chemically coded transfer can be envisioned'. The factors governing the site and sequence of successful transcription, however, remained unknown. The consensus was that every cell retains the full set of information, but that the inapplicable parts remain suppressed.

At the same time, various regulatory systems were discovered, beginning with Monod (1971) and extending into the present (e.g. for the homeobox: Holland and Garcia-Fernandez 1996; Ruddle et al. 1994). They show that regulatory genes produce molecules that can block the operator genes, behind which lie the respective structural genes whose transcription and subsequent translation yield complex functions and phenes. The consensus is that precisely these regulatory genes have increased most strongly from bacteria to humans, namely from 104 to 5 × 109.

The subdisciplines of microevolution then provided new insights into population dynamics, above all regarding the exchange of genetic material, and helped outline the speciation processes behind the divergence of populations. They also bolstered selection theory by revealing how the process selects for exaggeration as well as for mediocrity.

- (e2) Any credible opposition in the sense of Lamarckism has been silenced. Pragmatic reduction benefits from supplementation, but any insinuated ontological reductionism warrants opposition. All this has been accompanied by repeated calls for recognizing an 'epigenetic system' of gene interactions in the sense of an 'inner principle' ordering the flow of messages, among others by Waddington (1957), Haldane (1958), Whyte (1965), Riedl (1975), Wagnerx Wagner (1983, 1988), Alberch (1980).

This needs to be juxtaposed with a macro-evolutionism because micro-evolutionism neglects the following six major groups of facts lying beyond the species level: (i) complexity, (ii) regulation and (iii) the 'old patterns', (iv) the phylogenetic trees, (v) the structural and (vi) class hierarchies, i.e. the body plans and system groups (differentiated after Riedl 1975). A few examples for illustration:

(i) Complexity: The genome of an invertebrate, for example, typically consists of 4×10^8 base pairs transcribed in 1.3×10^8 triplets. This approximates the German-language Brockhaus Encyclopedia—with 50 letters in each of two columns, multiplied times 80 lines and 800 pages in each of the 24 volumes. Assuming that improving a triplet, a letter, would be based on chance, then only a single, random letter could be changed with each new edition (i.e. per 'individual'). Otherwise the mistakes would accumulate and the information content be eroded. Accordingly, about $1.3 \times 10^8/2$ attempts times 24 possible letters would be required,—i.e. about 1.5×10^9 or one-and-a-half billion reproduction steps or new editions—to improve the survival of the mutant in a population or the market success of the encyclopedia. Translated to a population of one million 'individuals', this event would require waiting for over one thousand generations.

Relinquishing the pearl-necklace model and examining a simple case of widespread polygeny cements the above conclusion. Let's again take the encyclopedia as an example and try to replace a single word, let's say 'and' with 'but'. It would require the above 1.5×10^9 attempts simply to achieve 'bnd', 'aud' or 'ant', which would not improve the text. All three letters would have to be exchanged simultaneously, requiring about $1.5 \cdot 10^{27}$ attempts—a clear impossibility. The conclusion must be that complex systems can only be adapted in larger, functional units.

(ii) The regulation problem is manifested in regulation errors. The synthetic theory provides no help in understanding how mutations in somatic cells of regeneration buds can allow correctly structured organs to arise at incorrect positions. This is precisely the case in heteromorphosis (Fig. 6.8), forwarded as an example here because they already prompted Darwin (see Sect. 6.3.2 (a2)) to assume 'inner mechanisms' to be at work.

System- or macro-mutations, unknown in Darwin's time, are even more numerous. Here, entire phene units are incorrectly positioned based on a similar process but involving mutations in the genome, i.e. inherited traits (Fig. 6.9). A similar phenomenon, albeit triggered by disturbances in the developmental process rather than being hereditary, involves 'phenocopies'.

All the structural genes contributing to a particular functional phene complex, both in somatic and germ cells, can probably be called up by a (single) regulatory gene. As elsewhere, errors in a process often shed light on the background conditions. Importantly, the same conditions or principles must also underlie the natural, error-free process. This is valid for all processes of 'degeneration', for the 'regulation' of damaged germ cells, for 'synorganization' (the joint transformation of phenes) and for 'Cartesian transformations' (harmonious phylogenetic transitions).

(iii) The problem of old patterns was already known to Darwin based on the phenomenon of 'spontaneous atavism'. A classical example is the three-toed mutant in horses, in which a stage characteristic of early horses recurs. The phenomenon is also known in humans, for example in the form of furry faces, the presence of a tiny tail, superfluous areolas, neck fistules as remnants of the milk crest or even of gill slits (compare Fig. 4.13).

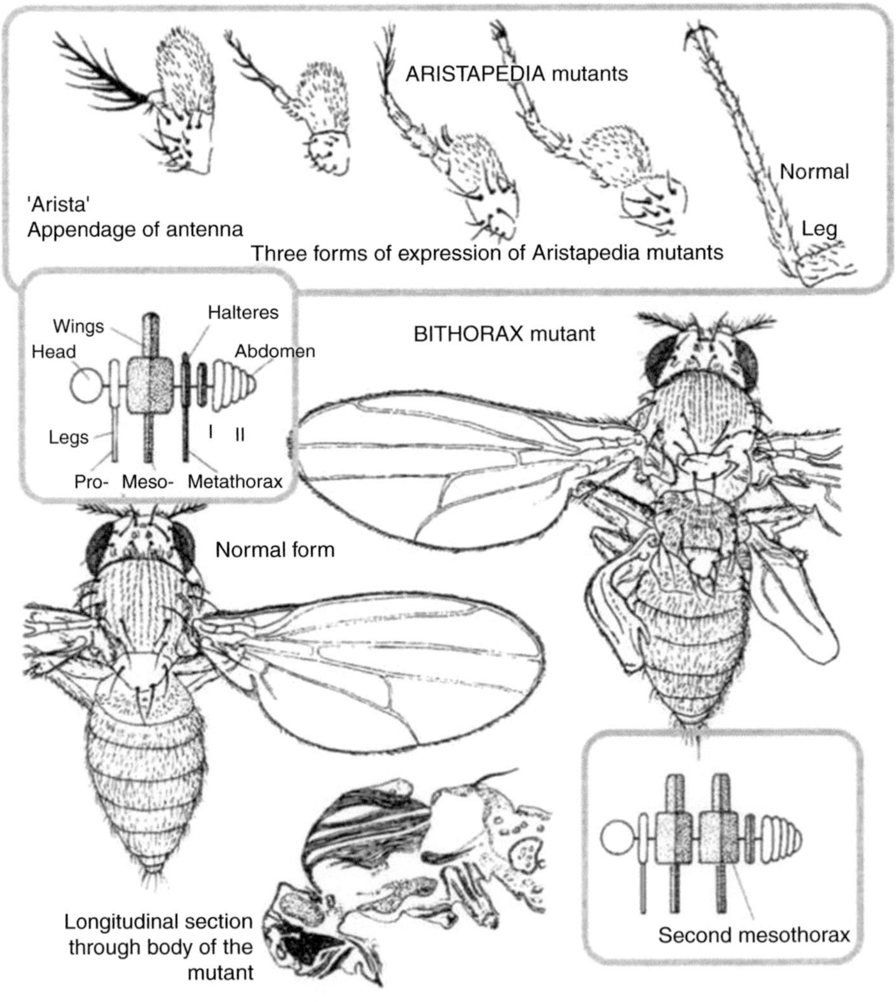

Fig. 6.9 Examples of system mutants in Drosophila: development of a leg instead of a normal appendage of the antenna, a redundant wing and a doubled thorax; inserts: schematic representation of the altered regions (after Riedl 1975)

The conclusion is that even obsolete phenes (interphones) exist in the epigenetic system as functional units and remain preserved as stations in the information transfer.

This also holds true for undisturbed processes, Why, based on Haeckel's law, must interphones be installed and passed through during ontogeny (e.g. the chorda in all vertebrates, the gill arteries in all mammals)? This helps explain why the 'induction patterns'—the pathways of information transfer in ontogeny—lead from the older to the younger phenes, and why these patterns are identical even in large taxonomic units (Fig. 6.20).

The next step involves going from the functional phenomena to examining those that must be understood historically and then to summarize the resulting demands.

(iv) The issue in phylogenetic trees involves the question why the developmental pathways of animal phyla (when setting time against transformation) become increasingly straighter and more elongate—as if they were headed toward a particular state (Fig. 6.10). The details support this: typogenetic phases of development are followed by typostatic phases, i.e. rapid transformations within a relatively short time period are followed by increasing constancy.

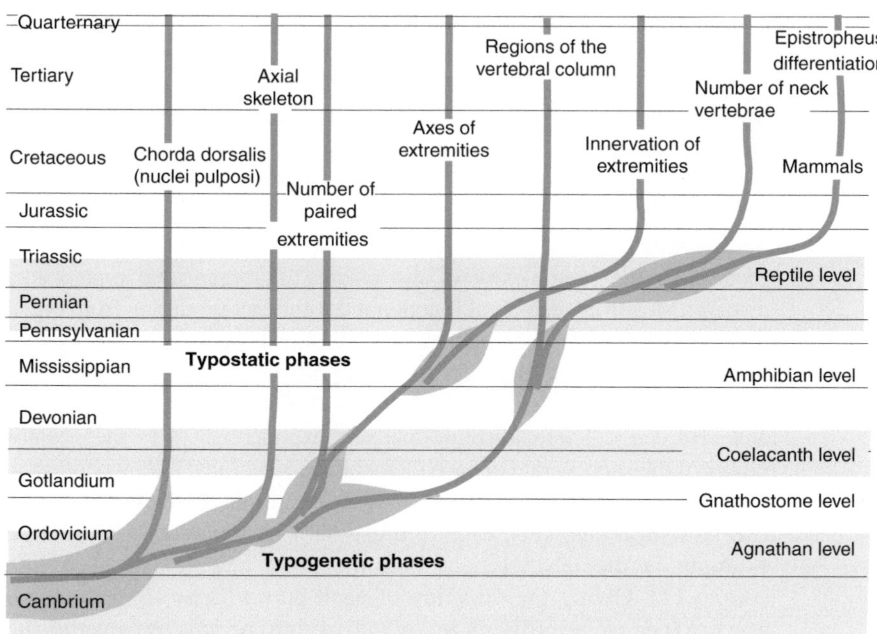

Fig. 6.10 Transition from typogenetic to typostatic phases, exemplified by several traits in the evolution of vertebrates. Note that typostases in the aftermath of typogeneses give rise to new traits that build upon the former. Left: geological periods indicated by lines; right: developmental stages delimited by fields

Closer examination reveals that new traits, rather than the old ones, gain new degrees of transformational freedom. These new traits build on the older ones, on those that set their foundations and define their functional delimitations.

(v) The issue in body plans is related to how we explain homologies. As determined in Sect. 4.3, homologies can be recognized based on those features that have resisted adaptive change. Their hierarchic organization also requires explanation.

(vi) The issue in the 'natural system' of organisms is whether that system might simply represent an imagined construct. This calls for substantiating the natural factors potentially behind a hierarchic pattern of similarities. The answer is to postulate 'constraints' arising from an 'inner selection principle'. Accordingly, the growing functional (and later also genetic) loading of the older traits incrementally restricts the chances of successful transformation (depending on the structural level in the hierarchy).

(f) The above demands can all be met by a systems theory of evolution (Riedl 1975; 1977; Wagner 1983; Wagner and Altenberg 1996). This approach considers the synthetic theory to be a necessary but not a sufficient explanatory model.

The first step is to introduce (fl) the contents of the theory, followed by (f2) the critiques.

- (f1) Basically, the theory envisions an inner selection principle that goes beyond blind mutability and short-sighted selection by the environment. This principle retains deep retrospection into the history of its own products. It assumes a 'recursive causality', a reciprocity, in which the chemically coded effects of the genes on the phenes retroact again from the phenes onto their own causes via a stochastic process (Fig. 6.7).

  The foundation is a symmetry involving a (i) phene- and a (ii) gene-initiated, two-stage model with (iii) corresponding outcomes showing a two-fold complementarity.

  (i) Examining the 'complexity of the phenes' in organismic organization reveals a wealth of reciprocal functional dependencies ranging from organology to the level of fine structures. All drastically reduce the chances that mutational changes of individual building blocks will be successful. They represent 'functional loadings, constraints or burdens' (German: 'funktionelle Bürden'). These loadings increase exponentially—to the negative power of the success rate of each element in a functional interdependency.

  First stage: The simplest case is the functional dependency of only two structural elements, each initially with its separate structures. An example is the development into the head and socket of a joint (Fig. 6.11). Giving the mutation of each part a high (10^{-6}) chance of success (i.e. every millionth reproductive step) reveals that overall success (i.e. of both parts) can be achieved only by undergoing complementary change. Much like rolling two dice, this translates into 10^{-12} of the cases. In a population consisting of 10^6 individuals, the success of a single part can be expected once per generation, but the success of the

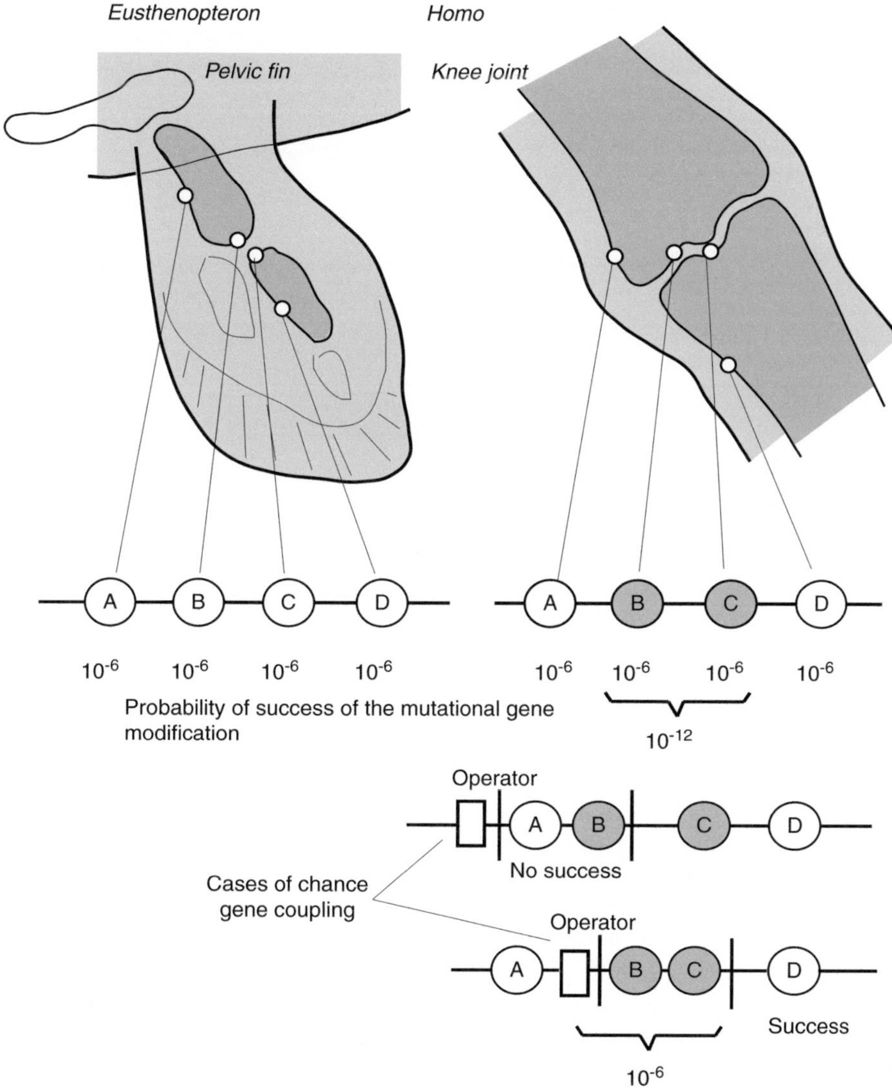

Fig. 6.11 Model of gene coupling due to a functional burden or loading exemplified by the development of a Devonian fish fin into a knee joint, as well as four genes (A to D) that code for the lengths and widths of both homologous supporting elements. The chances for successful change drop from 10^{-6} to 10^{-12} if the phenes are functionally linked—unless the respective structural genes are by chance coupled (Riedl 1975, 1977)

two parts independently of one another would require a million generations. The conclusion: more complex functional couplings would stand no chance of adaptation.

Second stage: If, by chance, both genes become coupled—for example under a regulator gene—then only the regulator need experience the successful mutation in order to change both of its subordinate structural

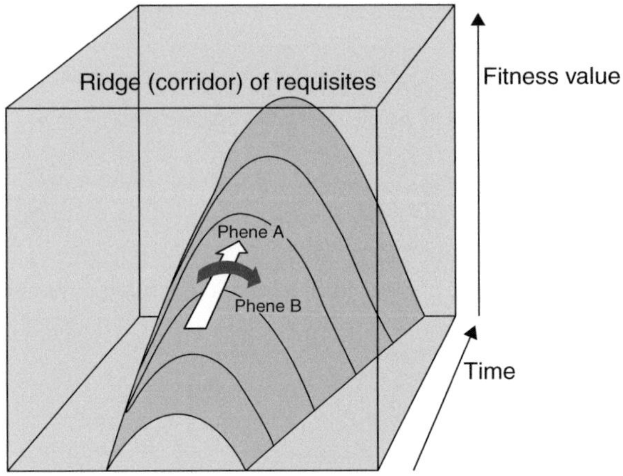

Fig. 6.12 Corridor model in the event of genetic loading, exemplified by a ridge to be scaled based on a population's fitness. Here, a pleiotropic gene controls two phenes, one in an advantageous (A), one in a disadvantageous manner (B). The balance of the two effects of the gene determines whether the population will climb the peak or slide off (see also Wagner 1988; Baatz and Wagner 1997)

genes. Giving the successful change of the regulator gene the same odds as that of each gene separately, namely 10^{-6}, increases the joint's chances of successful adaptation by a factor of one million.

The expectation therefore is that, among the many arising gene couplings, only some will be retained under the selection pressure of increasing functional loadings. Specifically, those that (by chance) couple structural genes coding for functionally dependent phenes; mutants with detrimental couplings will be eliminated.

(ii) A symmetric model to the above assumes, simply put, 'genetic loadings'. This has already been formalized (see Wagner 1983; Burger 1986; Wagner and Altenberg 1996; Baatz and Wagner 1997) (Fig. 6.12).

In a first phase, pleiotropies are assumed to already be present in the genome.

Second phase: A gene by chance codes for two phenes whose concordant change increases the fitness of the mutant. That mutant will quickly prevail in the population. If, in contrast, a gene codes for two phenes whose simultaneous change is disadvantagous to its bearer, then that mutant experiences negative selection pressure.

(iii) The correspondence of the processes is based on the fact that both models assume loadings arising either at the level of phenes or of genes. These loadings themselves again represent an interaction: new couplings of phene functions pull the epigenetic system in their direction, and pleiotropies in turn promote the chances for phenes to undergo special functional developments.

Both models describe an 'imitational epigenesis system': functional relationships at the phene level are either copied by the gene interactions, or the gene interactions promote certain functional developments. The result is a hierarchic pattern of gene interactions that strongly resembles the hierarchy of the homologs (Riedl 1977). Moreover, this mechanism—because it incorporates the historical aspect—also explains Waddington's (1957) postulated 'archigenotype'.

The complementary effects are valid for both models. In both cases the capacity to adapt complex systems in particular directions increases, albeit always at the expense of other developmental opportunities. The former makes the development of complex systems possible in the first place, but only at the expense of constraints. This yields the order behind the structural and class hierarchies in the organismic realm.

The parable of the two blind gamblers is a useful example. These two gamblers, Archaios and Neos (both species), play (compete) in front of the king (the environment). Each has two dice (two functions).

The first phase: The king rewards two sixes but communicates only success or failure (fitness) for each toss (reproduction). The players can do whatever they wish with the dice, but it must be done blindly (stochastics). Archaios knows that, on average, he will be successful in one out of 36 tosses and sticks with this strategy. Neos waits until he has a success and then ties the dice together in that position. He will be successful at least six times more often.

The second, complementary phase: the king changes the rules (fitness condition) and now rewards yellow-dice-six, red-dice-two. Archaios, the unspecialized player will continue to have the above, lesser degree of success. Neos, however, will never experience success again unless he can decouple the dice (which becomes increasingly unlikely if the game is continued with a larger number of dice).

The first phase contains the explanation for the adaptability of highly interlinked systems. It also harbors the family of open questions raised above under the items (i) complexity and (ii) regulation. The second phase explains conservatism based on the tiered structure of new constraints. It also explains the perceptible order of (iii) old patterns, of (iv) phylogenetic trees, of (v) body plans and of the directionality (vi) in the nature of the natural system.

Appended to this topic is the newer concept of 'generic properties' in the epigenetic process. This refers less to the genus-specific, the general or common features than to the associated consequences. For example the circumstance that relatively simple instructions from the genome can lead to highly differentiated results, patterns and structures due to the accompanying chemophysical environment (compare Newman and Comper 1990; Müller and Newman 1999). This issue is raised here because it is conceivable that the genome, which can achieve success only by considering those consequences, is steered in a particular direction by this process as well.

- (f2) Few critiques have been forwarded, but two basic positions can be distinguished.

  One (i) insinuates that the above principle incorporates Lamarckism, the other (ii) rebuffs the postulate of recursivity outright.

(i) The insinuation is very superficial and related to the expectation that, in the history of the theory, 'Darwinisms' could be countered only by 'Lamarckisms'. Closer examination might lead some to spy Lamarckism in the theory based on a potential effect of the phenes on the structure of the epigenetic system, through the back door so to speak. This misconception warrants explicit mention. In reality the system conditions operate in exactly the opposite direction, even counter to full adaptability to different environments.

Returning to the earlier example of the seven neck vertebrae in mammals shows that no fitness-boosting reduction (in dolphins) or increase (in giraffes) in number ever occurred (Fig. 6.13). Both groups of species had to

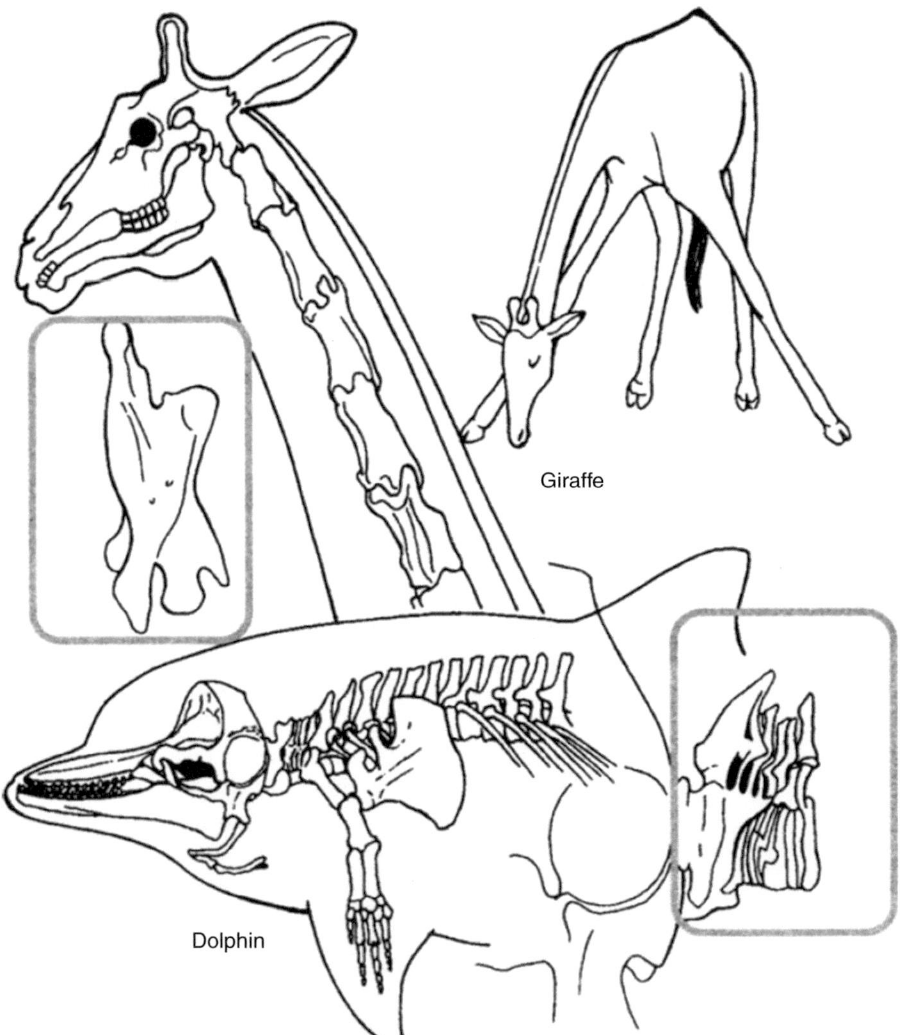

Fig. 6.13 Examples of functional burdens, illustrated based on the seven neck vertebrae in mammals. Note the technically suboptimal solution in both cases. The movement type of giraffes would have benefited from more, the dolphin from fewer neck vertebrae. Inserts enlarge one vertebra of the giraffe, the entire set for the dolphin (from Riedl 1994)

make do with seven neck vertebrae. Their only option was to extremely flatten or elongate them in order to adaptively approximate the respective environmental demands: giraffes would clearly have been more agile with more neck vertebrae and dolphins would have been able to better compensate the ram pressure on their heads with fewer.

This is merely an illustrative, functionally insightful example. In reality, all traits that are embedded in class hierarchies must be stabilized by constraints. This explains why we can recognize homologies based on those characters that resisted adaptation—a fact that cannot be overstated.

Of course, in assembling living systems, all homologies are functional and therefore underlie adaptive pressures. At the same time, they all represent compromises. This becomes increasingly apparent the older and the more heavily loaded or historically burdened they are. The spinal column can in fact be adapted (compare Chap. 4, Sect. 4.3.1 (c); Figs. 4.16 and 4.19). Nonetheless, no change in the construction principle itself can be successful—not even for turtles and trunkfish who, functionally, no longer need a backbone at all. The vertebrate eye has also clearly been adapted. Nonetheless, the key element that could stand improvement, namely turning the incorrectly oriented sensory cells toward the light, is unachievable.

Interestingly, the predictions that the theory made very early on (Riedl 1977) have come to be verified experimentally one after the other—yet without effecting a full breakthrough.

(ii) This probably reflects the postulate of recursiveness—albeit once again in an entirely superficial manner. This is because century-old dogmatics (the Weismann doctrine and the central dogma of olecular genetics) cannot easily be surmounted.

Examining the theory in greater depth reveals that no 'molecular solution to the phenomenon of life' is forthcoming. If the epigenetic system, as expected, proves to mirror the functional relationships of those phenes for which it codes, then molecule groups stand for complex anatomical units. These higher functions can only have been 'learned' from their products. Accordingly, the solution must once again be a bidirectional one.

6.3.3 The Explanatory Models in the Organic Realm

The organismic world—more so than the complex inorganic realm—cannot be understood without incorporating its history. Recall that all differentiation in the cosmos originates in the form of insertions. Namely as insertions between the constituents of the new system and the environment in which the new level arises (Chap. 4, Sect. 4.1.2 (b) etc.; Fig. 5.3). Accordingly, all organic systems can be envisioned as originating between inorganic ones.

The examples given for causal and explanatory relationships in physics, chemistry and geomorphology (Figs. 6.4, 6.5, and 6.6) showed that the ultimate causes accessible to us lie in the conditions of the macro- and microcosmos. The same also holds true for the entire organic realm. The macrocosmic side encompasses the gravitation that holds the organisms on the planet's surface along with the electromagnetic interactions that, in the form of irradiation by photons, provide the major energy source for the entire biosphere. They represent the environment in the broadest sense. The microcosmic side includes the strong and weak interactions, i.e. the conditions that create material and are then responsible (in the form of chemical bonds) for building up structures and for all further material and energy transfers in the organic realm. They represent the most fundamental constituents.

These two conditional aspects, which can be said to lie outside the organic realm, are decisive because organisms are open systems (see entropy problem in Chap. 4, Sect. 4.1.3 (a); Fig. 4.3). Their build up and maintenance of 'negentropy' or order—outside any physical equilibrium—is based solely on a throughput and consumption of order. Energy and matter must flow through them.

This is highlighted here to avoid redundantly expanding all the following examples of explanatory contexts to the underlying macro- and microcosmic conditions. The conditions at both these ends of the spectrum must be kept in mind.

Nothing in the organic realm can be understood without the historical perspective. I refrain from using the theory of evolution to introduce the explanatory models below because, in my opinion, that theory is incomplete, making it more opportune to begin with conventional examples (if only for didactic reasons).

The first examples are therefore taken from the fields of (a) physiology and (b) behavior, followed by (c) the twin problem behind explaining organismic structures and (d) the concept of evolution and its historicity, concluding with the next-higher (e) system of ecology.

Analogous to the presentation in the inorganic realm (Sect. 6.3.1), the dominance of cognition and the juxtaposition of structures/classes are givens. It is again useful to separate the (xl) symmetry of the explanation as a phenomenon from (x2) the remaining commentary.

(a) Explanatory models from the field of physiology are didactically useful because its theories, at least in the lower part of the double pyramid, approximate those in the better known inorganic systems of theories.

- (al) On symmetry: The first example—the 'cause of the human eye'—builds upon the subsystems detailing photochemical processes (Fig. 6.14). The 'all-trans-potential theory' is anchored in the fundamentals of photochemical processes, according to which a single incoming photon already changes the potential on the membrane of a photoreceptor cell. Clearly, the explanatory scheme of this theory is further rooted in the laws of chemistry and physics.

  A different type of theory is required in the direction of higher-level systems. Here, a theory of senses, of external or exteroreceptive senses and the excitability of protoplasm, extends to our expectations regarding the correspondence of an organic system with its environment. The prerequisites in this case are solar radiation reaching the Earth's surface, the emission of photons when hydrogen is

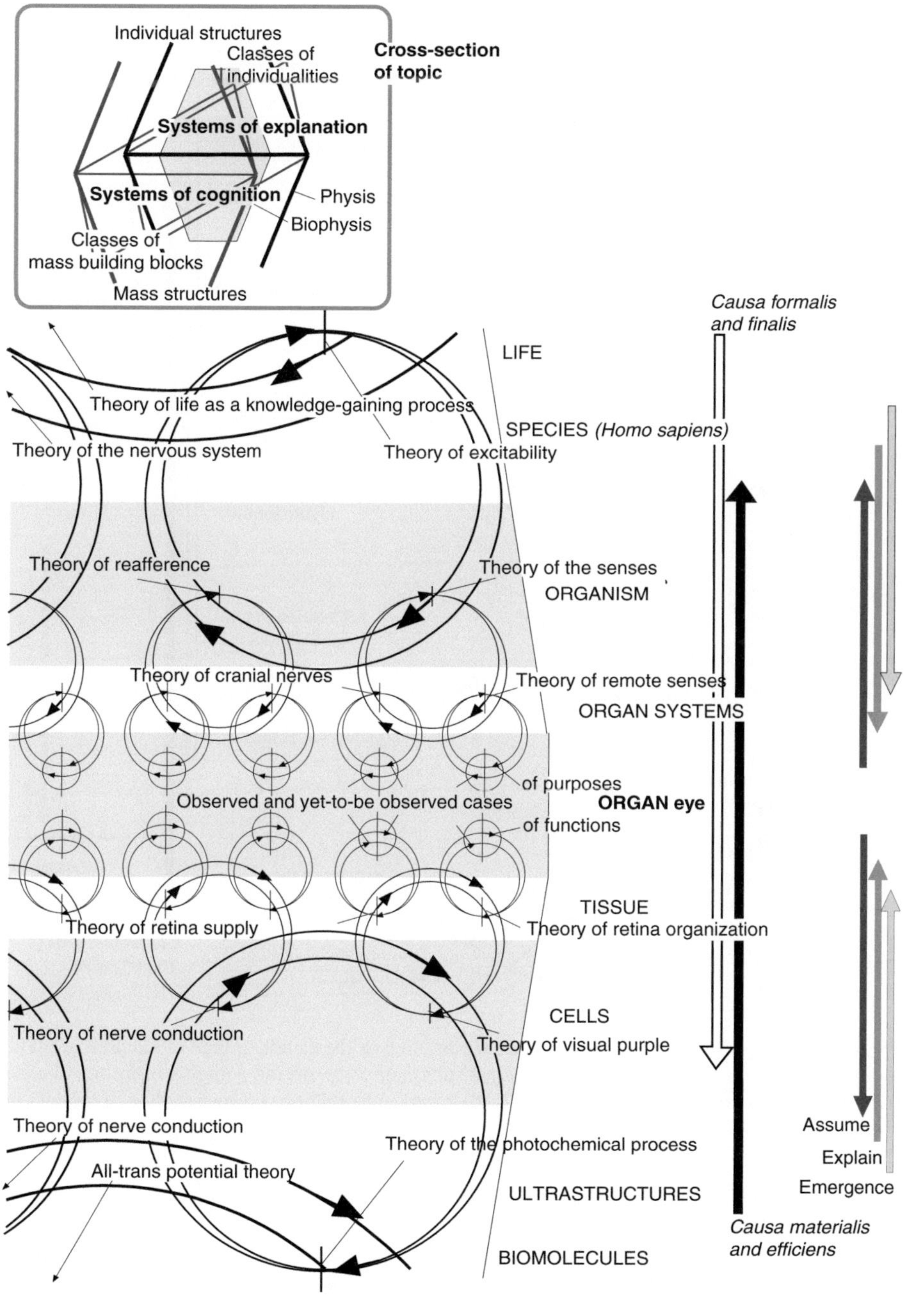

Fig. 6.14 Theory constellation in physiology, exemplified by the causes of our eye. Note that theories of physiology and molecular biology come into play toward the subsystems, those of behavior and learning processes toward the higher-level systems (after Riedl 1985). Insert (top left) highlights that solely phenomena of biophysics are considered in the framework of physis

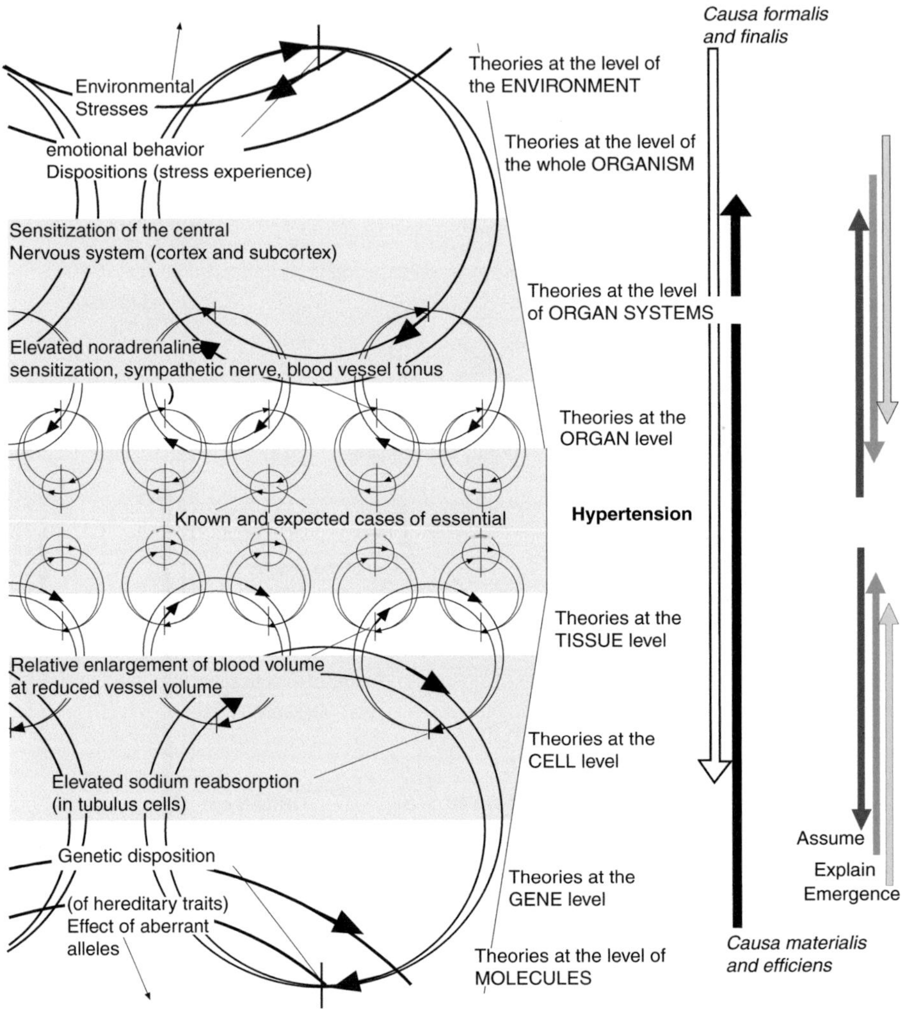

Fig. 6.15 Theory constellation in medicine, exemplified by the causes of high blood pressure. The bidirectional causal relationship recurs against subsystems toward the patient's make-up, against the higher-level systems toward the patient's environmental conditions (cross-section of topic as in insert in Fig. 6.14)

combusted to helium on the surface of the sun and, finally, the behavior of matter under high temperatures.

The second example (Fig. 6.15) stems from medicine, namely the 'cause of high blood pressure'. Here, the subsystems are based on genetic factors, which are attributable to mutative and hereditary dispositions, to structures of the base pairs, and ultimately to the laws of chemical bonds and those governing microphysics. The higher-level systems, in turn, lead from the stress experience itself to dispositions of society. The steps involve the pressures we are exposed to and our efforts to maintain a 'steady state', i.e. balanced sensibility. The task is to maintain the operational conditions of the system beyond any physical equilibrium—all this in an organismic and ultimately physical world replete with unpredictable disturbances.

- (a2) Textbooks in physiology and medicine typically provide explanations anchored in the pyramid of subsystems, i.e. those delving into genetics and biochemistry. Nonetheless, the role of the environment in the broader sense is central to this issue and to the approach taken in this book. In the high blood pressure example this involves the stresses humans are subjected to by external factors inherent to their social environment. In the eye it is the photons and their use in orientation: in our world this gives purpose to an eye, to the visual pigment rhodopsin, and to the application of photochemical processes (details in Riedl 1985).

  This also sheds light on the three pathways framework. The cognitive pathway begins in tangible dimensions. The explanatory pathway, however, which runs counter to it, represents an abbreviated and simplified repetition of the developmental pathway. Clearly, the possibilities inherent in photochemical processes and the utility of perceptive senses must have existed before the formation of an eye. The same holds true for genetic dispositions and the sensitization of the nervous system: they were no doubt present before high blood pressure existed.

  With regard to the four forms of causality, the structure, maintenance and operation of all physical drives arise basally from the functions of molecules involved in energy metabolism. Equally, all material causes are anchored in the cells and in cell reproduction. Conversely, the various form-giving selection criteria operate from the respective higher systems: the visual apparatus (eye–nerve pathways–visual cortex) on the eye, the eye on the retina, lens etc., and the retina on its structural layers. Importantly, the purposes of all these hierarchic levels are based on the advantages that perceptive senses lend for survival.

(b) The example from ethology involves the ritualized behavior of a duck species (Fig. 6.16), a behavioral type characteristic for many other species as well (Lorenz 1978).

- (b1) The symmetry of sufficient explanation is readily apparent in this discipline. Looking in the direction of the subsystems reveals that theories such as that of the 'special stimulus response' and of the 'autonomy of the nervous systems' are further attributable those governing neurophysiology, biochemistry etc. In contrast, in the direction of the higher-level systems, the framework references are the theories of hereditary behavioral programs along with those governing social behavior and the functions of socialization and species survival. Logically, signals and triggering mechanisms do not arise solely from their materials, but presuppose higher-level systems that lend them their legitimacy.
- (b2) This example again underlines that the cognitive process begins with the directly observable parameters and that, in the symmetry of the three pathways, the explanatory pathways correspond to those of the developmental pathways.

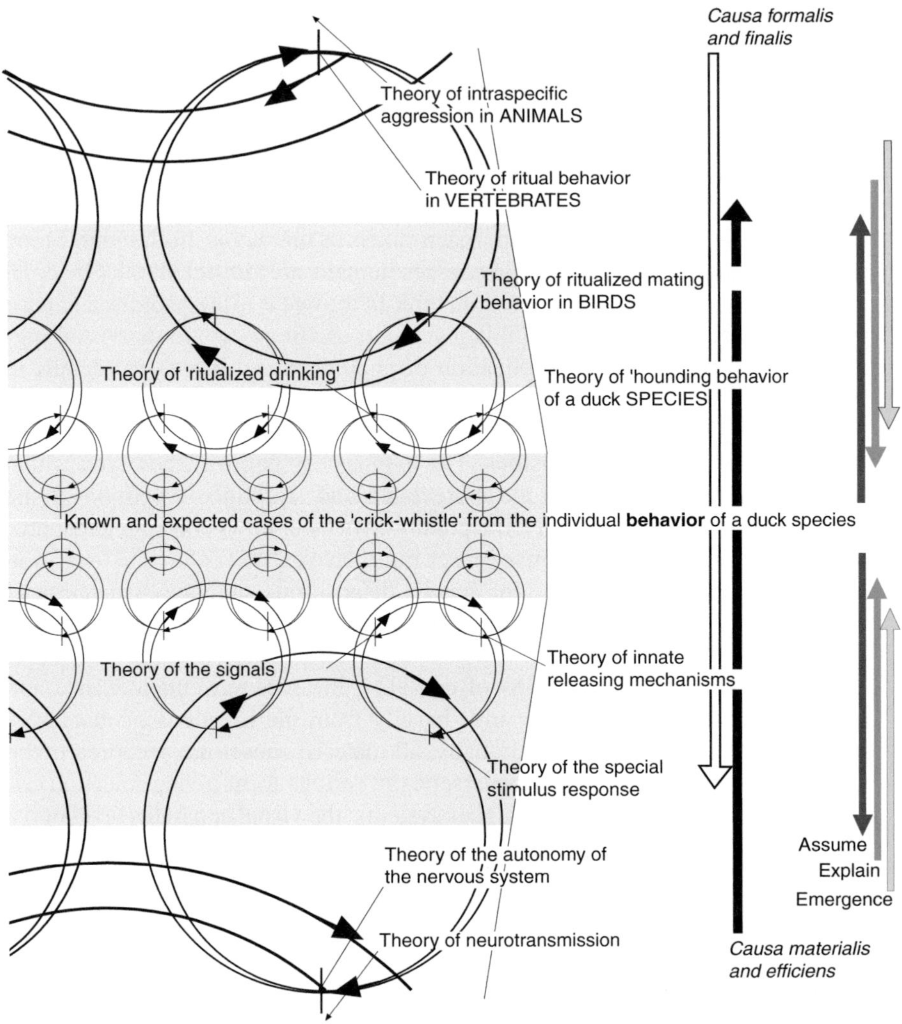

Fig. 6.16 Theory constellations in ethology, exemplified by the ritual behavior of a duck species. The theories extend into neurophysiology via the subsystems, into social behavior via the higher-level systems (compare insert Fig. 6.14; after Riedl 1985, modified)

This example also admirably highlights the four forms of causality, namely that the higher-level systems always provide the framework for understanding the purposes behind the subsystems, that the constituents develop in a preselective manner from the subsystems, and that the higher-level systems once again decide postselectively about the success of new systems based on their usefulness, their fitness.

(c) Explaining organismic components is a twofold task. Clearly, the explanation of an individual structure cannot be entirely dissociated from the evolution of the entire system. Nonetheless, the 'homology problem' for individual traits has become decoupled, warranting a separate presentation—especially because the overall problem is a conglomerate of these individual problems.

Two factors define both tasks. These essential foundations (Chap. 2, Sect. 2.1.3 (c); Fig. 2.1) are the correspondence versus the coherence conditions of selection. The former are driven by the mutagenic variability inherent in every organism, enabling adaptation to the environment and creating the diversity of species. The latter are responsible for the organization of an organism's structural components and for the order of living organisms (in structural and class hierarchies). The problem behind explaining identifiable structural elements lies less in their variability than in the reason for their constancy.

As determined earlier (Chap. 4, Sect. 4.3.1), homologs can be recognized based on those features that resist adaptive modification. This resistence, in turn, was explained by a symmetrical model of functional 'burdens' (Sect. 6.3.2 (f)). While these burdens or historical loadings do not necessarily reduce mutative modifications, they clearly do reduce the probability of functional success.

Again, the theory envisions two phases: coupling gene activities that code for functionally dependent phenes, and using pleiotropies of mutually dependent phenes. These initialize and drive the transformation and adaptability of complex systems. The results, in a second phase, cement the established couplings, leading to the above-mentioned 'organizational burdens', to the constraints that restrict or entirely exclude the probability of further successful adaptive changes (e.g. the chorda of vertebrates).

The next step is to collate the (cl) symmetry of causes into (c2) a systems theory.

- (c1) The hierarchic organization of the types of homologies itself calls for a symmetry of explanation: this is equally valid for the framework homologies and homodynamies as it is for the homonomies (Chap. 4, Sect. 4.3.1 (e)).

  The framework homologies themselves already reveal the tiered structure of their functional burdens. This is because successful changes in a system, for example of a vertebra, are initially tested against higher-level systems. In this case the framework is the function of the adjoining vertebrae, of a region of the spine, of the entire spinal column and ultimately the entire musculoskeletal system. Functional burdens can also be anchored in the subsystems of a vertebra. This holds equally true for the vertebral arch that surrounds the spinal cord as it does for the vertebral body, for the position, size and orientation of its articular surfaces, and ultimately for the organization and rigidity of the trabecular bone structure (Fig. 6.17).

  At the upper end of the scale, the organizational burdens are tested by the fitness of the individual vertebrate, at the lower end by the make-up of its genetic material. All the experience gained in comparative anatomy and systematics shows two things. First, for the entire range of such reciprocal interactions, only very harmoniously developing changes have been successful. Second, the fundamental architectural forms have become immutable: the columnar arrangement of the vertebrae from head to pelvis, the enclosure of the spinal cord, the arrangement of the spinal nerves, and finally the trabecular bone structure itself.

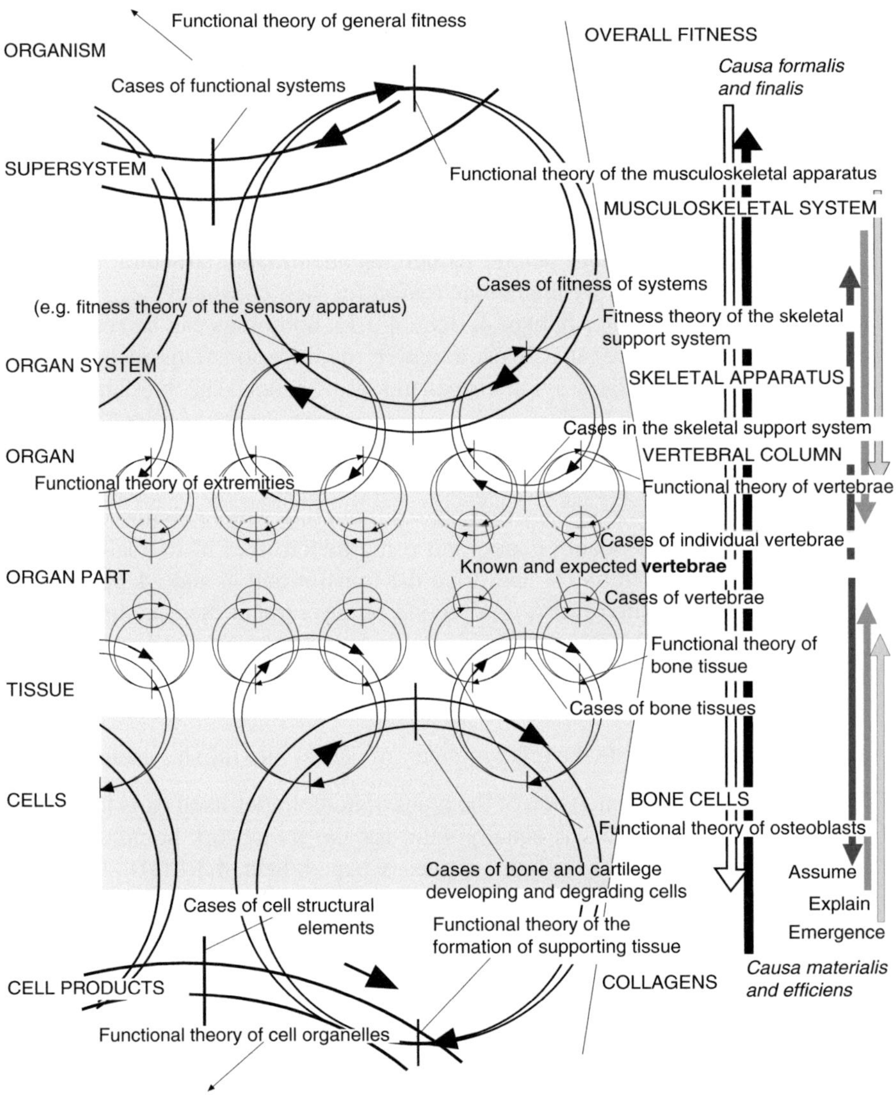

Fig. 6.17 Theory constellation in the concept of functional burdens, based on the vertebrae and vertebral column. Note the tiered system of conditions from cell differentiation to the fitness of the musculoskeletal system. Left: tier levels; right: the case representations (compare insert Fig. 6.14)

The homodynamies also form a hierarchic system of functional dependencies (see also Sect. 6.3.3 (d)): they represent the type of relationships for which genetic functions can be postulated, which also holds true for the explanations of homologies in general.

Standard building blocks—homonomies—are part and parcel of many individualized higher-level systems and, importantly, are also functionally responsible for all of these: the trabeculae not only for the articular surfaces of the vertebrae but everywhere where the musculoskeletal system is called upon to perform tasks. Finally, homonomous structural elements themselves consist of series of sub-homonomies. Thus, every trabeculum consists of bone cells, their organelles and further on down to the regulatory and structural genes that must be called up to produce bone material.

- (c2) System conditions: We still know little about how many of these functional coherences in the genome have already been copied as coupled structural plans, nor how many pleiotropies such coherent developments have already established and regulate. To remain with the above example, those mutations in the spinal column that have been successful (to the extent that they have not destroyed the organism) already demonstrate the existence of functional systemic couplings (short-tail and extra-vertebrae mutants; already highlighted by Waddington 1957).

In principle, this insight is not necessary to explain homology. It does, however, help to understand how such interlinked systems can be successfully modified. The explanation of homology initially requires recognizing the underlying historical loadings or functional burdens. Where these are insufficient to arrive at an explanation (see example presented in Fig. 4.13 and Chap. 4, Sect. 4.3.1 (b1)), the postulate of 'organizational burdens'—those that have become linked to or pre-dated the functional burdens—comes into play.

This system of organizational burdens also exhibits the same interplay between pre- and postselective selection. It is founded in the conditions of organization, in each case operating both from below (the material dispositions) and from above (the higher-level system of the bodyplan). The following section addresses the connection of the three pathways and the four forms of causality.

(d) We can elaborate evolution as an explanatory model based on the interplay of the individual conditions outlined above. The cases in which a mutation of a single structural gene successfully and directly affects a distinct phene are no doubt the exception. The much more likely scenario: interactions between the material dispositions and the bodyplan conditions led to a highly intertwined and hierarchically organized system of inserted interactions between genes and gene products.

As outlined earlier, this epigenetic system is expected to reflect—in simplified, symbolized form—the organizational degrees of freedom and the burdens behind the adaptability of complex structural elements and to define the given constraints. These are the homologies, from the most basal homonomies up to the bodyplan 'type'.

Explaining identifiable structural elements reflects an antagonism between two selection criteria, namely those behind adaptation to the environment and those that harmonize internal organization. The history of evolutionary theory (Sect. 6.3.2) shows that we now largely understand the adaptation process. The prime forces include mutability, the recombination of genes, population dynamics, the opening of new ecological niches, specialization processes and selection by the environment.

We know less about the establishment of order within the organismic realm. Its theoretical underpinnings still require further development.

The model for a theoretical framework explaining the orderliness of evolutionary products must combine two perspectives: both functional and genetic burdens may be at work here. This calls for delving into two families of theories, namely both morphological and genetic. We can justifiably postulate that the theories on the morphological level correspond with those on the genetic level and vice versa.

This perspective enables two predictions. On the one hand, we can expect morphological and genetic theory to exhibit symmetries despite their different methodological approaches. On the other hand, a range of familiar concepts still need to be (re-) formulated to fit the conventions of theory formation.

The core elements of morphological theories have already been discussed under the terms type, bodyplan, forms of homology and homonomies (Chap. 4, Sect. 4.3), touching upon the genetic theories (in Sect. 6.3.2). Stages in ontogeny provide a bridge between the two and are therefore addressed in more detail here. Somewhat simplified (see also Chap. 4, Sect. 4.2.3 (d)), evidence from comparative anatomy and systematics indicates the role of palingenetic traits, i.e. those that recapitulate phylogenetic processes.

This calls for (dl) outlining ontogenetic development. It again proves useful to separately treat (d2) the symmetry of the relationship and (d3) the context of the four causes and the three pathways.

- (d1) The ontogenetic perspective provides evidence that much of this development is predetermined by phylogenetic processes. This insight bases on a detailed differentiation of ceno- and palingenetic traits, whereby the cenogenetic ones go beyond the adaptations to larval or embryonic life to include all simplifications and predispositions. These underline the extent to which the epigenetic system has already adopted organizational constraints from the realm of the phenes.

  This can be illustrated by analyzing several well-known phenomena in ontogenetic development: (i) cleavage, (ii) gastrulation, (iii) induction, (iv) phenotypic epigenesis as well as (v) 'mobility and training'.

  (i) Cleavage refers to the division of the fertilized egg cell into homologizable, i.e. identifiable cell individuals, the so-called blastomeres (overview in Korscheit and Heider 1936). This process typically already defines the symmetries of the respective organism (Fig. 6.18) and clearly represents a predisposition. It is unlikely that ancient seas were initially populated by two- and later by four-celled stages. Multicellular organisms must have arisen before any three-axis organization of the later bilaterally organized creatures, or originated via plasmodial conditions, i.e. multinucleate conditions.

  The process involves a cenogenetic simplification (compare Chap. 4, Sect. 4.3.1 (d)) or symbolization. Nonetheless, a predisposed organization of body axes can only secondarily be related to readying an organism's fitness in the environment. The primary driver must be an organizational principle. The goal: to definitively and broadly distribute the information regarding the fundamental orientation of the future organism. Importantly,

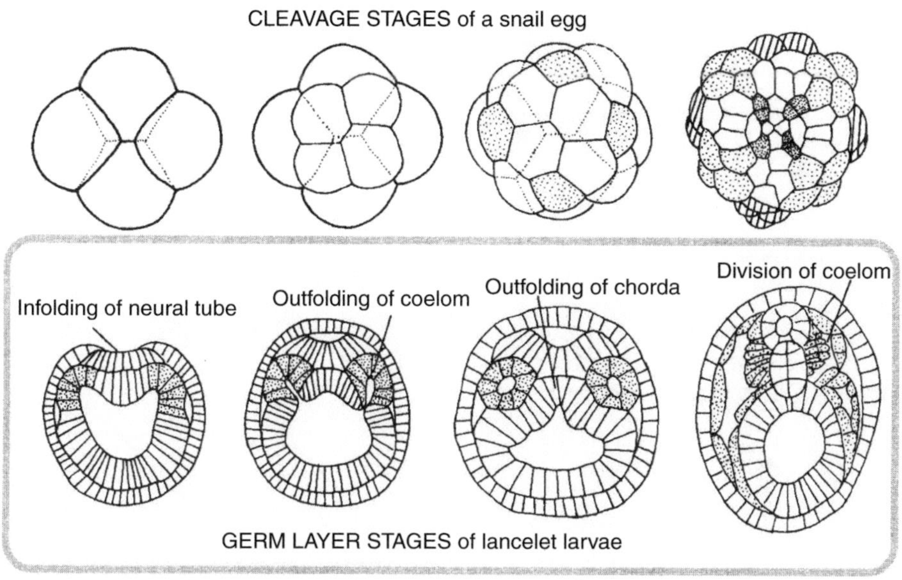

Fig. 6.18 Stages of cleavage and germ layer formation. Blastomeres with predetermined (presumptive) functions are shaded equally, as are the equivalent germ layers. Note the schematic nature of these architectures (after several authors; from Riedl 1975)

this order can only be 'learned' palingenetically from the bodyplan of the later organism. This is because multicellularity had to have arisen before the development of the three body axes (in bilateral organization: dorsal/ventral, anterior/posterior end, right/left).

The blastomeres of the so-called 'mosaic embryo' can form only one part of those three axes, for example the structural elements of either the anterior end, the dorsal side or the right body half. Accordingly, the alternative programs for the respective alternative blastomeres must be suppressed in the genome. This itself is yet another predisposition for future organization, a predisposition based solely and directly on that later organization and dictated by the functional higher-level system. The advantages afforded by that body structure in the environment play a more indirect role.

Based on the established theories on structural and regulatory genes, we need a theory on 'suppression' (Fig. 6.19) to explain how blastomeres are excluded from performing certain tasks.

(ii) This is even more convincing in gastrulation. Gastrulation refers to the organization of the 'presumptive' functions of entire layers of associated cells for specific structural regions of an organism (compare again Fig. 6.18). An outer group of cells, the 'ectoderm', can form only skin and skin derivatives, e.g. nerve channels, lens and hair. The deeper-lying blastomeres, the 'endoderm', can form only the gut and its appendages. Finally, the 'mesoderm' ultimately forms the entire locomotory system including the chorda primordium and the skeletal and muscle systems; it can arise in various ways, even from a single presumptive mesoderm cell.

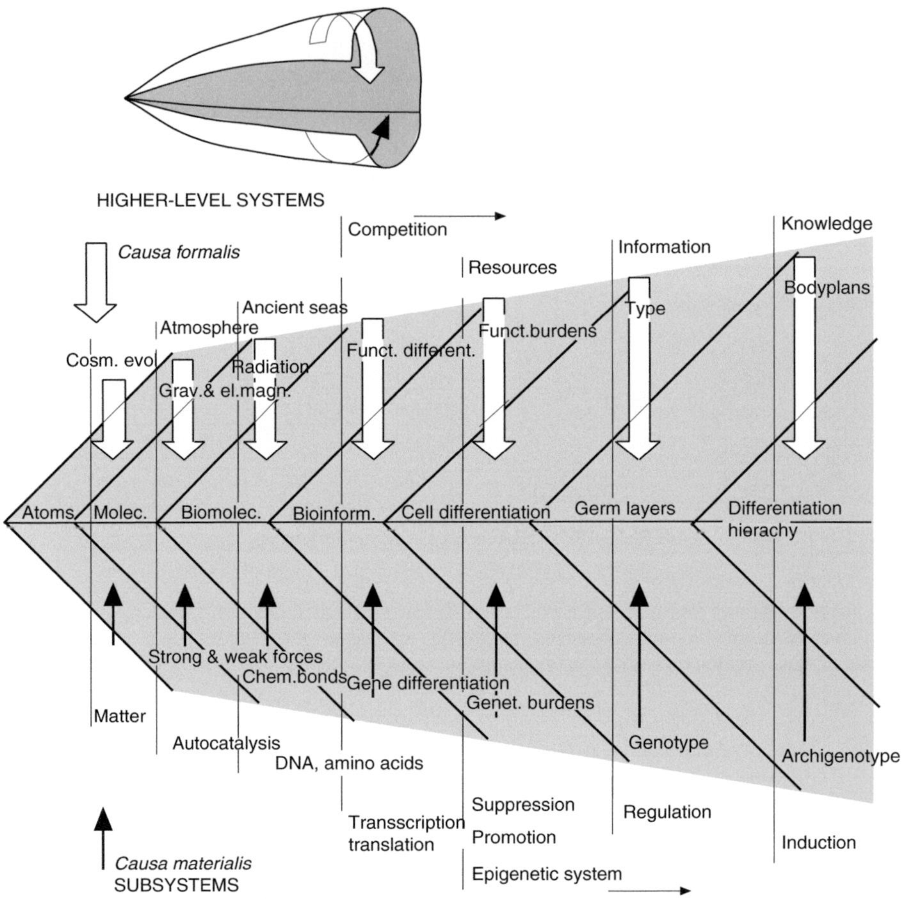

Fig. 6.19 The interrelationships in evolution between material and formal conditions, in each case going beyond the tier boundaries. Seven tiers of the hierarchic system are presented. They are additive over the time axis (in right-hand direction). The initial systems are labeled on the margins of the diagram. Systems directed toward organisms are entered along the axis. The phase-overarching *causae* are positioned between the two. The insert (top) depicts the actual convolute configuration of the scheme because the initial conditions also arose from one another. Compare overview in Figs. 4.2 and 5.3. (grav. & el. magn., strong & weak = the four physical forces)

This represents yet another mechanism that partitions construction plans and ultimately promotes the fitness of the complete organism. At the same time, however, as in house construction, it pre-sorts all the materials and structural plans for the future foundation, walls and roof. This approximates the broadest framework homologies such as those discussed earlier (e.g. the 'locomotory system').

Clearly, the locomotory system of vertebrates could not have arisen simply because two folds of the gut segregated to form the primary body cavity (Fig. 6.18, coelom), which subsequently unfolded lamella-wise into structures that represent the anlagen of the skeleton, connective tissue and musculature. Phylogenetically, they had to have differentiated cohesively. The fact that these germ layers are assigned such separate tasks must be dictated by the later differentiated product itself.

We need a theory (Fig. 6.19) that helps explain how a germ layer is assigned the specifics of its future functions. This calls for envisioning a regulatory system in which all the genes that are needed to produce that germ layer's specific products are switched on synchronously, whereas all others are suppressed. This must involve functional and/or genetic burdens (loading/constraints) that subsequently lead to organizational burdens as outlined in Sect. 6.3.2 (f1). These are unalterable principles of organization; their palingenetic expression already initializes the bodyplan.

(iii) The phenomenon of induction was addressed in Chap. 4, Sect. 4.3.1 (e), Sect. 4.3.3 (a) and Sect. 4.3.2 (e2). Numerous such transfers of information are known (Gilbert 1991). As observed by Riedl (1975), the direction is often from a phylogenetically older to a younger structural element—a logical constellation for any type of information transfer involving coordinated configuration processes. At construction sites the catchword is 'actual measurements'.

An astounding characteristic is that the pattern of this transfer remains the same across all the species of entire phyla (Fig. 6.20). Moreover, the message is understood by that full range of organisms, in the vertebrates for example from hagfish to chickens. This has justified (Chap. 4, Sect. 4.3.1 (e2)) the term homodynamy—homologous messages.

Little is known about the nature of the message transmitter or the manner of reception, somewhat more about the shape of the molecules being sent. In any case, this chemically coded information stems from gene products. This in itself represents a higher form of epigenetic action. Because it is hereditary, this action must involve the activation of special gene segments (even if they still await identification).

The bodyplan principle comes into full play in such induction flows (Fig. 6.20). The mechanisms must evidently be 'learned' from the bodyplan. After all, the anlagen or primordia of numerous presumptive organ groups are involved, and these communicate with each other, stepwise, about their later differentiations and functions. Accordingly, the chorda determines the segmentation of the dorsal mesoderm, which determines that of the vertebrae and the ganglionic ridge, which in turn together determine the arrangement of the spinal ganglia.

The selection principle has long shifted internally, inside the system itself. Similarly, in manufacturing an automobile, the respective tests, for

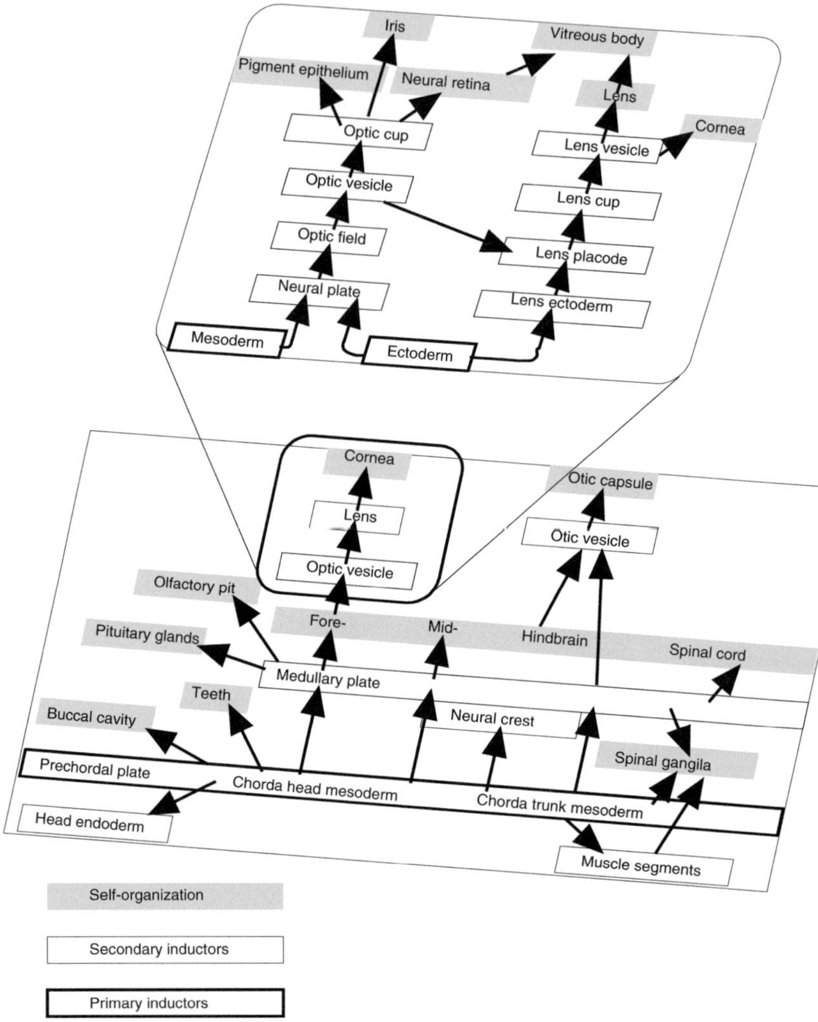

Fig. 6.20 Outline of a basal induction pattern in vertebrates. The development of the eye is extracted from the overview and detailed. The arrows indicate the direction of induction, whereby primary and secondary inductor blastemes as well as those involving self-organization are differently highlighted (after Riedl 1975)

example the fit of the pistons and cylinders, cannot be left to market forces. While automobiles do comply with the market, like in all complex systems the proven mechanical fits are a predetermined given.

(iv) Phenotypic epigenesis leads into the perceptible realm of systemic coordination and even goes one step further based on our knowledge about its genetic control. This heading encompasses a family of epigenetic phenomena showcasing the patterns of regulatory processes, in particular the

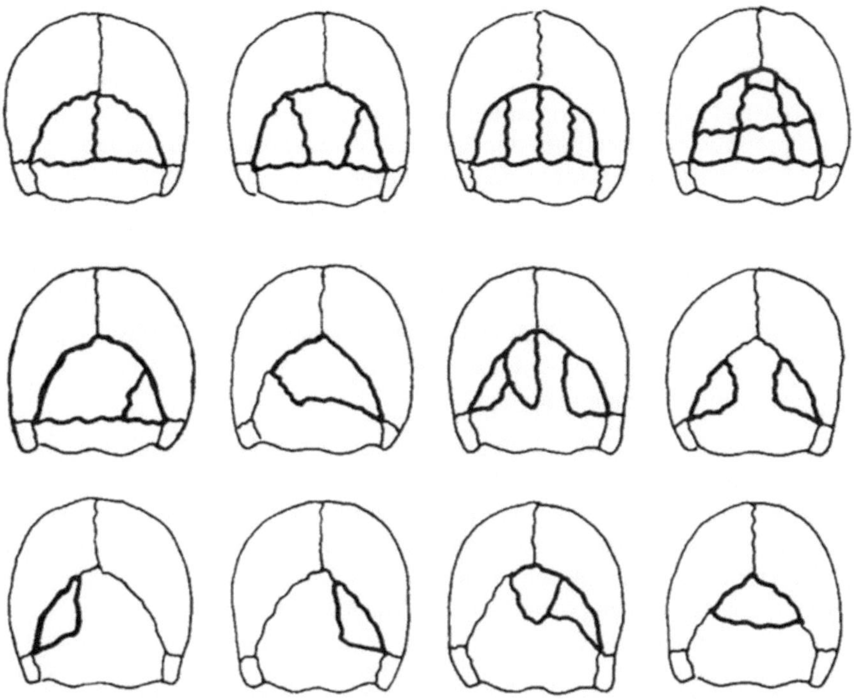

Fig. 6.21 Epigenetic degrees of freedom, exemplified by the variations of bone sutures (in bold) on the back of the human skull (from Hauser and De Stefano 1989)

relationship between degrees of freedom and fixations in the configuration of complex systems (Fig. 6.21).

Physical anthropology and human skulls provide a very comprehensive documentation of this (Hauser and DeStefano 1989). They reveal that the individual structural elements of a particular level, for example the bones of the neurocranium, communicate with each another. These bones never overgrow one another or leave gaps, but always abutt 'as programed'. Where exactly they meet, and where they form which sutures, then underlies controlled degrees of freedom.

(v) The final issue is the function of mobility and training. Recent experience shows their key role in organismic development. The catchwords are 'internal programs' and 'environmental demands', the latter again presupposing internal programs.

With regard to mobility, research shows that extremities, for example those of a chicken embryo (Müller 1994), start to move early on and that the correct formation of the joints is damaged when these movements are prevented by applying local anaesthetics. Equally, every mother is aware that her later-stage fetus can 'kick', and we know that unborn babies even suck their thumbs. The underlying controls remain to be deciphered.

Maturation requires training in the environment itself. Full development is possible only after dealing with external stimuli. If kittens are kept (see Chap. 3, Sect. 3.3.1 (b1)) in an environment lacking certain stimuli or pushed around in carts instead of being allowed to crawl about, the retina program's contour intensification, although prepared, fails to develop. This harks back to the so aptly named concept of the innate master (Angeborenen Lehrmeister) (Lorenz 1978): the final, jelled versions of innate programs are coupled to the environment.

Contour intensification itself, however, is merely a constructive abstraction in the potential circuitry of the retina. After all, the world does not consist primarily of 'contours': These merely play a set role in the 'technique' of gestalt perception as part of the overall organizational programs in the vertebrate bodyplan.

- (d2) The symmetry of sufficient explanation is presented here in a juxtaposition of morphological and genetic theories (because such symmetry is expected in both groups). The main demand placed on this explanatory principle is that it incorporate the solution to problems that the synthetic theory has left open (Sect. 6.3.2 (e2)). It must do justice to the 'systems theory of evolution' (presented in Sect. 6.3.2 (f1) as a supplement to the 'synthetic theory').

In examining the symmetry of causal relationships, structural research—as the name implies—has highlighted the phenes, from the cell organelles to the bodyplan and type. In contrast, the equivalent conditions postulated here for the genetic system are poorly understood. This reflects the technical prerequisites and the minimal methodological overlap in the two realms. Morphologists often do not know what to make of the insights generated by molecular genetics, and molecular geneticists tend to be unaware or wary of the morphological method.

Simply put, the phenes are plainly visible, but the theorems of morphology and systematics (Chap. 4, Sects. 4.3 and 4.4) provide the conceptual system behind them. These (state-of-the-art) theorems best describe the entities found in nature. The theoretical underpinnings discussed above are manifest throughout.

In genetics, the fundamental principles such as the transcription and translation of genetic information are known. The same holds true for certain basic principles of gene regulation, forms of operator genes, cases of gradient development, and the coding of more complex phenes—as exemplified by the homeobox genes. Insights into higher regulatory systems are currently restricted to induction processes (treated in Chap. 4, Sect. 4.3.1 (e), Sect. 4.3.3 (a) and paragraph 4.3.2 (e2)), to the epigenetic variation in the phenes, and to actions in embryos and juvenile stages. Theory plays an important role here as well. The basic tenet is a developmental process involving basic building blocks and leading to higher complexity. Feedback from higher-level systems is left aside (a paradigm that requires re-setting!).

The comparison of the two theoretical frameworks is once again best served by examining intermediate-level, tangible objects. The phene level is opportune because phenes represent easily perceptible structural elements: the balance between their stability and modifiability is the basis for all further theory development (Fig. 6.22).

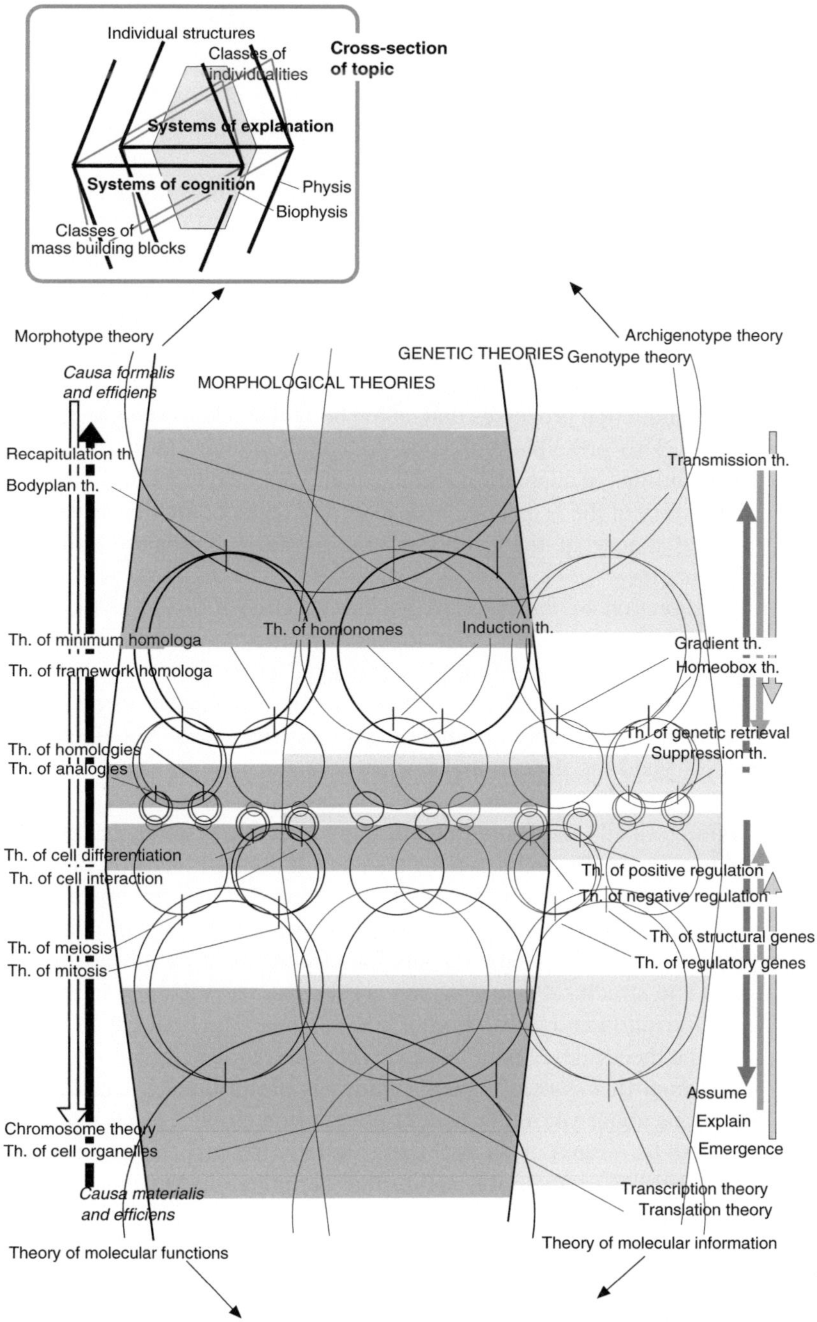

Fig. 6.22 Juxtaposition of morphological and genetic theories, incorporating the principles of functional and genetic burdens (historical loadings) to help explain organic order. Several examples are matched up tier-wise. Note (insert) that the theories of cognition and explanation are positioned behind one another

This calls for differentiating theories taking on (i) the higher-level systems versus (ii) the subsystems.

(i) In the direction of the higher-level systems the successive levels are the theories of homonomies, of the forms of homology, of body plans, and of the type. As outlined earlier, the general principle of the 'morphological type' follows from the forms and degrees of freedom of the body plans. The body plans, in turn, are derivable from the forms of homologies, and these from the differentiability of homonomies. Cases in point are the incremental individualization of arthropod extremities or of the teeth of reptiles to those of mammals.

Symmetrically, at the gene level, homonomies must reflect a principle of identical information retrieval. Equally, the homologies must reflect the series of induction processes that, at the bodyplan level, compose the archigenotypes. This principle embodies the complex interrelationships between the possibilities of adaptation and the fixation of adaptational limits, tailored to the history of the respective bodyplan. The cases of archigenotypes justify a general theory of the genotype, just as this reciprocally validates the archigenotypes.

(ii) In the direction of the subsystems, the successive levels in the realm of phenes are: the concepts of histology, of cytology (cells and tissues), of cell organelles, and of equal functions for equal ultrastructures such as membranes, fibers or tubuli. These concepts involve established systems and classifications, but probably not theories in the strict sense. Nonetheless, these concepts yield the definitional frameworks and prognoses we expect from theories.

Symmetrically to this, we expect to find genetic systems that govern this increasingly subdivided series of standard building blocks. A hierarchy of operon systems can be postulated. It is composed of theories governing positive and negative regulatory systems. These are based on the principles of the operon system, which in turn are founded in theories on the regulatory, operon and structural genes, which are themselves anchored in theories on the transcription and translation of genetic data. The final tier in this direction is the theory describing molecularly coded information.

Although this system of explanations is provisional, it enables making predictions about yet undiscovered phenomena (Riedl 1977). Interestingly, it can also be refuted or, as is already partially the case, be confirmed based on such new cases (overview and further literature in Akam et al. 1988; Raff 1996; Scharloo 1991).

- (d3) The next step is to incorporate the four *causae* and three pathways. The approach is based on the duality or bidirectionality of the structural conditions and the parallelism in the morphological and genetic theories (Fig. 6.22). The overall perspective in both theoretical frameworks again ranges from the most comprehensive theories on macro-phenomena at one end to those on micro-phenomena at the other.

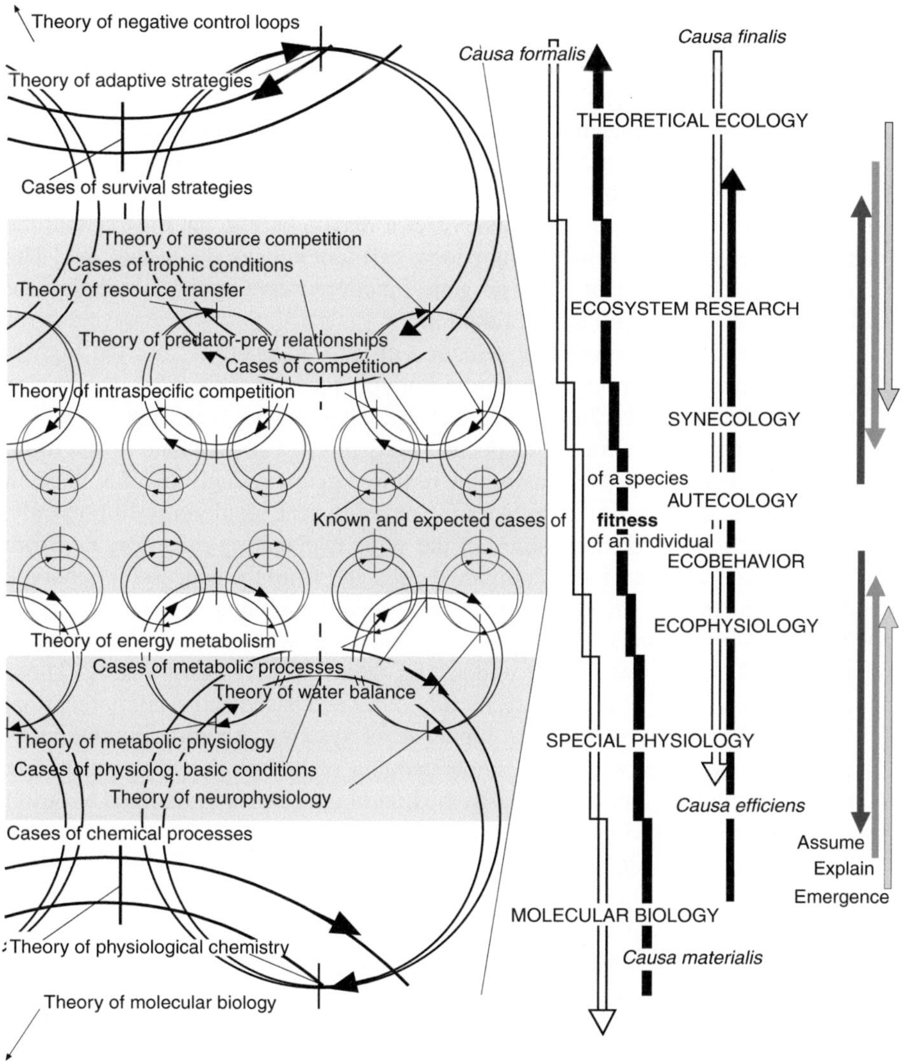

Fig. 6.23 Constellation of ecosystem-relevant theories. Note the pertinent disciplines (labeled on the right), (compare insert Fig. 6.14; after Riedl 1985, amended)

This rendition is restricted to the state of a fully formed, higher organism. The phylogenetic process is omitted in this drawing plane (Fig. 6.22). For that aspect, see Fig. 6.19. This constellation proves to be valid for all systems with a historical component—for the cosmos as a whole and for the physiological and behavioral examples given here (Figs. 6.14, 6.15, and 6.16). It also holds true for the ecological example (Fig. 6.23) and for all intentional systems (in Sect 6.4).

- The next step is to bring the explanatory model into the context of (i) the four forms of causality, followed by outlining its relationship to (ii) the three pathways.

(i) The forms of causality: The system is fueled by its access to energy in the form of photons and by molecular processes. They operate from below, from the cell organelles up through the entire system. Accordingly, they are responsible both for structuring the phenes and for driving the genetic processes. The same holds true for the materials. On the phene level it may appear trivial to state that the organelles are reflected in cell functions, cell functions in the tissues, etc. The justified postulate is that all basal gene functions serve as the foundation for (and extend through) all higher functions.

Vice versa, the form-giving selection criteria all operate from the higher-level systems. The principle behind a bodyplan selects the organs, the organs their tissues and cells. Equally clearly, the archigenotypes determine the induction patterns required for the respective body plans. The postulate is that those patterns played a role in releasing the required gene products and that these in turn must be the selection criteria for the required gene regulations. Importantly, all purposes are to be understood in the same explanatory direction, i.e. from the morphotype down to the individual cells and from the successful genotype down to the most elementary processes of gene regulation.

(ii) The first step in incorporating the three pathways is to examine the cognitive pathway. The histories of the underlying theories differ considerably. Those dealing with phene systems are naturally older and based on gestalt perception. This influenced the perceptible higher-level systems. Accordingly, Goethe's type concept (presupposing an understanding of homology) appeared before Owen's homology concept. Equally, the homology criteria formulated by Adolf Remane were followed only later the validations thereof by the present author. The prerequisites for formulating these terms, however, run in the opposite direction: the conditions behind degrees of freedom and constraints—still an 'inner' or 'esoteric principle' for Goethe—justify the forms of homology, the homologies in turn the bodyplan concept, and the body plans the development of the type concept. The succession in the subsystems runs, methodologically dictated, from the tissues and cells to the organelles and ultrastructures.

The genetic-ontogenetic theories are rooted in two approaches: those of embryology and those of molecular genetics. In the direction of the higher-level systems, the germ layer and induction theory were developed early on, in the late twentieth/early twenty-first century, to then be followed by Waddington's theory of the archigenotype. The time axis is reversed in the direction of the subsystems. Technological advances and growing confidence led from the concepts of transcription and translation to the operon systems, the gradients and homeoboxes. Nonetheless, the underlying prerequisites can again be read from the case examples up through to the principle itself.

The explanatory pathway leads from both ends of the double pyramid (Fig. 6.22) to the case examples of organismic organization. All homologies can be explained by bodyplan constraints, and these in turn by type theory. The latter enables predicting—depending on bodyplan—generally valid, typical basic structures, degrees of freedom and fixation. Equally, the specific suppressions can be explained by the respective inductions, which in turn are determined by their archigenotype (which itself reflects the general specifications of the genotype). The explanatory principle is now parked under the concept of functional burdens.

Towards the molecular end of the spectrum, the specifics of the homonomies can be explained by the dispositions of their substructures (down to the molecular level). Equally, the specifics of the regulatory systems can be attributed to the regulatory principle and to the ranking of the genes. The explanatory principle is now parked under the concept of genetic burdens.

As far as the developmental pathway (emergence) is concerned, it once again runs parallel to the explanatory pathway. All complex systems must, as entities, have become saddled with functional followed by genetic burdens. This means that—in synch with the adaptability of systems both simple and complex—the constraints that define orderly evolution were allocated from the onset. This is expressed in the morpho- and genotypes of all organisms and, from that level, has determined the degrees of freedom and fixation of the subsystems. Moreover, the disposition of the nucleic and amino acids was indisputably a bottom-up precondition for the evolution of the genes, and the disposition of the proteins a precondition for the evolution of all phenes.

(e) Explanatory models in ecology: the example selected here involves the interplay that determines species fitness. This primarily refers to an individual's reproductive success or to the potential stability and possible range expansions of a species.

Such a system goes beyond mere trophic, climatic and edaphic components (i.e. diet-related, climate-relevant and spatial conditions) to include organismic factors. The communities of species themselves become an environment for the member species. This is valid for the entire framework of producers, consumers and reducers, largely represented by the flora, the fauna and the microscopic organisms that put the transformed materials back into circulation.

This represents the most complex system in nature, in principle encompassing all representatives of all species along with all their functions, from behavior to molecular processes. It also lends itself to abstraction, to developing a theoretical framework of principles governing all living organisms. Information storage along with material and energy flows are the currency. They not only explain the operational conditions of the biosphere but also provide the backdrop behind evolution as a whole.

Again, we can separately examine (el) the symmetry of the explanatory structure and (e2) the relationship to the three pathways.

- (e1) The symmetry of the explanatory structures in ecology (at the upper end of complexity) is didactically equally valuable as at the lower end (the explanatory principle of physiology; Sect. 6.3.3 (a)). This because we again begin at an intermediate, visible dimension, whether it be a section of forest, a herd, a stream or a bay. The material cycles, for example in a boreal mixed forest or in the South Atlantic Ocean, are one step removed and grasped only later. The same holds true for the transmission of DDT along the food chain or the route heavy metals take through the groundwater.

This circumstance disciplined ecologists early on, forcing them to build on their experience at that intermediate level of observation before tackling the higher-level systems of aut- and synecology, ecosystem research and theoretical ecology, i.e. before separately focusing on species, communities and systems of communities along with their overarching principles (Fig. 6.23). These, in turn, then had to be reconciled with the law of entropy, with the interplay between Earth and the cosmos.

Working in the direction of the subsystems takes us through behavior and reproduction, energy and material flows, down to the biochemistry of energy gain. The successive tiers in both directions are indispensible to adequately explain ecological relationships.

- (e2) This constellation is also reflected in the three pathways. The cognitive pathway's point of departure lies in intermediate, graspable system dimensions. Theories on competition and subsequently to those outlining adaptive strategies were later steps (Fig. 6.23). These, in turn, led in one direction to theories on resources and in the other to their application in physiological chemistry. Conversely, the explanatory pathways proceeded 'inward' from those two endpoints. Thus, the abundance of species and of biocoenoses, as ecological relationships, are founded both in potential adaptive strategies and in chemophysical dispositions.

  The paths of the four forms of causality are equally distinctive. The driving forces originate from the molecular biology-based transfer of photon reception into energy storages and extend through the entire hierarchy of operation. Equally, all basic materials are formed at that lower level, even if they evolve into new system features from tier to tier, and as species, biocoenoses and ecological regions develop ever new qualities. Conversely, the form-determining selection criteria at every level recur onto the next higher level, and all purposes can be attributed to the ultimate strategy, namely survival in the respective communities.

6.4 The Principles of Understanding

This sixth book chapter began with the prospect of ordering the topics it seeks to explain based on the objects those topics encompass. This reflects a lack of awareness for their underlying origins and developments. It also represents an artificial partitioning inasmuch as human behavior itself is controlled by a tight interplay of conscious and subconscious processes.

The arrangement here broadly accommodates our conventions of differentiating between explanation and understanding, between nature and artefacts. Nonetheless, the borders are fluid and, as determined earlier, the methodological approaches in both are principally the same (with the caveat that the programs governing gastrulation and organismic behavior can come to gain an intentional character—see Chap. 5, Sect 5.3 for details).

The next step is to examine how the theories help practically explain the complex products made by humans.

The examination starts with (Sect. 6.4.1) human behavior, followed by (Sect. 6.4.2) those artefacts calling for a historical perspective and (Sect. 6.4.3) institutions whose explanation must forego that perspective.

6.4.1 The Explanatory Models of Human Behavior

This subchapter refers to behavior in the broad sense. Ethology teaches us that human behavior is anchored in phylogenetic predispositions. These predispositions, however, are superimposed by cultural phenomena, whereby consciousness, intentionality, language and culture play new roles. In the present context, the concept of behavior also encompasses intentions and actions. The physical products of such actions, namely artefacts, are treated at a later point.

The theories behind the attempts to explain such phenomena were already addressed in Chap. 5, Sects. 5.3.2 and 5.3.3, but now call for a causal pathway. That causality can be traced back as far as to energy transformation (albeit only from the theoretical perspective and via our own causal understanding as outlined in Sect. 6.2.3).

The text is best divided, following convention, into the (a) psychological and (b) sociological perspectives. This is because social psychology establishes a useful bridge. This division is justified on two levels and therefore (x1) the symmetry of the explanatory context is again presented before (x2) the relationship to the four causes and the three pathways.

(a) Researchers have repeatedly bemoaned the lack of comprehensive concepts in the explanatory models of psychology. They have had to make do with 'microtheories' to justify individual investigations (Foppa 1975). Moreover, the families of these theories are divided into two camps, the theriomorphic and culturomorphic approaches. The former derives our mental capacities entirely from those of mammals (therio), the latter from human culture. One monograph on learning theory (Bower and Hilgard 1983/84) even split the issue into two separate volumes along these lines.

- (a1) In the now familiar bidirectional explanation model (Fig. 6.24), the theriomorphic and culturomorphic programs are positioned at the lower and upper ends, respectively, of the theory pyramids.

  An adult coming to the aid of a child he or she does not know can serve as an example. In our culture, we initially explain that action—in the direction of the subsystems—based on theories governing the higher, ratiomorphic competence of our cognitive and social faculties. These then interface with

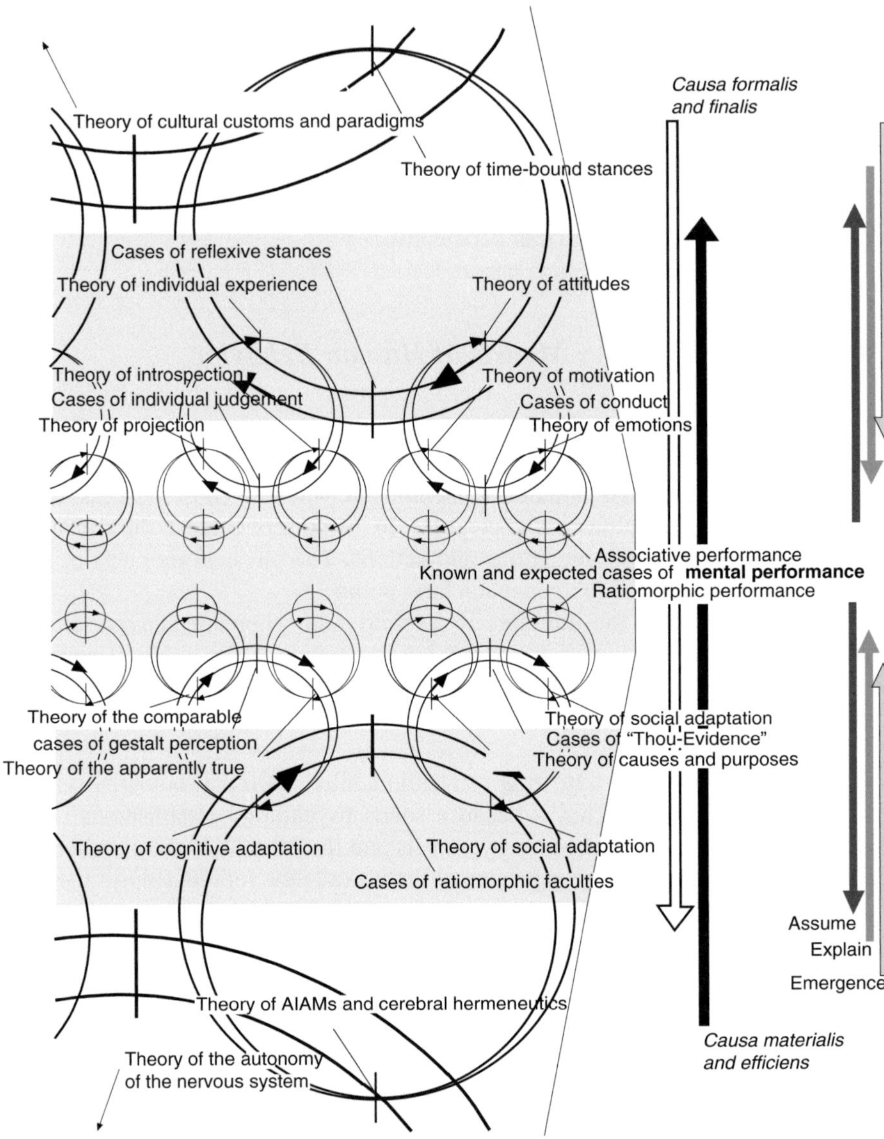

Fig. 6.24 Constellation of theories in psychology, highlighting the intersection of ratiomorphic and rational performance. The directional pathways of cognition, explanation and emergence are outlined. (AIAMs = instantaneous information processing mechanisms, compare insert Fig. 6.14; Thou-Evidence (German: Du-Evidenz) = ability to recognize and respect another person or a (higher) animal as an individual)

the theories on gestalt perception and emotional drives, then further with those governing the neuronal computations ultimately based on AIAMs ('Augenblicks-Information-Auswertenden-Mechanismen'/instantaneous information processing mechanisms) (Lorenz 1978) and on cerebral hermeneutics (Stent 1981).

In the direction of the higher-level systems, we initially rely on associative capacities such as assessing the child's activity, the surrounding conditions and the potential consequences. These in turn are evaluated in the context of our theories on individual experience and attitudes, which are ultimately crowned by cultural imprinting and human nature itself.

- (a2) Regarding the three pathways, the cognitive pathway is naturally based on evaluating the observed action. It continues to the higher levels based on interpreting the experiences and attitudes of the actor, which in turn helps draw conclusions about that person's cultural predispositions. Personal experience, the subject matter of individual- and social psychology, is a central element here. Toward the lower levels, the approach is again anchored in the action itself: it then leads from the various postulates in the theory on ratiomorphic faculties down to theories on the fundamentals of human perception.

  The explanatory pathways go in the opposite direction, with their starting point anchored in the most general of conditions. From above, society defines how attitudes develop, which in turn influences how actions are evaluated. From below, the AIAMs have phylogenetically laid the foundation for the types of perception and ratiomorphic faculties of living organisms. The developmental pathways (emergence) must have taken the same route.

  With regard to the forms of causality, the physical drives extend from the energy gained in cell metabolism through to all the actions of individuals and of the societies they populate. The same holds true for materials. The purposes, foremost that of survival, begin in the environment and extend, like the form-determining selection criteria, through the hierarchic levels down to the cellular processes. Recall again that all the intermediate tiers must have developed as insertions: their genesis requires the same attention afforded to the evolutionary processes illustrated in Fig. 6.19.

(b) Explanatory models in sociology can closely resemble those outlined above for psychology. Considering the action of a group instead of an individual (as in Fig. 6.24) enables directly tying into the above experience. The history of sociological theory is not pursued further here, even though it would be insightful (Riedl 1985). It suffices to state that the hermeneutic method was repeatedly addressed and that the interplay of hypotheses has received renewed recognition.

- (b1) The symmetry of these interrelationships is evident. Thus, the actions of a group can be explained, in the direction of higher-level systems, by the 'institutions' in which the group is embedded and the conditions those institutions (social position, city, nation or confederation of states) dictated. In the direction of the subsystems, group actions must be understood based on its members and their individual experiences, attitudes, associative competence and ratiomorphic faculties.

- (b2) The three pathways can be treated in the same context as the individual actions above. The cognitive pathways are rooted in examples of group actions, ultimately providing insight in two directions: the surrounding milieu and the competence of the individuals. Again, the explanatory and developmental pathways run counter to each other.

  Finally, the four causes—much like in the example of individual actions—extend in opposite directions through the entire system of expectations or theories (once again with due consideration of their tiered origins).

6.4.2 Explanation of Artefacts Exhibiting Genealogies

All complex systems have, as determined at length here, a historical component and in some cases even genealogies. This requires the presence of comparable entities exhibiting a connection between structural and class hierarchies (see Chap. 3, Sect. 3.4; Chap. 4, Sects. 4.3 and 4.4). In the present case, the classes of the artefacts are represented by cultures.

Artefacts are created by actions. Accordingly, the respective explanations require documenting and incorporating everything we know about individual and collective activity. Such an explanation is, in itself, insufficient because the act of creativity spawns new structures with new qualities, yielding new arrangements and forms of order. These very innovations need to be explained in order to convincingly elucidate the developmental process.

The explanatory structure behind cultural products provides insight into another aspect discussed earlier on and underlying all considerations: cognitive processes are rooted in phenomena involving intermediate-level, 'graspable' perceptibility. Nonetheless, starting out from any level ultimately ultimately yields the same symmetries, the same three pathways and the same four forms of causality. Philology provides the best example.

Let's start with the explanatory processes in (a) prehistory and archeology, continue with (b) philology and literature studies, and conclude with (c) history and (d) art history. The treatment is again subdivided into the (x1) symmetries and the (x2) causes and pathways.

(a) The explanatory models in prehistory and archeology arose jointly, as did the fields themselves. The differentiation was based on the availability or lack of written documents. The explanatory models, however, followed a very closely related approach. The distinction between external and internal interpretations, i.e. purpose and material, goes back to the turn of the twentieth century. The 'new archeology' of the 1970s then differentiated tiered systems and hierarchies of purpose and objects. This culminated in the context of discovery and the context of validation (Levin 1973), in my terminology the cognitive and explanatory pathways (details in Riedl 1985).

    Of course, not all prehistorians and archeologists immediately followed suit, for one because the process of induction remained controversial and because the differences between pragmatic and formal treatment remained to be fully resolved (examined elsewhere in this book). Overall, however, the systems context established itself.

(a1) Examining the symmetry of the explanatory models follows the traditional approach in these disciplines, namely by focusing on 'things'. More precisely, this refers to the theory behind a thing, behind an individual find, whether it be a hand axe, a clay pot, a figurine or an inscribed stele. In the direction of the higher-level systems, the theory behind a thing proves to be a member of a theory describing a group of things, for example a grave, a burial ground or a settlement. Ultimately, it proves to be part of the theory behind a particular culture. Toward the subsystems, the theories revolve around the composition, handling and manufacture of the particular item, the lowest tier again encompassing theories on the producer's knowledge and tools.

(a2) Anyone who has participated in an archeological dig can confirm that the discovery process, much like unearthing a fossil, is a downright physical experience—shovel-full by shovel-full or even brush stroke by brush stroke. With each step, an expectation, a theory, is discarded, modified, expanded or reinforced.

This nicely highlights the relationship between the three pathways. Things or items form the point of departure—even when that 'thing' is merely a burn mark or the infilling where a wooden pile once stood, even if these things are often immediately 'seen' in fields of similarity and interwoven with expectations and theories. The cognitive pathways invariably run through the entire hierarchy of the respective structure: upwards to the theory behind a culture, downwards to the theory behind the techniques someone may have used. All explanation arises bidirectionally from these encompassing theories. The developmental pathway is clear: the cultures, along with the individually acting persons, were always present before the artefacts, which inserted themselves between these two endpoints.

The relationship between the four causes and the structural hierarchy outlined above is clear. All physical drives arise from an actor and extend up to the workings of his or her culture. The same holds true for the materials, from the clay in the bricks to the city itself. The form-determining selection proceeds in the opposite direction: the concepts of a culture, a cultural sense, determine the type of infrastructure, the buildings themselves and their materials down to their decorative elements and the tools used. Equally, all purposes find their explanation in the next-higher systems: the material in the stone itself, the stone in the building, and the building in the settlement envisioned by a culture.

(b) The process of understanding in philology and linguistics has contributed significantly to modeling our approach to complex systems. Recall Boeckh's insights back in the late Goethe era, which served as a starting point for the present methodological discussion on hermeneutics (Chap. 4, Sect. 4.2.1 (a)). Boeckh's analyses went beyond the tiered structure of words, sentences, contexts, authors, literary genres and zeitgeist to include the reciprocity involved the process of grasping meaning.

- (b1) Boeckh commented on this symmetry in gaining understanding: "As the individuality of speech is expressed ... through the selection and (!) composition of the linguistic elements, both its sides must emerge in this dual relationship" (edition 1966, p. 126). Even the potential circularity is recognized, comprehended and dispelled: "The circle is dispelled here approximatively in that the purpose determining the class can already be partially recognized without complete knowledge of the individuality. This incomplete understanding of the class then provides insights into particulars of the individuality, giving the generic interpretation an improved foundation. Accordingly, both types of interpretation continue to intermesh reciprocally" (p. 131).

  Translated into the present terminology (illustrated in Fig. 6.25, tiers 5 and 6): Case examples of contexts enable developing a theory on style (based on the descriptive structure of the style), just as, conversely, the cases of styles yield a theory on context (based on the descriptive sense of the text syntax).

  Again, no theory on style stands in isolation. Rather, each must prove itself (along with its normative structures) against cases of the zeitgeist (tier 7), just as the theory on context (along with its normative sense) must demonstrate its validity based on examples of sentence meaning (tier 4).

  Although this verbalization of the interrelationships may sound somewhat confusing, it describes precisely how all of us approach a text. Perhaps a better option for understanding the deciphering process is simply to focus on an individual step in the respective illustration (Fig. 4.10). This helps avoid being distracted by the overall framework. Of course, this once again demonstrates how poorly language-based thought is suited for dealing with complex systems.

  Importantly, this interplay between the tiers is only one facet in dispelling the allegation of circularity. Chap. 4, Sect. 4.2.1 (d); and Fig. 4.6 outlined the value of recognizing that no theory is an island: each one is embedded in a hierarchy of theories.

  A case in point is the theory of word meaning (Fig. 6.25), which arises from cases of sentence meanings. Here, the word level naturally contains many words, each of which needs to be contradiction-free based both on sentence interpretations and symbol meanings. The theory of sentence meanings, in turn, which arises from cases of deciphered word meanings, yields many sentences at the sentence level. These must be equally contradiction-free and prove themselves within the context level.

  The first example was taken from the interplay between context and style (tiers 5 and 6) and demonstrated that these recur back to the zeitgeist and to sentence meanings (7 and 4). The second example links word and sentence meaning (tiers 3 und 4) and recurs back to context and symbol meaning (5 and 2).

  This once again highlights an important relationship touched upon earlier, but which philology illustrates more convincingly than any other discipline: the investigation can start out from any level, and only a contradiction-free overall system guarantees sufficient understanding of a written document. Texts are a case in point because at least five tiers are manifest—from the symbol to the context. The task of a reconstruction may be to initially deter-

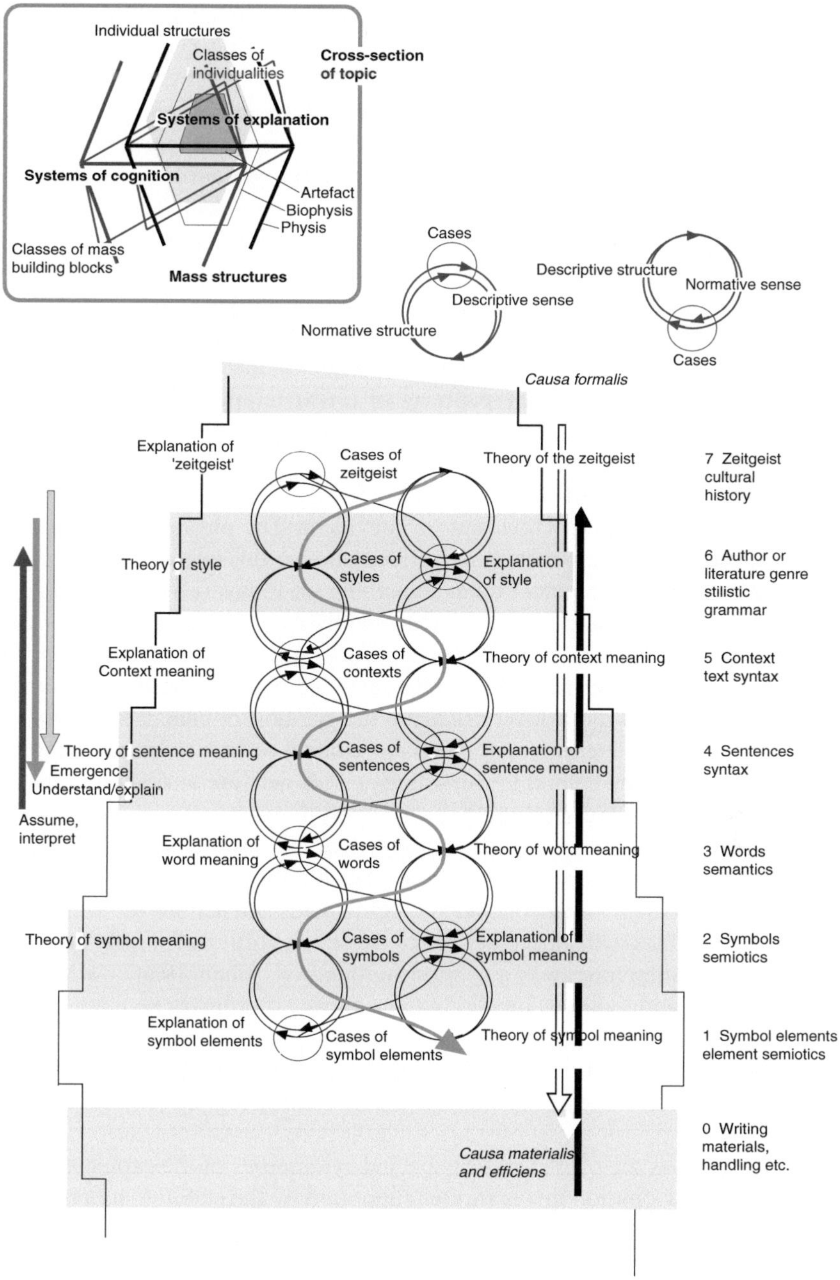

Fig. 6.25 Overview of the explanatory framework in philology, ordered according to normative and descriptive structures and the determination of sense, from the symbol elements (1) to the 'zeitgeist' tier (7). Determinations of the normative sense are designated by white arrowheads. Insert: note that artefacts are now included in the framework of biophysis and physis

mine the genre, author and zeitgeist. These levels are all accessible in everyday reading and correspondence.

- (b2) The genealogy of texts is based on the desire to express oneself, a phenomenon going back to the beginnings of language. In spoken language, the sound is positioned at one end of the scale, the person's message at the other end (context, author and time are still merged). The words themselves already represent insertions because they presuppose semantics and conceptualization. This is reflected in early logographic characters or hieroglyphics, which became differentiated into syllable- and alphabet-based texts.

  The three pathways are recognizable here as well. Accordingly, deciphering the hieroglyphics began centrally at the word level, revealing contexts in the upward direction and syntactic phonetic symbols in the downward direction. In everyday life, the explanatory pathway of a text is apparently determined by knowledge about the author and his or her context. The sense is presupposed. In the deciphering process, however, the explanatory pathway also progresses via the understood symbols.

  The relationship to the four causes is evident. The physical drives emanate from the base, from the writer's motivation, whether this be expressed in a chisel blow or in the effort to reflect upon something or, today, relying on and paying for a computer. The materials are represented by the physical substrate on which the symbols are placed. In the opposite direction, the formal conditions select and order the sentences according to the intended context; the sentences in turn order the words and the words once again the symbols or characters (designated in Fig. 6.25 as the determination of the 'normative sense' of the adjoining tier). Finally, the purposes behind the structures on all tiers are attributed to the context, to the intent of the message.

(c) The explanatory principle in the humanities has also expanded in recent decades. Most of us were taught history in an approach dubbed by critics as 'top-down history'. The experts then concluded that the dates and actions of military leaders and potentates were necessary but by no means sufficient to fully understand history. The alternative was a 'bottom-up history' (Ehalt 1984)—a history of small groups and families, involving the stances of 'ordinary people'.

  This further polarized history studies, which were already variously split into political institutions, pressure groups, nations and confederations of states, or into histories of mentalities, economies, regions and of the world (Fig. 6.26).

  - (c1) This sets the scene for the expected symmetries of the explanatory principle. In my opinion this is further supported by the stability at both ends of the hierarchic constellation: at the tip of the upper pyramid the longstanding metaphysical concepts such as Christianity or Islam, but also of Enlightenment and Scientism, at the end of the lower pyramid cognitive and social faculties along with family and clan.

    Importantly, cultural history demonstrates along a lengthy axis that all cultural institutions originated as insertions between those endpoints. This is equally valid for the armed forces and bureaucracies, fiefdoms, industries, departments

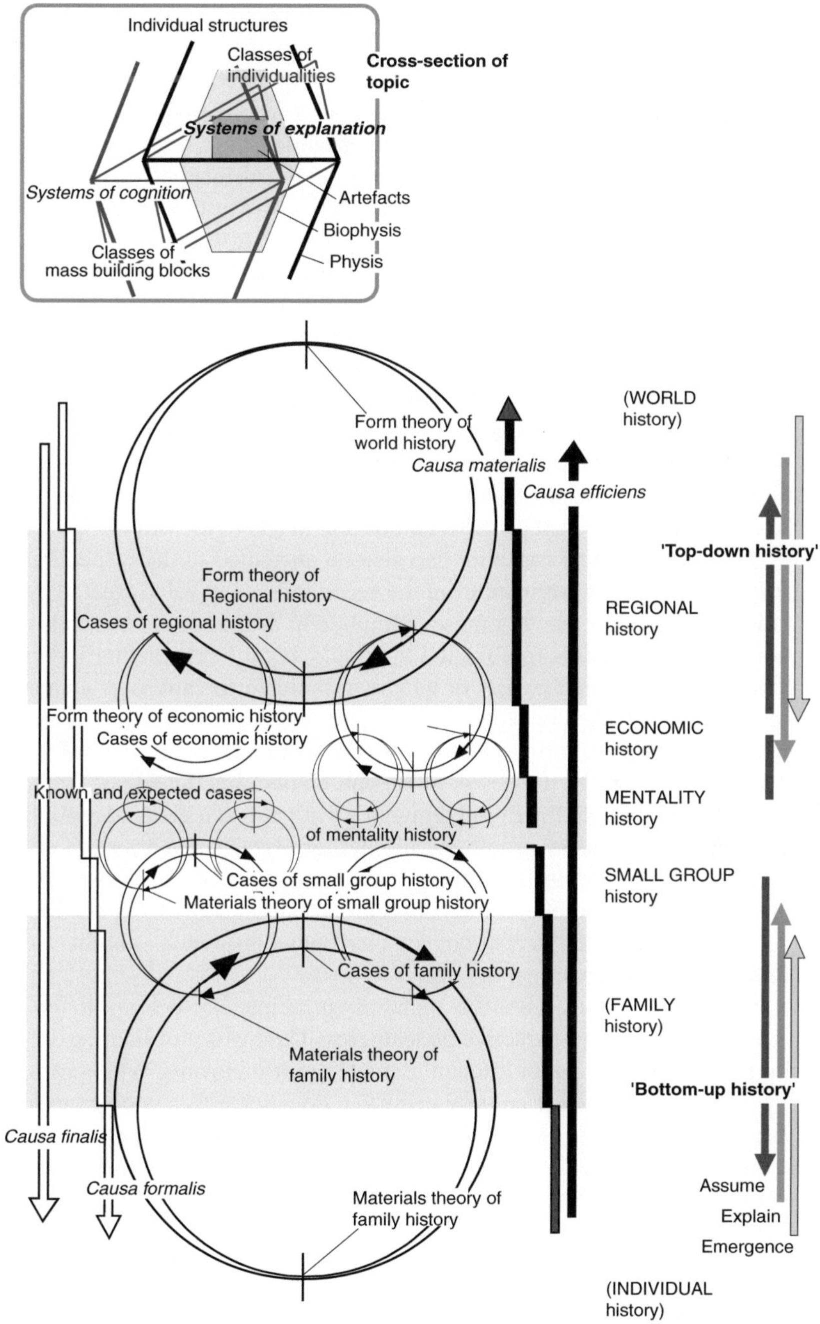

Fig. 6.26 Explanatory constellation in the humanities, a 'top-down history' and a 'bottom-up history', in an interplay between cases and theory and arising from a mentality- and small group history (after Riedl 1975, amended)

of veterans affairs or for art associations. And the preservational conditions of these insertions decrease toward the middle of this systemic constellation.

- (c2) What about the relationship between this hierarchic constellation and the three pathways? All tiers were actually always readily visible, and all knowledge gain was derived from them. The interrelationships, however, first became apparent based on the correlations underlying the visible perceptions.

  A case in point is the explanatory pathway. It reveals that all inserted institutions involve a two-fold causal relationship. Neither of the two extreme solutions provides an adequate explanation. Thus, the idealistic solution, which strives to attribute all causes to the metaphysical background of the historical past, fails. The same holds true for the materialistically inclined solution of psychohistory, which refers back solely to the faculties and needs of the people. In fact, neither of these two approaches is advocated in its pure form today.

  Again, the developmental pathway runs parallel to both explanatory pathways. The metaphysical concept, initially cloaked in mythos and magic and uniting the populations, was positioned at the cradle of human history just like the approach based on the faculties and needs of the actors.

  The four forms of causality can also be attributed to the respective tiers. All physical forces originate from the vigor and added value created by individuals, regardless whether these culminate in armies or in financial institutions. The same holds true for the materials, regardless whether individuals unite to form hunting parties or gangs, to build fortifications or a cathedral, or to join into armies or sects.

  The form-determining selection criteria operate from the higher-level systems, whether this be the power of the church defining the artistic framework of the Gothic period, the Enlightenment giving rise to legions of republics, or the petroleum industry promoting urban highways. The purpose of all these endeavors is rooted in the ultimate aim, namely survival—or at least in a presumed better, more legitimized, secure or fathomable life. And all this can be variously reflected in fashion, bank accounts, insurance, mobility or taxes for education and national defense.

(d) Finally, art history along with its explanatory principle also belongs to the group of artefacts exhibiting instructive genealogies. The history of literature, of seafaring, of weapons, even of kitchen utensils merit inclusion. While art history comprises only part of our cultural history, it does serve as a good example.

The discussion about the differences between its schools is in itself informative. The one side, wrote Sedlmayr, attempts to coin the descriptive terms based on the individual object, the other side strives to arrive at a definition by differentiating a few selected basic terms. The "need to overcome this rigid either/or of both positions" (Pächt 1977) was soon recognized. This insight actually closely resembles the discussion surrounding the problem of morphological type. Pächt holds that theses should be subjected to numerous controls involving repeated back-and-forth predictions and conclusions until they have passed the test.

This back-and-forth always uses case examples from a particular tier to form a theory governing the adjoining tier (Fig. 6.27). Accordingly, history encompasses the contemporary style, and that style in turn encompasses schools and masters,

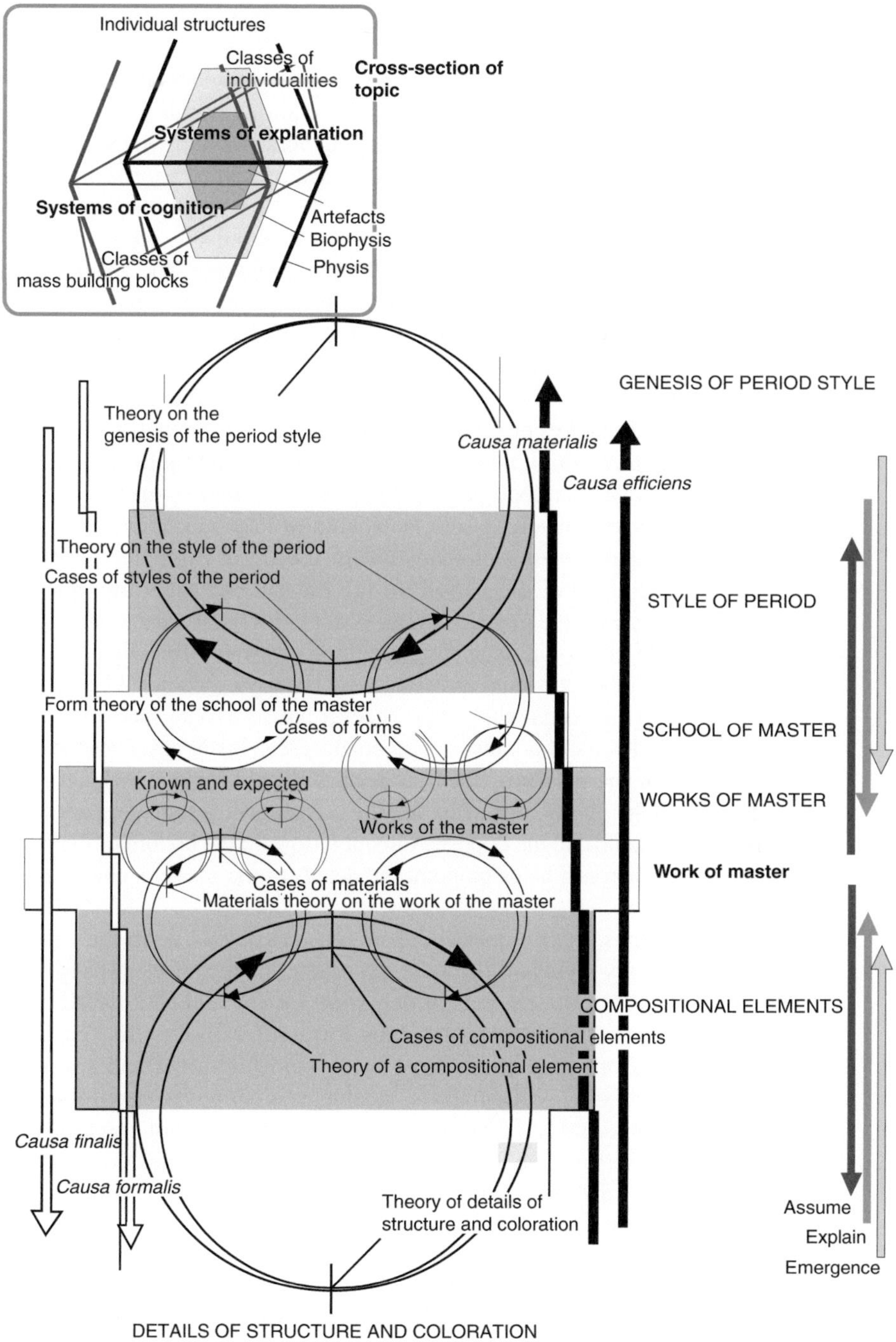

Fig. 6.27 Explanatory constellation in art history in an interplay between cases and theory, starting from the work of a master (from Riedl 1985)

compositional elements, color palettes and designs, along with techniques and details about handling.

- (d1) The symmetry in the process of gaining sufficient understanding here is already evident in the distinction between a psychology versus a sociology of art. This is sometimes referred to as bottom-up versus top-down esthetics.

  The rise and fall of the major epochs in art history can be approached both via historicism and via the subsequent 'art brut' and its repercussions. Both can both be understood based on the attitudes of groups as well as on the viewpoints of individual commissioning clients and buyers. The groups originally comprised the clergy or city councils, later gallery owners, art critics and the media. In contrast, the clients and buyers were churches, art patrons, citizens, dealers, collectors and public authorities.

  The two cliques are by no means independent of one another, yet their intentions often diverge considerably. Sandwiched between them are the actual actors, namely the artists, who, based on their strengths and vision, either consolidate or transcend prevailing styles. And artists—irrespective of their import—remain players in said meshwork of interdependencies.

- (d2) Art history is informative on many levels. It can, for example, mirror the soul and the intentions of the period. In the present context it introduces a measure of clarity into the general lawful order behind complex systems.

  This is already evident in examining the three pathways. The cognitive gain clearly stems from the individual artwork itself. The saying goes, "Whoever has seen a work of art has seen nothing, but whoever has seen many has seen something." An isolated work of art always triggers unease in historians. Developing a series of theories, and bolstering them by reciprocal confirmation, helps narrow down the school the master represents and the contemporary style (going toward the top of the pyramid). In the opposite direction (toward the bottom of the pyramid), this same approach helps pinpoint the details of the master's color palette and style.

  The explanations, in turn, proceed from those two ends, from the genesis of the contemporary style and from the genesis of the techniques. These two endpoints actually form the points of departure for any and all artwork.

  Regarding the relationship to the four forms of causality, the physical drives originate from the bottom-most level, from the artist and from the recruited paint mixers, stonemasons, adepts and shippers, regardless of whether these workers were remunerated in gold or with a hot bowl of soup by the church. The same holds true for the materials, whether they be pigments or stone, which then, via numerous steps, ultimately composed the ceiling fresco in the Sistine Chapel or the Medici tombs.

  The form-determining selection operates from the opposite direction. Thus, the contemporary style and its transformations form the master, the masters their respective schools, with the compositional elements determining the selection of materials and tools. Equally, all these levels recur back to the top tier, and all of them combined recur to the expression and statement made by a work in a particular style.

6.4.3 The Explanation of Institutions in Civilization

All technologies, agreements and even the most stolid products of civilization clearly have a history and often a genealogy. Their study, however, has not yet attained the level reached in the scientific disciplines outlined above. Rather than dabbling in specifics, the presentation here focuses on their current position in the systemic framework.

Two more easily communicable examples suffice: one from (a) economics and one from (b) jurisprudence. They can serve as a model for the treatment of other case studies. As above, the (x1) symmetries are presented first, followed by the (x2) causes and pathways.

(a) The example selected from the economic sciences involves industry, although transport, tourism or the monetary economy would serve equally well. Industry is perhaps most insightful in arriving at an adequate explanatory model, especially when the dimension between the production site and national industry is tangible.

The strong differentiation within the economic sciences has done little to promote a clear overview of the field. The broad distinction into the alternatives Business Administration versus Economics already indicates that explanations can emanate both from the whole and from its parts. Their respective solutions can diverge. 'The customer is king' continues to be a maxim—as if the causal relationships could be unraveled from a single standpoint. After all, back in the 1960s, Galbraith (1968) already demonstrated that industry needs to cultivate the market. This case study highlights all the hurdles (Chap. 3, Sect. 3.3.3) facing innate human intuition in a world that has simply become too complex.

The school of thought that best approximates the systems theoretical approach advocated here is that practiced at the University of St. Gallen. Its doyen, Hans Ulrich, declared as far back as 1981—and in opposition to the one-dimensional models—that one must "impartially test … new insights based on an understanding, hermeneutic or dialectic social science operating on another paradigm; this requires not only explanatory models in the sense of rationalism, but also insights that one might term 'models of understanding'" (pp. 19–21).

(a1) Applied to our expectation of symmetry in the explanatory model, we can presuppose a tiered framework of interrelationships that distinguishes between individuals, production sites, national industries, as well as state and global economies (Fig. 6.28). The automobile industry is a case in point. From the lower end, this assumes individuals with a disposition to purchase a vehicle. The cornerstones: economic wherewithal, willingness, an infrastructure with purchasable vehicles, open roads, gasoline stations and parking opportunities. The framework also presupposes that a number of individuals are willing to participate in producing those vehicles. This is trivial and yet a conditio sine qua non for an automobile plant.

At the opposite end, we require an industrialized civilization that promotes or at least tolerates such infrastructures. This, in turn, requires national indus-

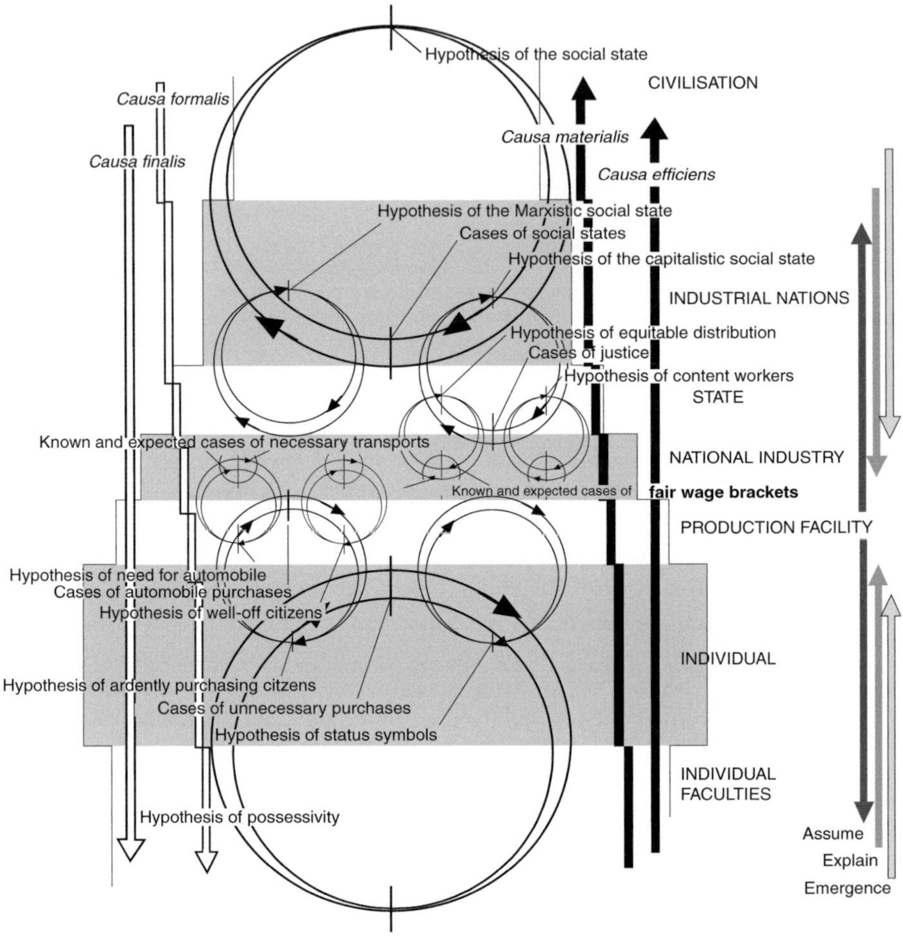

Fig. 6.28 Explanatory constellation -in economic theory in an interplay between cases and theory, starting from the theory of fair wage brackets, between production facility and national industry, and between the faculties of individuals and the conditions of a civilization (compare insert Fig. 6.27)

tries that produce steel and sheet metal or that can acquire these in exchange for other products. Other prerequisites: roads, fuel oil, etc. Finally, the population must support or at least approve of all this. The conditions at this end are equally decisive as those from the other end, but only both perspectives combined yield a full explanation.

(a2) With regard to the three pathways, the perceptions emanate from industry and production sites, which are decades old at least. Rather than being a strictly internal affair, however, their existence is attributable—toward the higher tiers—to factors involving ownership and the liquidity of the industrial complex within the state and to the interplay between industrial nations.

Conversely, toward the lower tiers, it involves the ownership and liquidity of local banks, savings associations and individual bank accounts.

Accordingly, both explanatory pathways must proceed from the structure of a society as well as from that of its citizens. The developmental pathway is clear: societies were present before their industrialization, citizens before technical schools, and both before automobile factories.

With respect to the four causes, all drives and all material production originate from the basis, from the activity of individuals. This also includes activities that produce robots or pump oil from the ground—essentially products of single individuals.

The selection conditions run in the opposite direction: industrial society acts on the public institutions of a state; the state on the potential national industries via infrastructures, taxation and subsidies; and industry on the market via supply, pricing and advertising. Ultimately, the market plays a role in determining what citizens feel they desire and can afford.

The presumed purposes of the whole machinery run a similar tack: from the generated needs to the production sites, to the capitalistic welfare state and to the hypothesis of a humane civilization nurturing a better, more secure and fulfilling existence.

(b) The second example stems from jurisprudence or the sociology of law. The foundation here is—hermeneutics. Symptomatically, this particular method, which drove the discussion on methodology in cultural history, is also being most explicitly applied at the extreme end of complex matters.

Recall the *hermeneutica sacra* and its twin *hermeneutica prophana* (Chap. 4, Sect. 4.2.1 (a)), which were developed early in the Renaissance for legal opinions on complex last wills and testaments. Modern legal theory has never quite discarded them. They remain present above and beyond the phase of 'natural justice or law'. That justice, rather than being based solely on nature, invokes a divinely founded order (German: Ordnung göttlicher Stiftung) and resurfaces as the 'positive law' advocated in the discussions of the Positivists today. Evaluating a case in criminal law, Hassemer wrote back in 1968, requires a "reciprocal enlightenment of the underlying meaning by examining the whole and its component parts" (p. 163). Some have expressed this as casting the eye back-and-forth between the topmost premise and actual life circumstances, others have raised it in the framework of the discussion on morphological type.

(b1) These considerations rely on recognizing the tiered structure of theoretical entities (Fig. 6.29) and on the symmetry of the rules of interpretation. This tiered system extends from word meaning to legal concepts, from legal principles and the legal system as a whole to ideological postulates and to the concepts of a nation representing a culture.

There is general agreement that comparisons must go beyond the legal cases themselves to encompass the evaluations underlying those cases. They should also incorporate the relevant legal principle along with the legal thinking behind each principle. The further consensus is that there is no lower- or uppermost

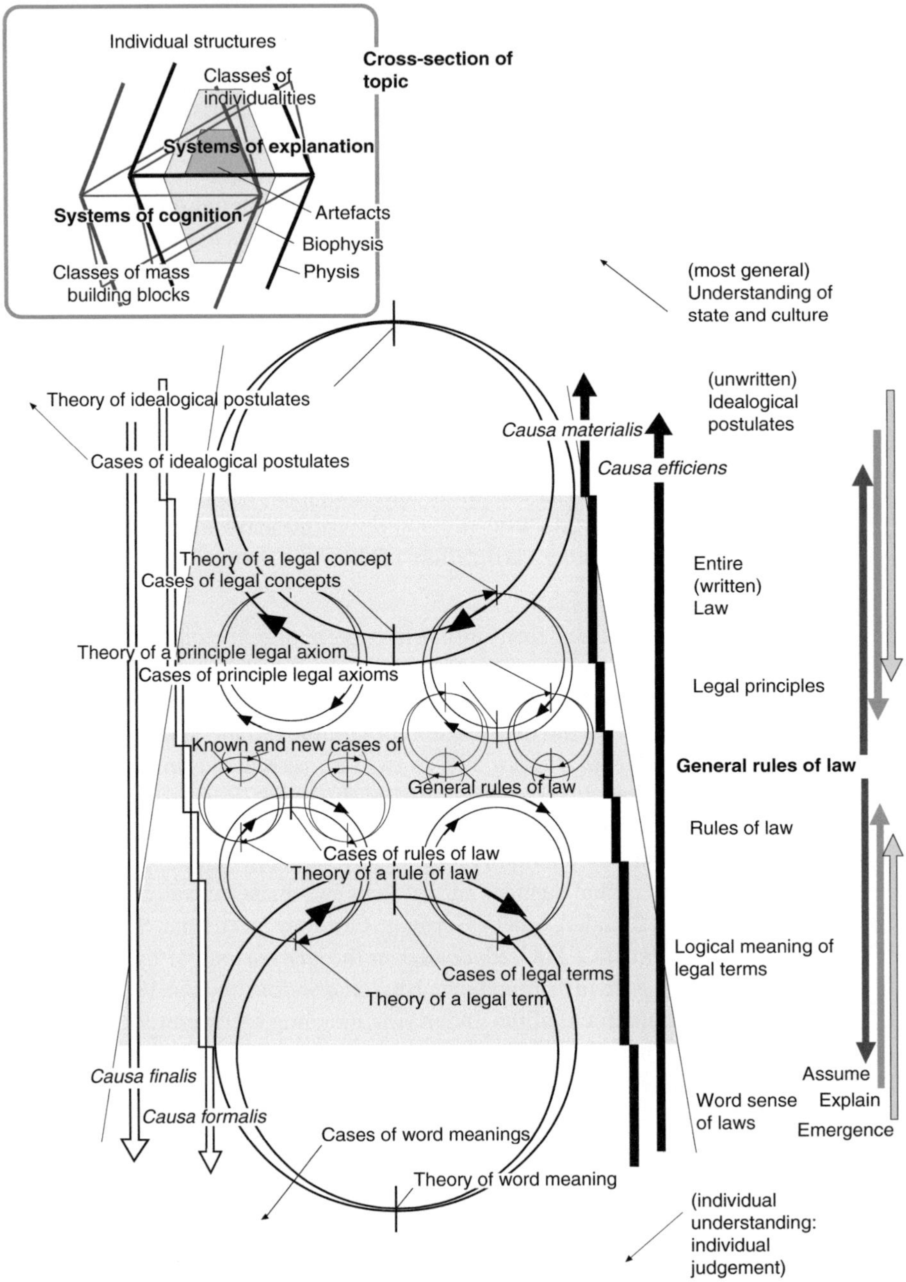

Fig. 6.29 Explanatory constellation in legal theory in an interplay between cases and theory, starting from the level of general rules of law between legal principles and rules of law, between idealogical postulates and word meaning

point of departure for deriving the tiered framework. This is because, beginning from the bottom, even the word meaning must be validatable based on the legal concept and legal principle. Equally, top-down, a teleological interpretation of ultimate purposes cannot always be the definitive principle behind an interpretation. In fact, the interpretation must itself refer back to ideological postulates and further to the legal system.

In the debate on ‘interpretational theories’, the sequence of the interpretational steps remains open. The combined experience gained in deciphering complex systems has shown that it matters little (i.e. is equally beneficial or detrimental) what level is first chosen to begin the process of interlinking with adjoining levels. The decisive factor is that the theories of every level confirm one another in a reciprocal manner throughout the overall system.

(b2) Deciphering a legal system (which is necessary if we wish to pursue the three pathways in the present context) is perhaps best illustrated based on an exotic legal framework. The same didactic strategy was already applied (in Sect. 6.4.2 (b)) in examining philology—we are simply too steeped in our own legal principles to see the forest for the trees. Accordingly, when Hammurabi’s Code was first scrutinized, the legal concepts or legal principles at the intermediate levels provided the basis for initially determining the word meanings in one direction, the idealogical postulates in the other.

The situation is similar in everyday modern life. In the case of traffic regulations, for example, citizens are first confronted with requirements and prohibitions: it is up to the driving schools to convey the nuances in the meaning of these terms and to outline the underlying postulates.

Again, the two explanatory pathways run counter to each other: the case examples are approached from both the word meanings and from the idealogical postulates. Equally, in written law, the expectation is that all legal concepts arose in a balanced interaction between the meanings of the words and the idealogical visions of the respective culture.

The final step is to examine the relationships of the four causes to the tiered structure of a legal system. The physical effort (the role of the *causa efficiens* in the tiers) is of less interest to legal practitioners. This although the development, implementation and maintenance of every legal system involved considerable physical efforts, ranging from ivory tower deliberations to academic and turbulent altercations, from simple agitation to any nmber of wars. All the drives originated in individuals, regardless of whether they were huddled in offices or united into armies—something for historians to muddle over.

The material aspect tells much the same story. In written law this refers to the letters in the alphabet, ergo different in Chinese than in our own language. Nonetheless, even within the European languages, differences in mentality and in the terms can manifest themselves as different constituents of a legal system. The fact that these terms form sentences and build up rules of law is self-understood. Importantly, these terms must prove themselves up into the realm of idealogical postulates. This issue is best left to linguists and cultural historians, but is equally a topic for comparative law.

The selective function of formal causes (*causa formalis*) is more relevant in everyday legal practice. Here, idealogical postulates determine the framework of the legal system. This, in turn, determines the legal principles, which themselves help derive general rules of law. These then function selectively in the choice of legal concepts and in the word meanings for their expression. This represents a crossroads between legislation and justice.

Finally, the *causa finalis* pertains to the purposes of law. Here, the structures of each and every tier draw explanation from the next higher tier, therefore ultimately from the idealogical postulates of a nation representing a culture. This broaches the issue of legitimacy. Natural law invoked divine insight, 'positive law' invoked the sovereign. The latter, the citizens as the sovereign, should be integrated into the legislative process in the form of a living democracy and in accordance with the laws governing complex systems. The legitimacy of an 'evolutive law' must be derived from the humanitarian mandate of the system as a whole.

Chapter 7
Overview and Outlook

Every aspect of human existence is complex, and we cannot escape that complexity. Even a single quantum of light that reaches us triggers complex reactions, be it in our eye or only on our skin. And the simplest fright, a falling rock for example, can churn up our entire nervous system. 10^{28} bits of information are encapsuled in the molecules contained in each and every human body. Expressed in bits, this somewhat more than contained in all of the libraries in human history, including all of the titles in all their editions.

Even our comparatively simple artefacts are complex. How much knowledge is required to handcraft a sharp-edged stone tool, not to mention a modern automobile? Nuclear fission is said to be based on a simple principle, but we can point our fingers at any number of highly complex weak links in nuclear power plants and concede the catastrophic health and political consequences in recent history. Our penchant for dangerous oversimplification is equally apparent in genetic issues, or should we really believe the gene manipulators who declare that our genes or 'nothing but' a chain of molecules.

It is worth repeating: Everything in our lives is replete with complexity—an inescapable destiny. Has this issue suddenly reappeared after some period of past neglect? The fact is that we have always suspected—and respected—complexity. Why else are we so in awe of the unfathomable, of the highly intricate and of historically evolved phenomena? Are the sciences perhaps at fault for simplifying our worldview?

In a nutshell, it makes a lot of sense to address complexity. This retrospective view begins by summarizing (Sect. 7.1) the world around us and human cognition, followed by (Sect. 7.2) the deceptions we fall victim to.

R. Riedl, *Structures of Complexity*,
https://doi.org/10.1007/978-3-030-13064-0_7

7.1 On the Unity of the World and of Cognition

The connection between our environment and human cognition may seem trivial at first sight. After all, what else can we perceive other than the world around us? And this mix even encompasses all the fantasies and follies swirling in our heads, which have become equally 'real' elements in the modern world.

This calls for (Sect. 7.1.1) looking back, from a distance, at why we have strayed from tackling complexity and for (Sect. 7.1.2) mustering key insights that could help deepen our understanding.

7.1.1 On Our Faculties, Language and Culture

The search for an approach to tackling our drift into simplification yields answers at four levels. The initial step is to consider our make-up, our faculties, the genetic programs governing how we perceive the world. Our cognitive apparatus has been formed based on our particular world of experience, but it is clearly adapted more for pure survival than for unraveling universal truths. The cruel selection that enforced this may well have abated as soon as early cultures learned to overcome the fundamental rigors of nature. Nonetheless, we remain adapted cognitively and socially to the world of early humans, to the simplest associative solutions to very simple everyday threats. These adaptations were subsequently completely overwhelmed by the demands of burgeoning culture.

Language—both body language and spoken language—then arose under the pressure of coherence conditions, not under correspondence conditions like our forms of perception (German: *Anschauungsformen*; Kant: 'inate intuition'). Language was developed to achieve mutual understanding and less to depict the world. The nascence of consciousness forced human cultures to choose between putting their trust in visual perception or in language-based thought. To this very day, our philosophical schools suffer from the charge to validate cognition in a contradiction-free manner. And, finally, the heroic effort to free human speech of its contradictions has completely disconnected pure logic from 'dirty reality'.

Such philosophizing has given rise to all the sciences. They, in turn, have created a 'compartmentalized world of thought' due to the unresolved phase transitions and to the way in which all complex phenomena emerge and develop. The sciences have introduced diverging terminologies and languages, constructed enormous systems of knowledge tailored to their very own slice of the cake. And they have acted as if that was all there was to it, often succumbing to the siren song of the doable and then being usurped by the powers to be.

This has driven our culture into deductive straits: Every schoolteacher knows that it is easier to teach and grade deductive outputs achieved by following rules (such as those of syntax and mathematics) than to evaluate inductive, creative achievements. This helps explain why we are governed by lawyers and why even innovative heads of state prefer law-abiding citizens.

This situation has gradually become more transparent, as has the accompanying dilemma. We increasingly recognize the loss of the 'longitudinal' connections, of the holistic view that does justice to the complexity of the world that sustains us. And we are beginning to recognize the full extent of the damage we have wreaked. At the same time, we are far from seeing who might be powerful enough to stand up and renounce the underlying power structures.

This calls for seeking the 'truth', for attempting to depict the true nature of the world, and for making recommendations about how we might once again better adapt to it.

7.1.2 The Systems of Cognition and the Structures of the World

This is the place to recapitulate the key cognitive prerequisites for grasping complexity.

Perceivable phenomena can experience various shifts in their explanation without themselves undergoing any change. Changes in the perceived phenomena themselves, however, require altering the explanation. Neglecting this circumstance means failing to recognize the priority of cognition. The result: cognition tends to be replaced by explanation.

Cognition is anchored in the *simul hoc* of gestalt perception and is a largely preconscious phenomenon. Explanation, in turn, involves the *propter hoc*, which has largely been added on as a conscious, experienced-based construct. Failure to recognize this difference can mean that currently unexplainable things threaten to slip off our cognitive radar screens.

Our modes of thought have all been honed by the underlying structure of the world and are therefore fundamentally equivalent. Neglecting this relationship dooms us to continue accepting the alleged contradictions between causalistic and hermeneutic worldviews as an unresolvable dilemma.

Individual chains of expectations, even single theories, must be embedded contradiction-free in hierarchies of theories and be confirmable in the broadest of contexts. Failing to recognize this relegates any claim to 'empirical truth' to a toothless phrase.

We perceive complex systems conceptually as hierarchies of structures and classes. Neglecting the difference between the two undermines any attempt to validate structural terms or to establish the order behind the class concepts.

Both structures and classes take on the form of double pyramids of aggregate (standard) versus individual (unique) systems. Their common base lies in the mesocosmic realm, which is more readily perceivable and tangible to us. Their tips point in the direction of the overarching theories of the micro- versus the macrocosmos—as defined by the respective 'state of the art'. Neglecting this insight blurs the borders between unique, historical phenomena on the one hand and repeatable, manipulatable phenomena on the other. This renders those borders open to manipulation.

All complex systems originate as insertions between the potential constituents (preselection) and the environment (postselection). No one-sided explanation can ever do justice to complex systems. Failure to recognize this hinders sufficient understanding.

Complex systems arise via phase transitions that involve the emergence of new features that are not contained in the constituents. The notion of explaining them based on the four physical forces or interactions is an empty postulate. That approach must be superimposed by the double duality ingrained in our understanding of cause and effect. Denying this blocks any differentiated perspective of complex causes.

The processes of cognition run counter to those of explanation. The explanatory pathways, however, repeat or recapitulate the developmental pathways of complex systems. Failure to recognize this relationship imparts a peculiar note of randomness to our take on natural laws.

7.2 On Naive and on Nasty Deceptions

Oddly enough, the naive deceptions marking cultural history have mostly been benign, regardless of whether they involve Neanderthals laying flowers in graves, arranging oneself with Zeus and Poseidon, or erecting columns in gratitude to God for having survived the plague. Our interactions with the unknown remained respectful and sacrificial in nature. We should not overlook the distinct possibility that many of today's scientifically founded expectations and theoretical frameworks may prove to be deceptions as well. As long as they prove harmless, however, we can relegate them to the bin of naive deceptions that have always attended human existence.

Today's animal protection, nature conservation and cultural preservation societies also take on this respectful stance. This reflects resistance against our technomorphic, commercialized civilization and our continued poor understanding of the overall frameworks. Such respectful attitudes toward the enigmatic help protects us from the nasty type of deceptions.

The truly nasty deceptions are an entirely different animal. As one might expect, they are rooted in a vicious circle of collective arrogance and ignorance. What precisely makes a deception evil? It turns out that we perceive some such deceptions as being less evil when, for example, only certain individuals or groups are granted unjustified advantages. When this takes on a collective character, however, when such preferences are generally damaging and put humanity as a whole at risk, then we perceive them as being evil. Where, then, lie the advantages or disadvantages for humanity?

Returning to the core issue, we can ask (Sect. 7.2.1) whether explanation can replace cognition and (Sect. 7.2.2), should this entail losses, then how can they be counteracted.

7.2.1 *Can Explanation Replace Cognition?*

"Representatives of a systematic group are similar because they are related and not … (related) because they are similar"—a quote taken from the systematist Ernst Mayr (1969, p. 68), who attributed it to the renowned paleontologist George Gaylord Simpson. The ultimate question to the reader: is this correct or is perhaps the opposite true? Even having fully digested the entire preceding text may not elict a spontaneous response. In fact, the answer requires some deeper thought. This inherent problem of our make-up, our program, will continue to accompany us in the future as well.

Naive deceptions already begin at the simplest level, namely at the fundamental argumentation behind this entire book—the differentiation between cognition and explanation. It is readily apparent that 'similar because related' involves explanation (i.e. based on a similar genetic make-up), whereas 'related because similar' involves a cognitive process (based on comparable forms). This raises the logical question of how an explanation can be forwarded without having previously recognized the similarity.

The notions that paleontologists and systematists have presented in a confusing or at least unacceptably simplified manner are equally muddled in comparative anatomy. This is reflected in the assertion by the anatomist Dieter Starck that: "Homology determinations therefore require knowledge about phylogeny" (1978, p. 11). Knowledge about phylogeny sets its sights on explanation. In contrast, the cognitive process involves homologies that, as underlined in this book, are based on perceiving fields of similarity and on high degrees of confirmation.

I have specifically cited leading authorities who have significantly advanced the field of evolutionary theory since the 1960 and 1970s rather than outsiders. And no one except myself (Riedl 1985) has as yet taken issue with that mainstream. Even in personal discussions, those very same experts opined that formative perception and recognizing fields of similarity are attributes of 'good old common sense' rather than being a topic of science. The suggestiveness of explanation is as strong as blind trust in perceptive intuition. Accordingly, the foundations of their respective sciences remain intuitional.

I still list this deception under the 'naive' category. But might it actually usher in and drive nasty deceptions? The answer depends on whether we believe that naive deceptions can promote a worldview capable of severely harming us in the future.

The underlying paradigms play a role here. Galbraith, for example, maintained back in 1968 that most economic theories are built on a mix of oracles, invocations and phantom apparitions. If true, that probably suffices to set the alarm bells ringing. Returning to the question whether explanation can replace cognition. Our society can in fact make do with this. More precisely, it can operate with such a paradigm and even achieve successes. But what about the losses?

7.2.2 On the Origin, Nature and Controllability of Losses

The cognitive process is tuned to the qualities of historical, irreversible phenomena and operates largely holistically, cybernetically, synthetically. It developed based on true complexity over the course of evolution. In contrast, the explanatory process is attuned to quantities of largely reversible subprocesses and operates mostly reductionistically, logically, analytically. It avoids complexity. It helped us circumvent the complexity that determines our lives.

This process is also driven by the suggestiveness of manipulation and experiment, and by the expectation that we already have an understanding of the complex backgrounds. Equally harmful is the notion that we need not wait for a sufficient understanding of complex phenomena. This approach initially paid off at the individual level, granted influence to groups, and lent power to the state. Those temporary group advantages in turn promoted an underlying technomorphic, commercially oriented mentality. This ultimately spawned the arrogance to think we can improve nature (environmental engineering) and regulate the world (globalization). Finally, the path toward nasty deception was also sown by the pursuit of convenient solutions and by shirking responsibility.

Complex systems require effort to grasp. "Accordingly, it is," as Lorenz (1973b, p. 4) expressed it, "much more expedient to pursue simple, tidy tasks involving self-recording instruments chock full of electronics. This yields a wealth of impressive data and enables conducting ever more impressive mathematical analyses."

The motivation behind this movement is the desire to bring the sciences—foremost biology, much like chemistry—closer to the scientific ideal embodied by physics. The mistake is the failure to recognize that physics itself is undergoing change. The idea was to avoid being degraded to a 'merely descriptive' endeavor and to achieve the status of an 'exact' science. The maze of confusion perpetuated in distinguishing between 'descriptive' and 'exact' is plainly evident.

Appeals, such as those by Bertalanffy (1968), Weiss (1971), and even passionate ones such as Lorenz' "The Fashionable Fallacy of Dispensing with Description" (1973b), have echoed unheard over the past two decades.

What has remained and proliferated is our interventions into complex systems without having adequately understood them beforehand. A case in point is civilization's treatment of the environment—our destruction of forests, degradation of the soil, plundering of the seas, and pollution of the stratosphere. Importantly, these activities were legitimized by the very perspectives fomented by science.

This book illustrates that sad state of affairs based on the reduction that evolutionary theory has experienced. In neglecting interdependencies, phase transitions and recursive causality, industry and politics have brashly plunged into poorly understood systemic frameworks. The underlying motivation: to manipulate any- and everything promising short-term profits.

The scope of the losses suffered to date is difficult to determine. It is entirely possible that states have already lost more 'gross national assets' than they have gained 'gross national products'. It is entirely possible that the nations on this planet already owe each other more than all of them own together (Riedl and Delpos 1996a). That, however, is another story.

This book sought to forward a theory of science that can teach us how to more professionally deal with recognizing and explaining complex systems. Academically scrutinizing the issue can help provide a blueprint for improved clarity and insight. The true litmus test, however, is actually translating this worldview into everyday life, to make it drive human endeavor and to ensure that it is echoed in educational policy. There is no shortage of targeted recommendations.

Importantly, educational policy is equatable with the universities and the degree to which they are influenced by industry, by national interests, by the scientific 'main stream' and on occasion by academies of science. The call is to uphold interdisciplinarity at these nodes, to foster synthetic approaches and promote competence. Phenomena that are not yet understood should once again become part of the curriculum and, should this call go unheeded, each one of us should make every effort to avoid being lulled into complacency. This would be a step forward in understanding phases, tiers and qualities. Mere training would blossom into real education.

Proceeding from that hub, 'teacher training' could metamorphose into true 'teacher education'. In laying down curricula, the disciplines need to be interlinked and the overall picture conveyed. The roles and significance of the various disciplines should be gauged based on their inductive content. This calls for promoting seeing and describing, far- and fore-sightedness, and topping it off with a dash of creativity. Wherever this call goes unheeded, each one of us should cultivate our own ethos.

These considerations are the springboard for ensuring that education encompasses the citizenry as a whole. After all, insights that evolve into majority opinion generate viable political will and practicable educational policy. What are the rewards? The capacity to arrive at a differentiated view of the world; the tools to delve cautiously into its complexity; the ability to recognize and respect our own limitations; the willingness to adapt ourselves anew to the world; and the humility to once again be mindful of uncharted territory.

Bibliography

Adanson M (1782) Histoire naturelle du Sénégal. Coquillages, Paris

Akam M, Dawsoon I, Tear G (1988) Homeotic genes and the control of segment diversity. Development 104:123–133

Alberch P (1980) Ontogenesis and morphological diversification. Am Zool 20:653–667

Altenberg L (1996) Genome growth and the evolution of the genotype-phenotype map. In: Banzhaf W, Eeckman F (eds) Evolution and biocomputation. Computational models of evolution. Springer, Berlin

Aristotle (1976) Metaphysics. Oxford Classical Text. Oxford University Press, Oxford

Ax P (1984) Das Phylogenetische System. Gustav Fischer, Stuttgart New York

Baatz M, Wagner GP (1997) Adaptive inertia caused by hidden pleiotropic effects. Theor Popul Biol 51:49–66

Bacon Fr (1620) Novum Organum scientiarum. Deutsch ab 1793. New edition, with other parts of the great instauration. Urbach P, Gibson J (eds) Open Court, 1994. La Salle, Chicago

Baron-Cohen S (1995) Mindblindness: an essay on autism and theory of mind. MIT Press, Cambridge Mass

Berlin B et al (1966) Folk taxonomies and biological classification. Science 154:273–275

Bertalanffy L (1968) General system theory; foundations, development, application. Braziller, New York

Boeckh A (1966) Enzyklopädie und Methodenlehre der philologischen Wissenschaften. I. Formale Theorie der philologischen Wissenschaften. Neuausgabe (von 1877). Wiss Buchgesellschaft,

Boerner P (1972) Johann Wolfgang von Goethe. Rowohlt, Reinbek/Hamburg

Boltzmann L (1979) Populäre Schriften. Ausgewählt und eingeleitet von Engelbert Broda. Vieweg u Sohn, Braunschweig

Borges JL (1966) Das Eine und die Vielen. Essays zur Literatur. Hanser, München

Bower GH, Hilgard ER (1983/84) Theorien des Lernens I/II. Klett-Cotta, Stuttgart

Brackman A (1980) A delicate arrangement. The stränge case of Charles Darwin and Alfred Russel Wallace. Times Books, New York

Braunfels W (ed) (1980) Die Kunst im Heiligen Römischen Reich. Band II. Die geistlichen Fürstentümer. Beck, München

Brehmer B (1980) In one word: not from experience. Acta Psych 45:223–241

Briggs J, Peat FD (1990) Die Entdeckung des Chaos. Eine Reise durch die Chaostheorie. Hanser, München Wien

Brockhaus Enzyklopädie (1966) 17. Aufl. in 20 Vol. Brockhaus, Wiesbaden

Brockman F (1968) Trees of North America. Golden Press, New York

R. Riedl, *Structures of Complexity*,
https://doi.org/10.1007/978-3-030-13064-0

Buffon G (1749–1804) Histoire naturelle générale et particulière, 40 Vol. Imprimerie Royale, puis Plassan, Paris
Bugnyar T, Huber L (1997) Push or pull: An experimental study on imitation in marmosets. Animal Behaviour 54:817–831
Bunge M (1982) Is chemistry a branch of physics? Zeitschrift f allg Wissenschaftstheorie 13(2):209–223
Burger R (1986) Constraints on the évolution of functionally coupled characters: A non-linear analysis of a phenotypic model. Evolution 40:182–193
Busch W (1982) Gesammelte Werke. Xenos, Hamburg
Callebaut W (1993) Taking the naturalistic turn; or how real philosophy of science is done. Chicago Univ Press, Chicago
Campbell D (1974) Evolutionary epistemology. In Schilpp P (ed) The library of living philosophers: 14 I and II. The philosophy of Karl Popper. Open Court, Lasalle, pp 413–463
Campbell D (1984) Evolutionary epistemology. In: Radnitzky G, Bartley III WW (eds) Evolutionary epistemology, rationality and the sociology of knowledge. Open Court, Lasalle
Capelle W (1968) Die Vorsokratiker; die Fragmente und Quellenberichte. Kröner, Stuttgart
Carnap R (1961) Der logische Aufbau der Welt. Meiner, Hamburg
Chaitin G (2012) Proving Darwin. Making Biology Mathematical. Vintage books, a division of Random House, Inc., New York
Chang Tung-sun (1952) A Chinese philosopher's theory of knowledge. A review of general semantics 9:203–226
Chinery M (1976) Insekten Mitteleuropas. Parey, Hamburg Berlin
Chomsky N (1970) Sprache und Geist. Suhrkamp, Frankfurt/M
Ciompi L (1997) Die emotionalen Grundlagen des Denkens; Entwurf einer fraktalen Affektlogik. Vandenhoeck u Ruprecht, Göttingen
Darwin Ch (1859) The origin of species by means of natural selection; or the preservation of favoured races in the struggle of live. Murray, London
Darwin Ch (1871) The descent of man and on selection in relation to sex, 2 Vol. Murray, London
Darwin Ch (1873) Das Variieren der Thiere und Pflanzen im Zustande der Domestikation (aus dem Englischen von J. Carus), 2 Vol. Schweizerbart, Stuttgart
Darwin E (1794) Zoonomia, or the laws of organic life. Johnson, London
Desmond A, Moore J (1991) Darwin. M. Joseph, London. Deutsch 1992: Südwest, München
Dilthey W (1883) Einleitung in die Geisteswissenschaften. Neuauflage, Bd. I–VII (1973). Vandenhoeck u Ruprecht, Göttingen
Dobzhansky T (1951) Genetics and the origin of species. Columbia Univ Press, New York
Dollo L (1893) Les lois de l'evolution. Bull Soc Belge Geol Pal Hydr 7:164-166
Dollo L (1922) Les Céphalopodes et Tirreversibilité de l'Evolution. (Das Gesetz von 1893) Les Lois de l'Évolution. Bull Soc Belge de Géol Paléont Etc, pp 164 (Procès-verbaux)
Dörner D (1975) Wie Menschen eine Welt verbessern wollten und sie dabei zerstörten. Bild der Wissenschaft, S 198–253
Driesch H (1908) Philosophie des Organischen. Wilhelm Engelmann, Leipzig
Ebeling W, Feistel R (1994) Chaos und Kosmos. Prinzipien der Evolution. Spektrum Akad Verlag, Heidelberg Berlin Oxford
Ebeling W, Freund F, Schweizer F (1998) Komplexe Strukturen: Entropie und Information. Teubner, Stuttgart Leipzig
Ehalt H (ed) (1984) Geschichte von unten. Böhlau, Wien Köln
Eibl-Eibesfeld I (1967, 1978) Grundriß der vergleichenden Verhaltensforschung. Piper, München Zürich
Eibl-Eibesfeld I (1984) Die Biologie des menschlichen Verhaltens. Grundriß der Humanethologie. Piper, München Zürich
Eigen M, Winkler-Oswatitsch R (1975) Das Spiel. Naturgesetze steuern den Zufall. Piper, München Zürich

Eisler R (1927–30) Wörterbuch der philosophischen Begriffe. Mittler u Sohn, Berlin
Engel A, König P, Singer W (1993) Bildung repräsentionaler Zustände im Gehirn. Spektrum der Wissenschaft 93/9:42–47
Engels EM (1989) Erkenntnis als Anpassung? Eine Studie zur Evolutionären Erkenntnistheorie. Suhrkamp, Frankfurt
Foppa K (1975) Lernen, Gedächtnis, Verhalten. Kiepenheuer u Witsch, Köln Berlin
Foucault M (1971) Die Ordnung der Dinge. Suhrkamp, Frankfurt/Main
Freytag-Löringhoff B v (1955) Logik. Ihr System und ihr Verhältnis zur Logistik. Kohlhammer, Stuttgart Köln
Fung Yu-Lan (1952) A history of Chinese philosophy. Vol. I: The period of the philosophers (aus dem Chinesischen von D. Bodde). Princeton Univ Press, Princeton
Futuyma JI (1986) Evolutionary biology. Sinauer, Sunderland, Mass, USA
Gadamer HG (1960) Wahrheit und Methode. Grundzüge einer philosophischen Hermeneutik. Mohr (Siebeck), Tübingen
Galbraith J (1968) Die moderne Industriegesellschaft. Droemer Knaur, München Zürich
Gell-Mann M (1994) Das Quark und der Jaguar. Vom Einfachen zum Komplexen. Die Suche nach einer neuen Erklärung der Welt. Piper, München Zürich (Orig. 1994: The quark and the jaguar. Freeman & Co., New York)
Gilbert SF (1991) Developmental Biology. Sinauer Ass Inc, Sunderland
Gipper H (1969) Bausteine zur Sprachinhaltsforschung. Neue Sprachbetrachtung im Austausch mit Geistes- und Naturwissenschaft. Schwann, Düsseldorf
Glass B, Temkin O, Straus W (1968) Forerunners of Darwin 1745–1859. John Hopkins Press, Baltimore
Gleick J (1988) Chaos – die Ordnung des Universums. Droemer Knaur, München (engl. Ausgabe 1987)
Gödel K (1931) Über formal unentscheidbare Sätze der Principia Mathematica und verwandter Systeme. Monatsschrift für Mathem u Physik 38, pp 173–198
Goethe JW v (1795) Erster Entwurf einer allgemeinen Einleitung in die vergleichende Anatomie, ausgehend von der Osteologie. II. Weimarer Ausgabe. Böhlau, Weimar
Goethe JW v (1817) Weimarer Ausgabe 1891, Nachdruck 1987, Bd. II/6. dtv, München
Goethe JW v (1824) Weimarer Ausgabe 1893, Nachdruck 1987, Bd. II/8. dtv, München
Goodal J (1965) Chimpanzees of the Gombe Stream Reserve. In: De Vore I (ed) Primate behavior. Holt, Rinehart & Winston, New York
Granet M (1971) Das chinesische Denken – Inhalt, Form, Charakter. Piper, München
Grant DA, Hake HW, Hornseth JP (1951) Acquisition and extinction of conditioned eyelid responses as a function of the percentage of fixed-ratio random reinforcement. J of Exper Psychol 42(l):l–6
Grant DA, Schipper LM (1952) The acquisition and extinction of conditioned eyelid responses as a function of the percentage of fixed-ratio random reinforcement. Journal of Experimental Psychology 43(4):313-320
Gregory W (1951) Evolution emerging, 2 Vol. Macmillan, New York
Habermas J (1970) Der Universalitätsanspruch der Hermeneutik. In: Bubner R et al (eds) Hermeneutik und Dialektik I. Mohr, Tübingen, pp 13–103
Haeckel E (1868) Natürliche Schöpfungsgeschichte. Georg Reiner, Berlin
Haken H (1978) Synergetics. An introduction. Nonequilibrium phase transition in physics, chemistry and biology. Springer, Berlin
Haldane J (1958) Theory of evolution before and after Bateson. J Genet 56:11–28
Hartmann N (1950) Philosophie der Natur. De Gruyter, Berlin
Hartmann N (1964) Der Aufbau der realen Welt. De Gruyter, Berlin
Hassemer W (1968) Tatbestand und Typus. Untersuchungen zur strafrechtlichen Hermeneutik. C Heymens, Köln Berlin Bonn München
Hassenstein B (1951) Goethes Morphologie als selbstkritische Wissenschaft und die heutige Gültigkeit ihrer Ergebnisse. Neue Folge d Jahrb d Goethe-Gesellschaft 12:333–357

Hassenstein B (1958) Prinzipien der vergleichenden Anatomie bei Geoffrey Saint-Hilaire, Cuvier und Goethe. Act Coli int Strasbourg Publ Fac lettr 137:155–168
Hauser G, DeStefano GF (1989) Epigenetics of the Human Skull. Schweizerbart, Stuttgart
Hebb D (1949) The Organization of behavior. A neuropsychological theory. Wiley, New York
Hemleben J (1970) Galileo Galilei. Rowohlt, Reinbek/Hamburg
Hempel C, Oppenheim P (1948) Studies in the logic of explanation. Philosophy of Science 15:135–175
Hennig W (1950) Grundzüge einer Theorie der phylogenetischen Systematik. Deutscher Zentralverlag, Berlin
Heschl A (1998) Das intelligente Genom. Über die Entstehung des menschlichen Geistes durch Mutation und Selektion. Springer, Heidelberg
Hillis D, Moritz C (1990) Molecular Systematics. Sinauer, Sunderland, Mass USA
Hinrichsen R, Schultz I (1988) Paramecium: a model system for the study of excitable cells. TINS (Trends in neurosciences) ll(l):27–32
Ho Mae-Wan (1984) Where does biological form come from? Rivista di Biologia 85:757–770
Horgan J (1995) Komplexität in der Krise. Spektrum der Wiss, pp 58–64
Holland PW, Garcia-Fernandez J (1996) Hox genes and chordate evolution. Developmental Biology 173:382–395
Hosseini A, Hogg DA (1991) The effects of paralysis on skeletal development in the chick embryo. I. General effects. J Anat 177:159–168
Huber L (1995) On the biology of perceptual categorization. Evolution and Cognition 1 (2):121–138
Huber L, Lenz R (1993) A test of the linear feature model of polymorphous concept discrimination with pigeons. Quarterly Journal of Expérimental Psychology 46B: 118
Huber L, Lenz R (1996) Categorization of prototypical stimulus classes by pigeons. Quarterly Journal of Experimental Psychology 49B:111–133
Hughes A, Lambert D (1984) Functionalism, structuralism, and 'ways of seeing'. J theoret Biol 111:787–800
Hume D (1739a) Dialogues concerning natural religion (posthumous). In: Essays and treatises on several subjects, Vol. 2. Tourneisen, Brasil
Hume D (1739b) Treatise on human nature (3 Vol.). Philosophical essays concerning human understanding (1748), später als: An enquiry concerning human understanding. (Mit dt.-engl. u. engl.-dt. Begriffsregister. Akademieverlag, Berlin 1965 XLI. Richter R (ed) Felix Meiner, Hamburg 1973)
Hu Shih (1953) The development of the logical method in ancient China. Paragon, New York
Jantsch E (1979) Die Selbstorganisation des Universums. Vom Urknall zum menschlichen Geist. Hanser, München Wien
Jaspers K (1957) Vom Ursprung und Ziel der Geschichte. Fischer, Frankfurt/M
Jaynes J (1993) Der Ursprung des Bewußtseins. Rowohlt, Reinbek/Hamburg. Orig. 1976: The origin of consciousness in the breakdown of the bicameral mind. Houghton Miffin, Boston
Kant IA (1781) Ausgabe B: 1787: Kritik der reinen Vernunft. Abgedruckt in: Kant IA, Werkausgabe Bd. III u. IV. Suhrkamp, Frankfurt/M
Kaufmann S (1993) The origins of order: self-organization and selection in evolution. Oxford Univ Press, New York
Koestler A (1968) Das Gespenst in der Maschine. Molden, Wien München Zürich
Koestler A (1972) Der Krötenküsser. Der Fall des Biologen Paul Kammerer. Molden, Wien München Zürich
Koenig O (1970) Kultur und Verhaltensforschung. Einführung in die Kulturethologie, dtv, München
Korscheit E, Heider K (1936) Vergleichende Entwicklungsgeschichte der Tiere, 2 Bde. Fischer, Jena
Kreiser L, Gottwald S, Stelzner W (1988) Nichtklassische Logiken. Akademie-Verlag, Berlin
Kreuzer F (1981) Leben ist Lernen. Von Immanuel Kant zu Konrad Lorenz. Ein Gespräch über das Lebenswerk des Nobelpreisträgers. Piper, München

Kreuzer F (1982) Offene Gesellschaft, offenes Universum. Franz Kreuzer im Gespräch mit Karl R. Popper. Deuticke, Wien
Kullmann W (1979) Die Teleologie in der aristotelischen Biologie. Aristoteles als Zoologe, Embryologe und Genetiker. Sitzungsver Heidelberger Akad d Wiss Philos-histor Klasse, 2. Abhandlung, pp 1–72
Kutschera F v (1972) Wissenschafts-Theorie, I und II. Grundzüge einer allgemeinen Methodologie der empirischen Wissenschaften. Fink, München
Lamarck J de M (1805) Considérations sur quelques faits applicables à la théorie du globe, observé par M. Peron dans ses voyages aux Terres australes, et sur quelques questions géologiques qui naissent de la connaissance de ces faits. Annales du Mus d'Hist Nat VI (59):26–52
Lamarck J de M (1809) Zoologische Philosophie. Deutsch v H Schmidt. (Französische Erstauflage 1809). Kröner, Leipzig
Lamnek S (1993) Qualitative Sozialforschung, 2 Vol. Psychologie. Verlags Union, Weinheim
Lenneberg E (1972) Die biologischen Grundlagen der Sprache. Suhrkamp, Frankfurt/M
Lévi-Strauss C (1968) Das wilde Denken. Suhrkamp, Frankfurt/M
Levin M (1973) On explanation in archaeology. A rebuttal to Fritz and Plog. American Antiquity 38(4):387–395
Lewin R (1993) Die Komplexitätstheorie. Wissenschaft nach der Chaosforschung. Hoffmann u Campe, Hamburg (Orig. 1992: Complexity. Macmillan, New York)
Linnaeus C (1758) Systema Naturae. Regnum Animale. Tomus I. Holminae, L. Salvii, 824 pp
Lorenz K (1941) Kants Lehre vom Apriorischen im Lichte gegenwärtiger Biologie. Blätter für Deutsche Philosophie 15:94–125
Lorenz K (1963) Haben Tiere ein subjektives Erleben? In: Meinke H (ed) Jahrbuch 1963 – Technische Hochschule München. Verlag Techn Hochschule, München, pp 57–68
Lorenz K (1965) Über tierisches und menschliches Verhalten. Piper, München
Lorenz K (1973a) (1983) Die Rückseite des Spiegels. Versuch einer Naturgeschichte menschlichen Erkennens. Piper, München Zürich
Lorenz K (1973b) The fashionable fallacy of dispensing with description. Naturwissenschaften 60(1): 1–9
Lorenz K (1974a) Die acht Todsünden der zivilisierten Menschheit. Piper, München Zürich
Lorenz K (1974b) Analogy as a source of knowledge. In: Les Prix nobels en 1973. The Nobel Foundation 1974b, pp 176–195
Lorenz K (1976) Die Vorstellung einer zweckgerichteten Weltordnung. Anz phil-hist Klasse, Österr Akad d Wiss 113 (So 2):39–51
Lorenz K (1978) Vergleichende Verhaltensforschung. Grundlagen der Ethologie. Springer, Wien New York
Lorenz K (1983) Nichts ist schon dagewesen: der Irrglaube an eine zweckgerichtete Weltordnung. In: Riedl R, Kreuzer F (eds) Evolution und Menschenbild. Hoffmann u Campe, Hamburg
Lorenz K (1992) Die Naturwissenschaft vom Menschen. Eine Einführung in die vergleichende Verhaltensforschung; Das,Russische Manuskript'. Aus dem Nachlaß herausgegeben von Agnes von Cranach. Piper, München Zürich
Luhmann N (1972) Rechtssoziologie (Vol 1 u 2). Rowohlt, Hamburg
Luria A (1976) The mind of a mnemonist. Regenery, Chicago
Lyell C (1830) Principles of geology. Murray, London
Mahner M, Bunge M (1997) Foundations of biophilosophy. Springer, Berlin
Mainzer K (1994) Thinking in complexity. The complex dynamics of matter, mind, and mankind. Springer, Heidelberg
Malsburg C, Schneider W (1986) A neuronal cocktail-party. Biological Cybernetics 54:29–40
Malthus TR (1817) An essay on the principle of population. Murray, London
Mandelbrot B (1983) The fractal geometry of nature. Freeman, New York
Marr D (1982) Vision; a computational investigation into the human representation and processing of Visual information. Freeman, San Francisco

Maupertuis PLM de (1745) Venus Physique, contemant deux dissertations, l'une sur l'origine des hommes et des animaux; et l'autre sur l'origine des noirs. La Haye
Mayerthaler W (1996a) Linguistik und evolutionäre Erkenntnistheorie. In: Riedl R, Delpos M (eds) Die Evolutionäre Erkenntnistheorie im Spiegel der Wissenschaften. Wiener Univ Verlag, Wien
Mayerthaler W (1981) Wie lernen Kinder sprechen? Die Brücke. Kärntner Kulturzeitschrift, 7 Jahrg, pp 24–27
Mayerthaler W (1982) Das Hohe Lied des Ding- und Tunworts bzw. Endstation „Aktionsding". Papiere zur Linguistik, Heft 27/2:25–61
Mayerthaler W (1996b) Linguistik und EE. In: Riedl R, Delpos M (eds) Die Evolutionäre Erkenntnistheorie im Spiegel der Wissenschaften. Wiener Universitätsverlag, Wien, pp 294–305
Mayr E (1942) Systematics and the origin of species. Columbia Univ Press, New York
Mayr E (1965) Numerical phonetics and taxonomy theory. Syst Zool 14:73–95
Mayr E (1969) Principles of systematic zoology. McGraw-Hill, New York
Mayr E (1988) Eine neue Philosophie der Biologie. Piper, München Zürich
McNeill D, Frieberger P (1994) Fuzzy Logic. Die, unscharfe' Logik erobert die Welt. Droemer Knaur, München
Medawar P, Medawar J (1986) Von Aristoteles bis Zufall. Piper, München
Mittelstrass J (ed) (1984) Enzyklopädie der Philosophie und Wissenschaftstheorie. Bibliogr Inst, Mannheim
Möbius F, Möbius H (1978) Bauornamente im Mittelalter, Symbol und Bedeutung. Edition Tusch, Wien
Mohr H (1981) Biologische Erkenntnis; ihre Entstehung und Bedeutung. Teubner, Stuttgart
Monod J (1971) Zufall und Notwendigkeit. Philosophische Fragen der modernen Biologie. Piper, München Zürich
Morgan C (1923) Emergent evolution. Williams and Norgate, London
Morowitz H (1970) Entropy for biologists. Acad Press, New York
Müller GB (1994) Evolutionäre Entwicklungsbiologie: Grundlagen einer neuen Synthese. In: Wieser W (ed) Die Evolution der Evolutionstheorie. Spektrum, Heidelberg, pp 155–193
Müller G, Newmann S (1999) Generation, integration, autonomy: three steps in the évolution of homology. In: Homology, Novaris Symposium 222. Wiley, Chichester, pp 65–79
Newmann S, Comper W (1990) 'Generic' physical mechanisms of morphogenesis and pattern formation. Development 110(1):1–18
Nicolis G, Prigogine I (1987) Die Erforschung des Komplexen. Auf dem Weg zu einem neuen Verständnis der Naturwissenschaften. Piper, München Zürich
Oberhummer H (1993) Kerne und Sterne: Einführung in die nukleare Astrophysik. Barth, Leipzig
Odum HT (1971) Environment, power and society. Wiley and Sons, New York London Toronto
Oeser E (1979) Wissenschaftstheorie als Rekonstruktion der Wissenschaftsgeschichte. Bd 2 Vol. Experiment, Erklärung, Prognose. Oldenbourg, Wien
Owen R (1848) On the archetype and homologies of the vertebrate skeleton. Assoc Rep 1846:169–340
Pächt O (1977) Methodisches zur kunsthistorischen Praxis. Prestel, München
Patterson C (1982) Morphological characters and homology. In: Joyseyand K, Friday A (eds) Problems of phylogenetic reconstruction. Academic Press, London for the Systematic Society, pp 21–74
Peitgen H-O, Richter P (1986) The beauty of fractals. Springer, New York Heidelberg
Piaget J (1969) Nachahmung, Spiel und Traum. Die Entwicklung der Symbolfunktion beim Kind. Klett, Stuttgart
Piaget J (1973) Der Strukturalismus. Walter, Olten
Piaget J (1975) Das Erwachen der Intelligenz beim Kinde. Klett, Stuttgart
Piaget J (1978) Das Weltbild des Kindes. Klett, Stuttgart
Piaget J, Inhelder B (1958) The Growth of Logical Thinking from Childhood to Adolescence. Basic Books, New York

Pittendrigh C (1958) Adaptation, natural selection and behavior. In: Roe A, Simpson G (eds) Behavior and evolution. Yale Univ Press, New Haven, pp 390–416
Plate L (1925) Die Abstammungslehre. Tatsachen, Theorien, Einwände und Folgerungen in kurzer Darstellung. Gustav Fischer, Jena
Platon (1997) Phaidros oder vom Schönen. Rowohlt, Reinbek/Hamburg
Pöltner G (1993) Eine Auseinandersetzung mit der Evolutionären Erkenntnistheorie. Kohlhammer, Stuttgart
Popper K (1973a) Objektive Erkenntnis. Hoffmann u Campe, Hamburg
Popper K (1973b) Logik der Forschung. Mohr, Tübingen
Popper K (1994) Ausgangspunkte. Hoffmann u Campe, Hamburg
Popper K, Eccles JC (1977) The seif and its brain. Springer, Berlin
Prigogine I, Stengens I (1990) Dialog mit der Natur: Neue Wege naturwissenschaftlichen Denkens. Piper, München. Orig. 1984: Order out of chaos. Man's New Dialogue with Nature. Heinemann, London
Quine WV (1953) From a logical point of view. Harvard University Press, Cambridge
Raff R (1996) Shape of life: Genes, development, and the evolution of animal form. Univ of Chicago Press, Chicago London
Remane A (1971) Grundlagen des natürlichen Systems, der vergleichenden Anatomie und Phylogenetik. Koeltz, Königstein/Taunus
Remane J (1989) Die Entwicklung des Homologiebegriffes seit Adolf Remane. Zool Beitr NF 32(3):497–503
Rensch B (1973) Gedächtnis, Begriffsbildung und Planhandeln bei Tieren. Parey, Hamburg Berlin
Riedl R (1973) Energie, Information und Negentropie in der Biosphäre. Naturwiss. Rundschau 26(10):413–420
Riedl R (1975) Die Ordnung des Lebendigen. Systembedingungen der Evolution. Parey, Hamburg Berlin
Riedl R (1976) Die Strategie der Genesis. Naturgeschichte der realen Welt. Piper, München Zürich
Riedl R (1977) A Systems-analytical approach to macroevolutionary phenomena. Q Rev Biol 52:351–370
Riedl R (1978/79) Über die Biologie des Ursachen-Denkens – ein evolutionistischer, systemtheoretischer Versuch. Mannheimer Forum 78/79:9–70
Riedl R (1980) Biologie der Erkenntnis. Die stammesgeschichtlichen Grundlagen der Vernunft. Parey, Hamburg Berlin
Riedl R (1982) Das gespaltene wissenschaftliche Weltbild unserer Zeit. Aulavorträge Hochschule St. Gallen für Wirtschafts- und Sozialwissenschaften, St. Gallen 18:3–24
Riedl R (ed) (1983) Fauna und Flora des Mittelmeeres. Parey, Hamburg Berlin
Riedl R (1985) Die Spaltung des Weltbildes. Biologische Grundlagen des Erklärens und Verstehens. Parey, Hamburg Berlin
Riedl R (1986) Information aus biologischer Sicht. Biblos 35:14–26
Riedl R (1987a) Begriff und Welt. Biologische Grundlagen des Erkennens und Begreifens. Parey, Hamburg Berlin
Riedl R (1987b) Denkordnung als Abbild der Naturordnung (Wiederabdruck aus: „Evolution und Menschenbild" in: Riedl R (ed) Kultur – Spätzündung der Evolution?) Piper, München Zürich, pp 210–227
Riedl R (1991) Schrödingers Negentropie-Begriff und die Biologie. In: Zeitschr f Wissensch forschung Literas Univ Verlag, Wien 6, pp 53–65
Riedl R (1992a) Wahrheit und Wahrscheinlichkeit. Biologische Grundlagen des Für-Wahr-Nehmens. Parey, Hamburg Berlin
Riedl R (1992b) Bewußtsein, systemisch und evolutionär gesehen. In: Guttmann G, Langer H (eds) Das Bewußtsein. Multidimensionale Entwürfe. Wiener Studien zur Wissenschaftstheorie, Band 4, pp 135–162
Riedl R (1994) Mit dem Kopf durch die Wand. Die biologischen Grenzen des Denkens. Klett-Cotta, Stuttgart

Riedl R (1995a) Deficiencies of adaptation in human reason. A constructivistic extension of evolutionary epistemiology. Evolution and Cognition l(l):27–37
Riedl R (1995b) Goethe and the path of Cognition: An aniversary. Evolution and Cognition 1(2):114–120
Riedl R (1996) Cognition of evolution; Can causal explanation overrule Cognition? Evolution and Cognition, New Series 2(2):88–107
Riedl R (1997) From four forces back to four causes. Evolution and Cognition, New Series 3(2):148–158
Riedl R (1998) Dealing with complex systems; to decipher language and organisms. In: Van de Vijver G, Salthe S (eds) Evolving Systems. Reidel, Dordrecht, pp 267–278
Riedl R, Ackermann G, Huber L (1992) A ratiomorphic problem solving strategy. Evolution and Cognition 2(1):23–61
Riedl R, Delpos M (eds) (1996a) Die Ursachen des Wachstums. Unsere Chancen zur Umkehr. Kremayr & Scheriau, Wien
Riedl R, Delpos M (eds) (1996b) Die EE im Spiegel der Wissenschaften. Wiener Universitätsverlag, Wien
Riedl R, Huber L, Ackermann K (1991) Rational versus ratiomorphic strategies in human Cognition. Evolution and Cognition 1:71–88
Riedl R, Wuketits FM (eds) (1987) Die Evolutionäre Erkenntnistheorie. Parey, Hamburg Berlin
Rieger R, Tyler S (1979) The homeology theorem in ultrastructural research. Am Zool 19:635–664
Rieppel O (1990) Structuralism, functionalism, and the four Aristotelian causes. Journ of the Hist of Biology 23:291–320
Ringel E (1997) Selbstbeschädigung durch Neurose. Psychotherapeutische Wege zur Selbstverwirklichung. Fischer, Frankfurt/M
Ritter J (ed) (1971) In: Ritter J, Gründer K (eds) (1976) Historisches Wörterbuch der Philosophie. Schwabe, Basel Stuttgart
Roberts M (1985) Die Spinnen von Großbritannien und Irland. Bauer, Keltern
Ruddle FH, Bartels JL, Beriüey KL, Kappen C, Murtha MT, Pendelton JW (1994) Evolution of HOX genes. Annu Rev Genet 28:423–442
Sandkühler H (ed) (1990) Europäische Enzyklopädie zur Philosophie und Wissenschaft. Meiner, Hamburg
Scharloo W (1991) Canalization: Genetic and developmental aspects. Annu Rev Ecol Systems 22:65–93
Schischkoff G (1991) Philosophisches Wörterbuch. Kröner, Stuttgart
Schmidt SJ (ed) (1987) Der Diskurs des radikalen Konstruktivismus. Suhrkamp, Frankfurt/M
Schrödinger E (1957, 1977) Was ist Leben? Lehnen, München. Original-Ausgabe (1944) What is life? Mind and Matter. Cambridge Univ Press, London
Schwabl H (1958) Weltschöpfung. In: Paulys Realencyklopädie der klassischen Altertumswissenschaften. Suppl Band IX, Stuttgart, Druckenmüller, pp 1–142
Schweitzer F (ed) (1997) Self-organization of complex structures. From individual to collective dynamics. Gordon and Breach, Amsterdam
Scribner S (1977) Modes of thinking and ways of speaking: Culture and logic reconsidered. In: Johnson-Laird P, Wason P (eds) Thinking: readings in cognitive science. Cambridge Univ Press, Cambridge
Searles HL (1968) Logic and scientific methods. Ronald Press, New York
Sexl R (1982) Was die Welt zusammenhält. Physik auf der Suche nach dem Bauplan der Natur. Deutsche Verl-Ges, Stuttgart
Shannon C, Weaver W (1949) The mathematical theory of communication. Univ of Illinois Press, Urbana
Simon H (1965) The architecture of complexity. General Systems 10:63–76
Simpson G (1952) The meaning of evolution. Yale Univ Press, New Haven
Simpson G (1961) Principles of animal taxonomy. Columbia Press, New York

Singer W (1995) Time as coding Space in neurocortical processing: a hypothesis. In: Gazzaniga M (ed) The cognitive neurosciences. MIT Press, Massachusetts
Sjoelander S (1995) Some cognitive breakthroughs in the evolution of Cognition and consciousness and their impact on the biology of language. Evolution and Cognition 1 (1):3–11
Sneath P, Sokal R (1973) Numerical taxonomy: the principle and practice of numerical classification. Freeman, San Francisco
Snow CP (1959, reprint 1986) The two cultures: and a second look. Cambridge University Press, Cambridge
Sokal R, Sneath P (1963) Principles of numerical taxonomy. Freeman, San Francisco
Spencer H (1850) First principles of a system of synthetic philosophy, 2nd ed. Appleton, New York
Stark D (1978) Vergleichende Anatomie der Wirbeltiere auf evolutionsbiologischer Grundlage, Vol. I. Springer, Heidelberg New York
Stegmüller W (1979) Rationale Rekonstruktion von Wissenschaft und ihrem Wandel. Reclam, Stuttgart
Stegmüller W (1983) Erklärung, Begründung, Kausalität, Band 1, Teile A–G. Springer, Berlin Heidelberg New York
Stegmüller W (1987) Hauptströmungen der Gegenwartsphilosophie, 3 Vol. Kröner, Stuttgart
Stein B, Meridith A (1993) The merging of the senses. MIT Press, Cambridge
Stent G (1981) Cerebral hermeneutics. Journal Social Biol Struct 4:107–124
Stirner M (1845,1866) Der Einzige und sein Eigentum. Reklam, Stuttgart
Thenius E (1972) Die Kreidezeit. In: Grzimeks Tierleben; Ergänzungsband: Entwicklungsgeschichte der Lebewesen. Kindler, Zürich
Thirring W, Stöltzner M (1994) Entstehung neuer Gesetze in der Evolution der Welt. Naturwissenschaften 81:243
Thom R (1975) Structural stability and morphogenesis. Benjamin, Reading MA
Uexküll J v (1937) Umwelt und Innenwelt der Tiere. Z f Tierpsychol I, pp 33–34
Ulrich H (1981) Die Betriebswirtschaftslehre als anwendungsorientierte Sozialwissenschaft. In: Geist M, Köhler R (eds) Die Führung des Betriebs. Poeschel, Stuttgart, pp 1–25
Urey H (1952) The planets. Univ of Chicago Press, Chicago
Vester F (1975) Das Überlebensprogramm. Fischer TB, Frankfurt
Viollet-Le-Duc E (1875) Dictionnaire raisonné du mobilier français de l'époque carlovingienne à la renaissance (Vol. 6). Libr imprim réunis (Morel), Paris
Völk H (ed) (1993) Facetten der Astronomie. Barth, Leipzig
Vollmer G (1979) Evolutionäre Erkenntnistheorie. Hirzel, Stuttgart
Vorländer K (1990) Geschichte der Philosophie, 3 Vol. Rowohlt, Reinbek/Hamburg
Waddington C (1957) The strategy of the genes. Allen and Unwin, London
Wagner GP (1982) The logical structure of irreversible systems transformation. A theorem concerning Dollo's law and chaotic movement. J Theoret Biol 96:337–346
Wagner GP (1983) On the necessity of a systems theory of evolution and its population biologie foundation. Comments on Dr. Regelmann's article. Acta Biotheoretica 32:223–226
Wagner GP (1988) The influence of variation and of developmental constraints on the rate of multivariate phenotypic evolution. J Evol Biol 1:45–66
Wagner GP (1989) The biological homology concept. Annu Rev Ecol Syst 20:51–69
Wagner GP, Altenberg L (1996) Complex adaptations and the evolution of evolvability. Evolution 50:967–976
Wagner GP, Kratky K, Ackermann G (1992) A probabilistic model for the discrimination between periodic and non periodic series of events. Evolution and Cognition 2(1):23–61
Wainwright S (1988) Axis and circumference. The cylindrical shape of plants and animais. Harvard Univ Press, Cambridge Mass, London
Wallace AR (1855) On the law which has regulated the introduction of new species. Ann and Mag of Nat Hist 16:184–196
Wallace AR (1866, 1889) The darwinism. Macmillan, London (Deutsch von Braus D (1891) Vieweg, Braunschweig)

Wallace AR (1889) Darwinism. Macmillan, London
Watson JD (1968) The double helix. A personal account of how the double helix was discovered. Atheneum, New York
Webster G, Goodwin B (1984) A structuralist approach to morphology. Rivista di Biologia 77:503–534
Weinberg S (1977) Die ersten drei Minuten. Der Ursprung des Universums. Piper, München
Weiss P (ed) (1971) Hierarchically organized systems in theory and practice. Hafner, New York
Weizsäcker C-F v (1982) Die Einheit der Natur, dtv, München
Whyte L (1965) Internal factors in evolution. Braziller, New York
Wieser W (1989) Vom Werden zum Sein. Energetische und soziale Aspekte der Evolution. Parey, Berlin Hamburg
Williams P, Fitch W (1989) Finding the minimal change in a given tree. In: Fernholm B, Bremer K, Jörnvall H (eds) The hierarchy of live. Elsevier, Amsterdam
Wimmer M (1995) Evolutionary roots of emotions. Evolution and Cognition l(l):38–50
Wimmer M, Ciompi I (1996) Evolutionary aspects of affective-cognitive interactions in the light of Ciompi's concept of 'Affect-Logic'. Evolution and Cognition 2(l):37–59
Wygotski L (1976) Psychologie der Kunst. (Aus dem Russischen übertragen von Barth H) VEB Verlag d Kunst, Dresden

Index

A

Ackermann, 29, 54
Adanson, 166
Adaptation, 10, 18, 19, 23, 26, 31, 34, 62, 94, 122, 123, 140, 205, 215, 248, 249, 251, 254, 262, 265, 271, 306
Added value, 99, 185
AIAM, *see* Augenblicks-Information-Auswertenden-Mechanismen
Akam, 282
Alberch, P., 256
Albertus Magnus, 226
Algorithm, 31, 40, 54, 55, 57, 58, 60, 64, 93
Altenberg, L., 260, 262
Anagenesis, 37, 66, 97, 101, 178
Analogy, 56, 96, 117–120, 128
Anaximander, 16
Anaximenes, 226
Apomorphies, 166
A posteriori, 18
Appetence, 206
A priori, 17, 39, 64, 195
Archigenotype(s), 262, 282, 284
Architectural style, 5, 117, 197
Aristoteles, 192
Aristotelianism, 226
Artefacts, 3–6, 64, 66, 79, 86, 94, 103, 108, 122, 123, 141, 142, 150, 158, 161, 165, 185, 210, 233, 286, 287, 305
Association, 28, 53, 54, 91, 93, 133, 301
Associative learning, 28, 31, 52, 169, 201
Assumption(s), 2, 17, 40, 92, 96, 101, 183, 187, 199, 212, 213, 216, 231, 245, 249
Athenagoras, 24
Augenblicks-Information-Auswertenden-Mechanismen (AIAM), 205, 206, 288, 289
Augustine, 224
Autocatalysis, 247
Averros, 226
Avery, 256
Axiom, 198, 243

B

Baatz, 262
Bacon, 114
Baer, 253
Baron-Cohen, 216
Batson, G., 9
Baumann, 222
Bauplan, 146
Behaviorists, 216
Berkeley, 24
Bénard cells, 5
Berlin, 165
Bertalanffy, 36, 101, 310
Bidirectional, 99, 153, 265
Bifurcation(s), 80, 178, 232, 237
Biologism, 11, 12
Biosphere, 103, 185, 231, 285
Blanchard, 24
Blind gambler (parable), 263
Body language, 20, 21, 197, 306
Boeckh, 107, 292
Boltzmann, 103, 195
Boole, 162
Borges, J.L., 165
Bosanquet, 162

R. Riedl, *Structures of Complexity*,
https://doi.org/10.1007/978-3-030-13064-0

Bosch, 64
Bower, G.H., 287
Boundary(ies), 67, 70, 71, 86, 106, 141, 142, 153, 154, 158, 162–164, 174, 210
Boyle, 162
Brackman, 250
Bradley, 24, 162
Braunfels, W., 151
Brecht, 232
Brehmer, 70
Bricks, 105, 148, 208, 239
Bridge effect, 60
Bridgman, 162
Brockhaus, 144, 257
Bruno, 226
Bugnyar, T., 28
Buffon, 165, 248
Bunge, 37
Burdach, 8
Burger, 262
Busch, 118

C

Callebaut, 11, 220
Campbell, 10, 33, 122, 225
Carnap, 200, 201
Cartesian transformation, 258
Catastrophe theory, 247
Categories, 17, 18, 47, 140, 170, 202, 251, 309
Causa efficiens, 48, 148, 191, 213, 227, 236, 240, 303
Causa finalis, 148, 191, 226, 231, 237
Causa formalis, 148
Causa materialis, 148, 191
Causal chains, 183, 184
Causalistic, 2, 11, 12, 14, 41, 307
Chance, 35, 64, 91, 139
Chance analogy, 119
Chaos, 2, 4, 6, 36, 105, 230
Chaos theory, 13, 247
Character boundary, 174
Chinery, M., 168
Chomsky, 125
Christianity, 17, 226, 227, 231, 294
Ciompi, 206
Circularity, 106, 114, 143, 200
Cladistics, 166
Cladogram, 167
Class hierarchies, 45, 47, 78, 92, 94, 149, 153, 165, 209, 211, 241, 265, 271
Classes of systems, 78
Classification, 47, 82, 173
Coevolution, 243
Cognitive apparatus, 16, 18, 66, 169, 219, 225, 306
Cognitive dualism, 105, 193, 226, 228
Cognitive pathway, 43, 47, 241, 269, 286, 290, 291
Cognitive pyramid, 42
Cognitive symmetries, 148, 240
Coherence, 23, 271, 306
Coincidences, 136
Common sense, 168, 186, 199, 200, 245, 309
Communication, 18, 38, 58, 67, 96, 114, 187, 212
Comper, 263
Composition, 54, 58, 65, 86, 136, 237
Comte, 227
Conceptual reversibility, 229
Conceptualization, 187, 229, 294
Condillac, 24
Conditioned response, 201
Conditioning, 28, 53, 54, 59, 201
Confirmed prognoses, 55
Consciousness, 15, 16, 28, 33, 34, 36, 47, 53, 74, 80, 97, 183, 231, 287, 306
Consecutive coincidence, 54, 57, 138
Constancy, 32, 59, 64, 129, 170
Constituents, 2, 12, 37, 42, 82, 84, 97–100, 115, 182, 208, 229, 232, 245, 308
Constraints, 5, 23, 34, 52, 100, 211, 263, 265, 277, 285
Construction, 11, 33–36, 64, 92, 97, 105, 118, 149, 193, 234, 235, 237, 240
Constructivism, 34
Context meaning, 293
Continuum, 18
Contour, 280
Contour intensification, 59, 65, 280
Contradiction, 201
Contradiction-free, 23, 89, 108, 174, 187, 292, 307
Convergence, 167, 220
Copernican revolution, 221, 223
Correspondence, 19, 21, 23, 41, 123, 129, 132, 137, 187, 195, 201, 210, 266, 271
Correspondence conditions, 21, 306
Couturat, 162
Creationism, 248
Creator God, 231
Crick, 256
Culturomorphic, 287

Cuvier, 107
Cybernetics, 197

D
Daimler, 133
d'Alembert, 194, 227
Darwin, Ch., 223, 224, 250, 252, 254, 258
Darwin, E., 222, 249
Darwinism, 248, 252, 254
Dawidson, 24
Deception(s), 305, 308–310
Declaratio, 198
Declarative sentence, 72
Deduction, 29
Definition, 1, 3, 5, 8, 42, 49, 116, 153, 158, 168, 186, 189, 199, 202
Degrees of freedom, 5, 136, 145, 212, 279
Delpos, M., 1, 11, 18, 102, 186, 311
Demiurge, 222, 248
DeMorgan, 162
Descartes, 24, 162, 187, 195, 224, 228
Descendence, 223
Descriptive structure, 292
Desmond, 250
DeStefano, G.F., 279
Determination, 23, 72, 99, 125, 144, 153, 156, 170, 178, 212, 225, 228, 232, 272, 299, 309
Dewey, 162
Diagnosis, 170, 234
Diagrammatic type, 146
Dichotomy, 222
Diderot, 226
Differential diagnostic, 82, 172–174, 176, 177
Differentiation, 41, 42, 44, 66, 67, 78, 92, 93, 99, 100, 105, 118, 122, 212, 214
Dilthey, 17, 115, 196
Disappointment, 29, 108, 197
Discontinuity, 176
Discriminatory power, 82
Disposition, 23, 50, 187, 208, 234, 253, 286
Dobzhansky, T., 255
Dogmas, 201, 203, 227, 265
Dollo, 169, 237
Dörner, 184
Dostojewski, 218
Double pyramid, 43, 47, 202, 209, 210, 307
Driesch, H., 215, 255
Driving forces, 85, 102, 148, 193, 230, 231, 286
Dualism, 105, 195
Du-Evidenz, 216

E
Ebeling, 2
Eccles, 35
Educational policy, 311
Ehalt, 232
Ehrenfels, 9
Eibl-Eibesfeldt, 19, 217
Eigen, 221
Einstein, 190
Eisler, 23
Elimination, 26, 89, 184, 203, 204
Emergence(s), 2, 4, 36, 66, 67, 82, 96, 101, 200, 203, 229, 239, 241, 244
  phenomena, 2
  philosophy, 37
Empedocles, 248
Empiricism, 16, 23, 71, 191, 194, 195, 201
Empiricist, 23
Energy levels, 101, 243
Energy transformation, 203
Energy-information relationship, 100
Engel, A., 59
Engels, 18, 33
Enlightenment, 32, 84, 227, 294, 296
Entelechy, 181, 215
Entropy, 36, 37, 58, 100, 102
  export, 2, 4
Epigenetic (system), 136, 141, 146, 264
Epistemology, 3, 18, 123, 181, 220
Escher, 67
Esthetics, 6, 120, 298
Ethology, 11, 40, 58, 70, 132, 133, 135, 146, 174, 217
Eucken, R., 228
Evolutionary epistemology (EE), 10, 26, 33, 36, 39, 122, 202
Evolutive law, 304
Existential philosophy, 17
Explanandum, 199
Explanatory pyramid, 42
Extrapolation, 42, 56, 71, 202
Extra-subjective reality, 3, 10, 12, 17, 19, 31, 34, 47, 53, 55, 91, 92, 183

F
Falsificationism, 31
Fate, 194, 238, 255
Feistel, 2
Feuerbach, 24, 195
Field boundary, 174
Fields of similarity, 39, 63, 73, 75, 77, 78, 128, 137, 144, 150, 156, 181, 291, 309

Final cause, 86, 205
Fitness, 55, 98, 99, 205, 263, 272
Fixation, 145, 208, 279
Foppa, 287
Forces, 14, 39, 58, 70, 97, 197, 199, 220, 241, 247, 306
Formal causes, 204, 304
Formal conditions, 197, 206, 230, 233, 235, 239, 276
Forms of causality, 48, 82, 225, 226, 232, 239, 269, 273, 284, 290, 296, 298
Forms of perception, 16, 18, 19, 21, 29, 74, 191
Förster, 34
Foucault, M., 165, 166
Four causes, 192, 193, 226, 301, 303
Francé, 255
Fractals, 2, 67
Framework homologies, 129, 130, 136, 140, 164, 271, 276
Free will, 215, 216
Frege, 162
Freiberger, 38
Fulguration, 38, 155
Functional analogy, 119, 128
Functionalism, 9, 10
Function-structure dualism, 105
Fuzzy logic, 38

G

Gadamer, 106
Galbraith, 299, 309
Galilean, 226
Galileo, 42, 114, 118, 194, 226
Garcia-Fernandez, 256
Gastrulation, 274, 275
Gell-Mann, 2
Gene, 26, 28, 169, 305
Genealogical, 79, 170
Genealogies, 120, 296
Gene coupling, 261, 262, 271
Gene interactions, 141, 256, 262
Generalization, 28, 40, 48, 57, 108, 120, 212
Genes, 252, 255, 273
Genetic theories, 281
Genotype, 262, 281, 282, 285
Geoffroy Saint Hilaire, 107
Germ line, 255
Gestalt perception, 7, 9, 13, 14, 38, 47, 58, 78, 163, 181, 284, 307
Gestalt theory, 9
Gilbert, 277
Glasersfeld, 34
Glass, 248
Gleick, 247
Goal-oriented world order, 191
Gödel, 23
Goethe, 8, 107, 126, 127, 226, 284
Goodmann, 200
Goodwin, 10
Grant, D.A., 29
Gravitational forces, 70, 97, 241, 247
Gregory, W., 9, 15, 130, 147
Gross national assets, 310

H

Habermas, 106
Haeckel, E., 10, 17, 35, 140, 223
Haken, 4, 229, 247
Haldane, J., 256
Hammurabi, 303
Hartmann, 16, 92
Hassemer, 301
Hassenstein, 123
Hauser, G., 279
Hebb, 59
Hegel, 195, 228
Heidegger, 17
Heider, 274
Heisenberg, 35
Hempel, 107, 200, 201
Hemisphere preference, 7
Hemispheres (of the brain), 7, 91, 127
Hennig, W., 166
Hermeneutic circle, 106, 115, 116, 212
Hermeneutic method, 114, 289
Hermeneutica prophana, 107, 301
Hermeneutica sacra, 107, 301
Hermeneutics, 41, 42, 57, 106–108, 113, 115, 123, 136, 200, 217, 232, 288, 289
Heschl, 59
Heteromorphoses, 251, 252, 257
Hierarchy, 5, 80, 82, 93, 123, 130, 140, 141, 148, 150, 170, 174, 207, 209, 286
  of bodyplans, 147
  of framework homologs, 140
  of homologs, 131, 133
  of mass structures, 211
  of rule-governed relationships, 209
  of system groups, 174
  of theories, 292
Higher-level system, 94, 208, 209, 231, 244, 245, 247, 266–273, 275, 280, 282, 284, 286, 289, 291, 296
Higher-level theory, 57, 220
Hilgard, 287

Hillis, 169
Hinrichsen, 52
Hinst, 24
Historicity, 2, 4, 5, 45, 221, 228, 232, 237–239, 245, 266
Hobbes, 24, 194
Holbach, 194
Holistic, 1, 8, 48, 50, 307
Holland, P.W., 286
Homeobox gene, 157, 280
Homodynamy, 140–142, 277
Homology(ies), 128, 129, 135, 138, 140, 142, 155, 166, 168, 177, 265, 273, 309
 determination, 309
 problem, 166, 168, 270
 theory, 118
Homonomy, 140, 141, 153
Hooker, 254
Horgan, 2
Huber, L., 28, 76, 184
Hughes, 10
Humboldt, 108
Hume, 39, 147, 181, 182, 222, 249
Hylozoists, 226
Hypothesis, 29, 52–57, 60, 125, 183–185
Hypothesizing, 57

I
Idealism, 25, 126, 127, 195, 226–228
Idealistic philosophy, 17, 126, 231
Idealistic reductionism, 228, 231
Identification key, 82
Ideology, 31, 301, 303
Imprinting, 74
Individuality, 86, 108, 114, 153, 292
Induction (in ontogeny), 264
Information, 28, 29, 36, 67, 91, 93, 96, 103, 228, 305
Information-assessment-mechanism, 205
Information content, 103, 135, 157
Information–energy equivalence, 102
Information-processing apparatus, 85
Information transfer, 256, 259, 277
Innate hypotheses, 19, 39, 53, 70, 122
Innate master teacher, 59, 127
Inner intelligence, 255
Inner mechanisms, 257
Inner organization, 251
Inner selection principle, 260
Inorganic realm, 2, 3, 12, 14, 37, 79, 240, 245
Insertions, 94, 232, 294, 308
Instruction(s), 93, 94, 105, 141, 263
Intentionality, 287
Interdependence, 48, 122
Interdisciplinarity, 1, 311
Intermediate level, 47
Intermodality, 74
Internal selection, 255
Introspection, 197, 216
Intuition, 17, 18, 188, 309
Intuitionistic, 128
Invariant(s), 28, 40, 58, 247
Ionian natural philosophers, 191, 220
Ionian physiologists, 248
Irreversibility, 2, 4, 237
Isology, 140, 142

J
James, W., 162
Jantsch, E., 225, 243
Jaspers, 17, 225
Jaynes, J., 189

K
Kammerer, 255
Kant, 15, 17, 24, 39, 162, 187, 195, 220, 224, 225, 306
Karneades, 24
Kaufmann, S., 247
Kepler, 44, 194, 195, 226
Knowledge gain, 16, 17, 26, 38, 72
Koenig, 124, 217
Koestler, A., 93, 255
Koffka, 9
Kreiser, 38
Kreuzer, 31
Kullmann, 193, 226
Kutschera, 186, 187, 196, 199

L
Lamarck, 82, 179
Lamarckism, 251, 254, 256, 263
Lamarckists, 249, 250
Lambert, 10
Lamettrie
Lamnek, 115
Language-based thought, 39, 58, 66, 70, 72, 292, 306
Laplace, 220
Lateral inhibition, 59

Law times application, 34
Lawfulness, 54, 55, 64, 139, 189, 202
Law(s), 2, 4, 11, 14, 32, 34–36, 47, 49, 89, 107, 140, 182, 189, 200, 210, 217, 221, 230, 232, 237, 243, 246, 302
Legitimation, 187, 250
Leibniz, 24, 162, 195
Lenneberg, 125
Lenz, 76
Le Roy Ladurie, 232
Leukipp, 191
Level of order, 105
Lévi-Strauß, 189
Leukipp, 191
Levin, M., 290
Lewis, 162
Linguistic universality, 38, 67
Linné, 165
Loading, 23, 261, 262, 281
Lockc, 194
Logic, 1, 7, 12, 23, 31, 36, 38, 181, 306
Logical conclusion, 38, 72
Logical operation, 55
Longitudinal theories, 13, 26, 32, 37, 39, 51
Lorenz, 3, 10, 19, 31, 38, 58, 59, 92, 96, 118, 122, 128, 148, 155, 183, 231, 269, 310
Lorenz, K., 18
Luria, A., 38
Lyell, 249

M
Mach, 195
Macro-evolution, 257
Macro-mutation, 257
Mae-Wan Ho, 10
Maimonides, 226
Maintenance, 5, 37, 266, 269, 303
Mainzer, 3
Malebranche, 228
Malsburg, 59
Malthus, 250
Mandelbrot, 247
Marr, D., 61
Mass building blocks, 44
Mass hierarchies, 80
Material, 235
  causes, 85, 148, 204, 240
  conditions, 208, 231, 235
Maupertuis, 222, 248, 249
Max Weber, 196
Maxwell, 103
Maxwell's demon, 103
Mayerthaler, 36, 38, 72
Mayr, 80, 309
McNeill, 38
Medawar, J., 37, 229
Medawar, P., 37, 229
Medawars, 229
Memory, 26, 32, 41, 53, 62, 64, 73, 127, 186
Mendel, 254
Meridith, 74
Metamorphoses, 118, 126, 145, 146, 311
Metaphor, 56, 65, 93, 117, 120, 146, 155
Metaphysics, 16
Methodology, 2, 6, 8, 49, 110, 227
Microevolution, 255, 256
Micro-evolutionism, 257
Mill, 24, 162, 195
Milieu, 5, 97, 99, 232, 243
Miller, 245
Mills, 201
Mittelstrass, 23
Möbius, 151
Models of understanding, 299
Mohr, 3, 229
Monod, J., 256
Moore, J., 250
Morgan, 37, 66
Moritz, 169
Morphological distance, 167
Morphological method, 280
Morphological theories, 274
Morphological treatment, 13
Morphological type(s), 44, 107, 144, 146, 282, 301
Morphology, 2, 3, 8–10, 77, 92, 126, 127, 138, 152, 165, 181, 219, 240, 245
Morphotype, 281, 284
Müller, 263, 279
Mutability, 260, 273
Mutant, 257
Mutation(s), 10, 28, 32, 33, 102, 257
Mutual dependence, 122

N
Nägeli, 254
Natural law(s), 35, 138, 187, 200, 250, 304, 308
Natural order, 223
Natural system, 114, 123, 166, 177–179, 260
Negentropy, 102
Neodarwinism, 9, 10, 215, 248, 255
Neolamarckism, 255
Nested hierarchies, 80

Neurath, 24
Neuthomism, 224
Newton, 42
Nicolis, 2
Non-equilibrium thermodynamics, 2
Normative sense, 292, 293
Normative structures, 293
Nouns, 5, 38, 67, 72, 105
Nuclear forces, 13
Numerical taxonomy, 80

O
Oberhummer, 243
Object constancy, 40, 62, 65
Objective reality, 34
Oeser, 3
Oken, 8
Old Darwinists, 254
On intermodality, 74
Onditioned response, 201
Ontogeny, 35
Ontological reductionism, 50, 230, 245, 247, 256
Ontology, 92
Open systems, 4
Operational, 54, 71, 236, 285
Operon, 282, 284
Oppenheim, 107, 200
Optimization, 23, 174
Order, 2, 4–6, 8, 9, 28, 32, 34, 35, 49, 50, 52, 59, 74, 78, 85, 87, 93, 97, 101, 102, 117, 120, 122–124, 141, 174, 179, 183, 194, 231, 237, 245, 251, 261, 266, 290, 294, 307
Organology, 99, 206
Orldview, 12
Our worldview, 1
Owens, 8, 9, 128

P
Pächt, 296
Pangenesis theory, 249, 251
Paradigm(s), 2, 3, 23, 32, 100, 102, 103, 105, 113, 122, 219, 220, 309
Parler, 233, 234
Parmenides, 191, 227
Parsimony, 169
Patterson, 143
Pawlow, 24
Peirce, 162
Peitgen, 247
Perception, 14, 144
Phase, 2
Phase transition(s), 2, 4, 5, 36–38, 67, 69, 70, 100, 101, 113–115, 238, 239, 247, 306, 308, 310
Phene(s), 152, 252, 255, 259, 262, 285
  concept, 156
Phenetic, 143, 144
Phidias, 235
Philosophia perennis, 224
Physicalism, 227
Piaget, 9, 38, 58, 62
Pittendrigh, 205
Plate, 254
Plato, 227
Pleiotropies, 252, 262, 271, 273
Poincaré, 162
Pöltner, 18
Polygeny, 257
Polymorphism(s), 6, 154, 155
Popper, 26, 31, 115, 117, 200, 227
Population dynamics, 229
Positional criterion of homology, 155
Positional relationship, 116, 130, 146, 147, 155, 259
Positional trait, 136
Positive law, 301, 304
Positivists, 155, 301
Post hoc, 39, 147
Post-selection, 204, 237
Pragmatic reductionism, 50, 228, 229, 231, 240, 247
Pragmatic turn, 225
Pragmatism, 256
Precondition(s), 6, 16, 26, 42, 128, 182, 183, 195, 206, 232, 233, 235, 285
Preconscious, 11, 307
Predisposition(s), 69, 183, 234, 235, 249, 275
Pre-selection, 204
Preservational conditions, 37
Prigogine, 2, 5, 229
Primordial sea, 237, 245
Primordial soup, 98
Probability, 91, 120, 138, 139, 179, 271
Probability-based, 140
Probability theory, 29
Probando, 198
Process, 39
Prognoses, 23, 32, 34, 35, 41, 54, 91, 108, 113, 138
Program(s), 18, 54, 57, 62, 64, 65, 92, 96, 117, 309

Projection, 94, 197, 216
Proof, 114
Propter hoc, 39, 53, 147, 179, 181, 182, 199, 307
Protocol sentence, 155
Psychoanalysis, 218
Purposeful, 185, 228, 237
Pursuit of influence, 182, 230
Pursuit of truth, 15, 23
Pythagoras, 191, 193, 227

Q
Quantities, 230, 310
Quantum forces, 148, 236, 241, 242
Quantum physics, 245
Quine, 200

R
Raff, 282
Randomness, 47, 179, 308
Ratiomorphic apparatus, 163, 168, 186, 189
Ratiomorphic faculties, 125, 289
Rationalism, 23
Rationality, 16, 224
Reading direction, 167, 178, 179
Reading problem, 178
Reafference principle, 148
Realism, 33, 62
Reality, 3, 10, 12, 17, 19, 22, 25, 31, 33–35, 38, 52, 53, 62, 86, 92, 123, 146, 152, 183, 187, 198, 210, 306
Reason(s), 11, 15, 18, 19, 33, 53, 66, 93, 118, 181, 182, 191, 212, 227, 248
Rebut, 11
Recapitulation, 189, 194, 212
Reciprocal dependence, 122
Reciprocal enlightenment, 12, 83, 113, 301
Recognition, 2, 11, 41, 51, 57, 63, 75, 223, 232
Recursive causality, 5, 184, 260, 310
Recursivity, 113
Reductionism, 2, 50, 195, 228, 230–232, 237, 240, 245, 247, 256
Redundancy content, 102
Reflex, 26–29, 53–55, 57, 59, 60, 122, 205
Regulation, 257, 258, 263, 280, 284, 303
Regulatory gene, 256, 257, 275
Regulatory systems, 256, 282
Reichenbach, 162
Remane, A., 129
Remane, J., 167
Rensch, 183
Rescher, 200
Retrieval (of memory), 74
Reversibility, 228
Richter, 247
Riedl, 3, 5, 18, 26, 34, 54, 70, 103, 119, 243, 264, 295, 297, 309
Rieger, 143
Rieppel, 10
Ringel, 217
Ritter, 23
Roberts, M., 168
Roger Levin, 2
Romanticism, 228
Ruddle, F.H., 157, 256
Russell, 202

S
Sandkühler, 23
Sartre, 17
Scharloo, W., 282
Schelling, 195, 228
Schiller, 126
Schipper, 29
Schischkoff, 189
Schleiermacher, 108
Schlick, 24
Schmidt, 34, 223
Schneider, 59
Scholasticism, 226, 227
Schrödinger, 2, 6, 36, 102
Schultz, 52
Schwabl, 220
Schweitzer, 2
Scientism, 227
Second law of thermodynamics, 100
Scribner, 38
Searles, 162
Sedlmayr, 296
Selection criteria, 18, 20, 284, 286, 296
Selection pressure, 19, 66, 187, 262
Self-organization, 4, 172, 278
Semantics, 15, 67, 71, 72, 142, 294
Sexl, 35
Sextus Empirikus
Shannon, 103
Simon, 2
Simpson, 255, 309
Simul-hoc, 39, 47, 53, 60, 64, 78, 106, 117, 177, 179, 196, 307
Simultaneous coincidence, 53, 57–69
Singer, 59
Sjoelander, 74
Sneath, 167
Snow, 50, 196

Sokal, 80, 167, 168
Solution strategy, 54
Soul, 17, 127, 191, 193, 222, 224, 226
Space-time continuum, 18
Spencer, 250
Spender, 195
Spinoza, 162
Spiral processes, 31, 115
Spontaneous atavism(s), 124, 251
Standard building blocks, 45, 46, 84, 86, 88, 93, 122, 140, 210, 272, 282
Starck, 309
Stegmüller, 116, 196, 200, 220
Stein, 74
Stengers, 229
Stochastic processes, 251
Stöltzner, 221
Stölzner, 49, 221
Straus, 248
Strawson, 24
Structural criterion, 132, 155
Structural hierarchy, 43, 64, 109, 110, 206
Structuralism, 8–10
Subjective experience, 216
Subjective reality, 34
Substantiation, 127, 129, 198
Subsumption, 107, 113, 127, 139, 200, 202, 217, 232, 241, 243, 245
Subsystems, 42, 136, 231, 267
Subtheories, 220
Symmetry levels, 41, 140
Synapomorphy, 166
Synergetics, 2, 67, 195, 247
Synoptic, 1, 3, 6, 7, 9
Synorganization, 258
Syntax, 15, 71, 114
Synthetic theory of evolution, 255
System, 13
  categories, 86, 88, 89
  groups, 174, 176, 177
  mutants, 258
  of theories, 9, 10, 42
  properties, 4, 239
Systematics, 78, 80, 89, 109, 126, 138, 165, 169, 173, 206, 210, 219

T

Tarsky, 24
Teleology, 181, 214
Teleonomy, 181, 205
Telos, 214
Temkin, 248
Tertium non datur, 38
Thenius, E., 161
Theory development, 113
Theory of cognition, 17, 201, 225
Theory of evolution, 10, 11, 13, 26, 37, 49, 178, 219, 220, 248, 252, 266
Theory of ideas, 17
Theory of mind, 216
Theory of science, 38, 40, 72, 311
Thermodynamics, 4, 36, 100
Thou evidence, 216, 288
Three pathways, 82, 83, 85, 86, 88, 107, 211, 213, 241, 269, 274, 284, 286, 287, 289, 290, 294, 296, 298, 300, 303
Time and space, 18
Thirring, 221
Thomas von Aquin, 226
Thompson, 9
Tool use, 183
Transcendental, 17, 26, 195
Transitional criterion, 133
Transitivity, 66
Transmission, 123, 124, 178, 286
Trends, 5, 76, 158, 160, 162
Truth, 7, 15, 16, 42, 54, 97, 306, 307
Turgot, 227
Tyler, S., 143
Type of order, 105
Typostatic phases, 259

U

Uexküll, 152
Ulrich, H., 299
Unreal conditional clause, 201
Urey, 221, 245

V

Validation, 14, 16, 17, 41, 113–115, 138, 181, 182, 187, 188, 198, 238
Venn, 162
Verbs, 5, 72, 105
Vester, 184
Viollet le Duc, E., 161
Volk, 243
Vollmer, 3, 10, 122
Vorländer, 23

W

Waddington, 9, 262
Wagner, 29, 143, 262

Wainwright, 63
Wallace, 250, 252
Wallacism, 254
Watch-maker, 93, 94
Watson, 256
Wave–particle duality, 105
Weak forces, 239, 247
Weaver, 103
Webster, 10
Weighting, 54, 81, 106, 138–140, 157, 163, 164, 166–174, 193, 199
Weinberg, 97
Weismann, 255, 256, 265
Weismann doctrine, 255, 256, 265
Weiss, P., 9, 310
Weizsäcker, 72
Wertheimer, 9
Whitehead, 162
Whyte, 256
Wieser, 102, 103
Williams, 169
Wimmer, 206
Winkler-Oswatitsch, 221
Windelband, 196
Winkler-Oswatitsch, 221
Windelband, 196
Wolff Fuß, 194
Worldview, 12, 16, 31, 33, 34, 51, 220, 222, 228, 241, 305, 307, 309, 311
Wuketits, 3, 18
Wygotski, 232

Z
Zenon, 193

MIX
Papier aus verantwortungsvollen Quellen
Paper from responsible sources
FSC® C105338

If you have any concerns about our products,
you can contact us on
ProductSafety@springernature.com

In case Publisher is established outside the EU,
the EU authorized representative is:
Springer Nature Customer Service Center GmbH
Europaplatz 3, 69115 Heidelberg, Germany

Printed by Libri Plureos GmbH
in Hamburg, Germany